Lecture Notes in Physics

378

C. Bendjaballah O. Hirota
S. Reynaud (Eds.)

Quantum Aspects of Optical Communications

Proceedings of a Workshop
Held at the CNRS, Paris, France
26–28 November 1990

Springer-Verlag
Berlin Heidelberg New York London Paris
Tokyo Hong Kong Barcelona Budapest

Editors

Cherif Bendjaballah
Laboratoire des Signaux et Systèmes CNRS-ESE
Plateau de Moulon, F-91192 Gif-sur-Yvette Cedex, France

Osamu Hirota
Tamagawa University
6-1-1 Tamagawa Gakuen, Machida, Tokyo, Japan

Serge Reynaud
Laboratoire de Spectroscopie Hertzienne
Université Pierre et Marie Curie
BP 74, F-75252 Paris Cedex 05, France

ISBN 3-540-53862-3 Springer-Verlag Berlin Heidelberg New York
ISBN 0-387-53862-3 Springer-Verlag New York Berlin Heidelberg

Printed in Germany

Printing: Druckhaus Beltz, Hemsbach/Bergstr.
Bookbinding: J. Schäffer GmbH & Co. KG., Grünstadt
2153/3140-543210 – Printed on acid-free paper

Preface

This volume in the series Lecture Notes in Physics contains the proceedings of the workshop on Quantum Aspects of Optical Communications (**QOC '90**) held at the CNRS–Paris (France) from 26 to 28 November, 1990. This conference was attended by about 80 participants from all around the world. A limited number of grants were offered to students presenting papers in the area of QOC.

The purpose of this workshop was to bring together researchers working in the various fields of QOC. Six sessions were organised in order that most of the topics involved in QOC were covered: quantum mechanics for optics, quantum measurements and information, quantum stochastic processes, quantum detection and photon channels, squeezed and nonclassical states, and devices for quantum communication. All the sessions were conducted by eminent scientists and began with invited talks each of 40 minutes. They were followed by contributed papers each of 20 minutes. There were about 40 papers in total.

There were also interesting discussion during and after the sessions on mathematical and physical implications of the lifting operation and its application to information dynamics, quantum nondemolition measurements, quantum entropy and information transfer, the standard quantum limit and its possible breaking, coherent squeezed light and its realization, quantum noise reduction in squeezed light, the phase operator in quantum optics, recent progress in the designing of the interferometer for the detection of gravitational waves and the concept of the instrument and its mathematical description.

The editors would like to thank the contributors for providing their papers on the first day of the conference or shortly after for a few of them. The editors would also like to thank all those people who helped before, during and after the meeting.

Finally, the following foundations are gratefully acknowledged for their support scientific, moral and financial:

CNES, CNRS, SFO, SFTI, from France, and
FAT, Fujitsu, Hitachi, NEC, Nomura, from Japan.

We look forward to the next QOC.

Paris, January 1991

C. Bendjaballah
O. Hirota
S. Reynaud

Contents

Session VI: Devices for Quantum Communication P. Chavel, Chairman

Session I

– Quantum Mechanics for Optics –

Chairman : G. Lindblad

Quantum Limits of Measurements and Uncertainty Principle

MASANAO OZAWA

Department of Mathematics, College of General Education
Nagoya University, Nagoya 464, Japan

1. Introduction

In the theoretical considerations on quantum aspects of optical communications, one of the most important programs is to establish limits of measurements which are subjected to the laws of quantum mechanics in a rigorous and unified manner. In such a program, it is natural to expect that the uncertainty principle will play a central role. However, the recent controversy [28,5,21,22], which arose in the field of gravitational-wave detection, on the validity of the standard quantum limit for monitoring of the free-mass position [4,6] revealed a certain weakness of our understanding of the Heisenberg uncertainty principle.

Historically, Heisenberg established the uncertainty principle from his analysis of *gedanken* experiments for certain processes of successive measurements [10]. Thus his uncertainty principle is often stated [25, p. 239] in a form that a measurement of one variable from a conjugate pair disturbs the value of the conjugate variable no less than the order of $\hbar$/(accuracy of the measurement). Nevertheless, we have not established a general theory of this kind of uncertainty principle, as pointed out by several authors [3,13].

Our uncertainty relation in current text books was first proved by Robertson [24] by a simple mathematical reasoning with use of the Schwarz inequality. However, it is often pointed out that Robertson's uncertainty principle does not mean the Heisenberg uncertainty principle. Robertson's uncertainty principle is only concerned with state preparations as in the statement that any state preparation gives an ensemble of objects in which the product of the standard deviations of conjugate variables is greater than $\hbar/2$ [3,12].

In this paper, we shall show how the Robertson uncertainty relation gives certain intrinsic quantum limits of measurements in the most general and rigorous mathematical treatment. In Section 2, fragments from our previous work on mathematical foundations of quantum probability theory are given (see, for the detail, [15,16,17,18,19,20]). In Section 3, some basic properties of root-mean-square error of measurement, called precision, introduced in [21] is examined and, in Section 4, a general lower bound of the product of precisions arising in joint measurements of noncommuting observables is established. This result is used to give a general proof of the uncertainty relation for the joint measurements which has been found by several authors [2,11,27,1]. In Section 5, we shall give a rigorous condition for holding of the standard quantum limit (SQL) for repeated measurements.

Lecture Notes in Physics, Vol. 378
C. Bendjaballah, O. Hirota, S. Reynaud (Eds.),
Quantum Aspects of Optical Communications

For this purpose, we shall examine another root-mean-square error, called resolution, introduced in [21] and prove that if a measuring instruments has no larger resolution than the precision then it obeys the SQL. As shown in [21,22,23], we can even construct many linear models of position measurement which circumvent the above condition. In Section 6, some conclusions from the present analysis will be discussed.

2. Foundations of quantum probability

Let $\mathcal{H}$ be a Hilbert space. Denote by $\mathcal{L}(\mathcal{H})$ the algebra of bounded linear operators on $\mathcal{H}$, by $\tau c(\mathcal{H})$ the space of trace class operators on $\mathcal{H}$ and by $\sigma c(\mathcal{H})$ the space of Hilbert-Schmidt class operators on $\mathcal{H}$. A positive operator in $\tau c(\mathcal{H})$ with the unit trace is called a *density operator* and $\mathcal{S}(\mathcal{H})$ stands for the space of density operators on $\mathcal{H}$. Denote by $\mathcal{B}(\mathbf{R}^d)$ the Borel σ-field of the Euclidean space $\mathbf{R}^d$. A map $X : \mathcal{B}(\mathbf{R}^d) \to \mathcal{L}(\mathcal{H})$ is called a *probability-operator-valued measure* (POM) if it satisfies the following conditions (P1)–(P2):

(P1) For any sequence $\langle \Delta_i \mid i = 1, 2, \ldots \rangle$ of disjoint sets in $\mathcal{B}(\mathbf{R}^d)$,

$$X(\bigcup_{i=1}^{\infty} \Delta_i) = \sum_{i=1}^{\infty} X(\Delta_i),$$

where the sum is convergent in the weak operator topology.

(P2) $X(\mathbf{R}) = 1$.

A linear transformation $T : \tau c(\mathcal{H}) \to \tau c(\mathcal{H})$ is called a *positive map* if $T(\rho) \geq 0$ for all $\rho \in \mathcal{S}(\mathcal{H})$. We shall denote the space of all positive maps on $\tau c(\mathcal{H})$ by $\mathrm{P}(\tau c(\mathcal{H}))$. A map $\mathbf{X} : \mathcal{B}(\mathbf{R}^d) \to \mathrm{P}(\tau c(\mathcal{H}))$ is called an *operation-valued measure* if it satisfies the following conditions (O1)–(O2):

(O1) For any sequence $\langle \Delta_i \mid i = 1, 2, \ldots \rangle$ of disjoint sets in $\mathcal{B}(\mathbf{R}^d)$,

$$\mathbf{X}(\bigcup_{i=1}^{\infty} \Delta_i) = \sum_{i=1}^{\infty} \mathbf{X}(\Delta_i),$$

where the sum is convergent in the strong operator topology of $\mathrm{P}(\tau c(\mathcal{H}))$.

(O2) For any $\rho \in \tau c(\mathcal{H})$,

$$\mathrm{Tr}[\mathbf{X}(\mathbf{R}^d)\rho] = \mathrm{Tr}[\rho].$$

An operation-valued measure $\mathbf{X} : \mathcal{B}(\mathbf{R}^d) \to \mathrm{P}(\tau c(\mathcal{H}))$ is called a *completely positive operation-valued measure* (CPOM) if it satisfies the following condition (O3):

(O3) For any $\Delta \in \mathcal{B}(\mathbf{R}^d)$, $\mathbf{X}(\Delta)$ is a completely positive map on $\tau c(\mathcal{H})$, i.e.,

$$\sum_{i,j=1}^{n} \langle \xi_i | \mathbf{X}(\Delta)(|\eta_i\rangle\langle\eta_j|)|\xi_j\rangle \geq 0,$$

for all $\Delta \in \mathcal{B}(\mathbf{R}^d)$ and for all finite sequences $\xi_1, \ldots, \xi_n$ and $\eta_1, \ldots, \eta_n$ in $\mathcal{H}$.

The *transpose* ${}^tT : \mathcal{L}(\mathcal{H}) \to \mathcal{L}(\mathcal{H})$ of $T \in \mathrm{P}(\tau c(\mathcal{H}))$ is defined by the relation

$$\mathrm{Tr}[{}^tT(a)\rho] = \mathrm{Tr}[aT(\rho)],$$

for all $a \in \mathcal{L}(\mathcal{H})$ and $\rho \in \tau c(\mathcal{H})$. In this case, ${}^tT$ is also positive in the sense that $T(a) \geq 0$ for any $a \geq 0$ in $\mathcal{L}(\mathcal{H})$. For any operation-valued measure $\mathbf{X}$, the relation

$$\hat{\mathbf{X}}(\Delta) = {}^t\mathbf{X}(\Delta)1 \qquad (\Delta \in \mathcal{B}(\mathbf{R}^d)),$$

determines a POM $\hat{\mathbf{X}}$, called the POM *associated* with $\mathbf{X}$. Conversely, any POM X has at least one CPOM $\mathbf{X}$ such that $X = \hat{\mathbf{X}}$ [16, Proposition 4.1]. POM's are called "measurements" in [12] and operation-valued measures are called "instruments" in [8,7]. Our terminology is intended to be more neutral in meanings in the physical context.

Suppose that a Hilbert space $\mathcal{H}$ is the state space of a quantum system $\mathbf{S}$. A *state* of $\mathbf{S}$ is a density operator on $\mathcal{H}$ and an *observable* of $\mathbf{S}$ is a POM $A : \mathcal{B}(\mathbf{R}) \to \mathcal{L}(\mathcal{H})$ such that $A(\Delta)$ is a projection for all $\Delta \in \mathcal{B}(\mathbf{R})$. A state of the form $|\psi\rangle\langle\psi|$ for a unit vector $\psi \in \mathcal{H}$ is called a *pure state* and ψ is called a *vector state* of $\mathbf{S}$. A finite set $\{A_1, \ldots, A_n\}$ of observables is called *compatible* if $[A_i(\Delta_1), A_j(\Delta_2)] = 0$ for all i, $j = 1, \ldots, n$ and Δ_1, $\Delta_2 \in \mathcal{B}(\mathbf{R})$. The *joint probability distribution* of a compatible set $\{A_1, \ldots, A_n\}$ of observables in a state ρ, denoted by $\Pr[A_1 \in \Delta_1, \ldots, A_n \in \Delta_n \| \rho]$, $(\Delta_1, \ldots, \Delta_n \in \mathcal{B}(\mathbf{R}))$, is given by the following *Born statistical formula*:

$$\Pr[A_1 \in \Delta_1, \ldots, A_n \in \Delta_n \| \rho] = \mathrm{Tr}[A_1(\Delta_1) \cdots A_n(\Delta_n)\rho].$$

The symbol ρ in the left-hand-side denotes the state for which the probability distribution is determined. For pure states $\rho = |\psi\rangle\langle\psi|$, the symbol ψ will be sometimes used instead of ρ in this and similar expressions, and then we have

$$\Pr[A_1 \in \Delta_1, \ldots, A_n \in \Delta_n \| \psi] = \langle\psi|A_1(\Delta_1) \cdots A_n(\Delta_n)|\psi\rangle.$$

By spectral theory, the relation

$$\hat{A} = \int_{\mathbf{R}} aA(da).$$

sets up a one-to-one correspondence between observables A and self-adjoint operators $\hat{A}$.

In this paper, by a *measurement* we shall mean generally an experiment described as follows. Let $\mathbf{P}$ be a quantum system, called a *probe system*, described by a Hilbert space $\mathcal{K}$. The system $\mathbf{P}$ is coupled to the system $\mathbf{S}$ during a finite time interval from time t to $t + \tilde{\tau}$. Denote by $\hat{U}$ the unitary operator on $\mathcal{H} \otimes \mathcal{K}$ corresponding to the time evolution of the system $\mathbf{S} + \mathbf{P}$ from time t to $t + \tilde{\tau}$. The time t is called the *time of measurement* and the time $t + \tilde{\tau}$ is called the *time just after measurement.* At the time just after the measurement, the systems $\mathbf{S}$ and $\mathbf{P}$ are separated and in order to obtain the outcome of this experiment a compatible sequence $\langle M_1, \ldots, M_n\rangle$ of observables of the system $\mathbf{P}$ are measured by an ideal manner. The observables $M_1, \ldots, M_n$ are called the *meter observables.* In order to assure the reproducibility of this experiment, the probe system $\mathbf{P}$ is always prepared in a fixed state σ at the time of measurement. Thus the physical process of a given measurement is characterized by a 4-tuple $\mathcal{X} = [\mathcal{K}, \sigma, \hat{U}, \langle \hat{M}_1, \ldots, \hat{M}_n\rangle]$, called a *measurement scheme*, consisting of a Hilbert space $\mathcal{K}$, a density operator σ on $\mathcal{K}$, a unitary operator $\hat{U}$ on $\mathcal{H}\otimes\mathcal{K}$ and a compatible sequence $\langle\hat{M}_1,\ldots,\hat{M}_n\rangle$ of self-adjoint operators on $\mathcal{K}$. Every measurement scheme $\mathcal{X}=[\mathcal{K},\sigma,\hat{U},\langle\hat{M}_1,\ldots,\hat{M}_n\rangle]$ determines a unique CPOM $\mathbf{X}:\mathcal{B}(\mathbf{R}^n)\to\mathrm{P}(\tau c(\mathcal{H}))$, called the CPOM of $\mathcal{X}$, by the following relation

$$\mathbf{X}(\Delta_1\times\cdots\times\Delta_n)\rho=\mathrm{Tr}_{\mathcal{K}}\left[(1\otimes M_1(\Delta_1)\cdots M_n(\Delta_n))\,\hat{U}(\rho\otimes\sigma)\hat{U}^{\dagger}\right], \tag{2.1}$$

for all $\rho\in\tau c(\mathcal{H})$ and $\Delta_1,\ldots,\Delta_n\in\mathcal{B}(\mathbf{R})$, where $\mathrm{Tr}_{\mathcal{K}}$ stands for the partial trace operation of $\mathcal{K}$. Then the CPOM $\mathbf{X}$ satisfies the following *Davies-Lewis postulates* (DL1)–(DL2) (cf. [8]):

(DL1) Measurement probability: If the state of the system $\mathbf{S}$ at the time of measurement is ρ, then the probability distribution $\Pr[X\in\Delta\|\rho]$, $(\Delta\in\mathcal{B}(\mathbf{R}^n))$, of the outcome variable X of the measurement is

$$\Pr[X\in\Delta\|\rho]=\mathrm{Tr}[\mathbf{X}(\Delta)\rho].$$

(DL2) State reduction: If the state of the system $\mathbf{S}$ at the time of measurement is ρ, then the measurement changes the state so that the state ρ_Δ, at the time just after measurement, of the subensemble $\mathbf{S}_\Delta$ of the systems selected by the condition $X\in\Delta$ is given by

$$\rho_\Delta=\frac{\mathbf{X}(\Delta)\rho}{\mathrm{Tr}[\mathbf{X}(\Delta)\rho]},$$

for any $\Delta\in\mathcal{B}(\mathbf{R}^n)$ with $\Pr[X\in\Delta\|\rho]\neq 0$.

Given an operation-valued measure $\mathbf{X}:\mathcal{B}(\mathbf{R}^n)\to\mathrm{P}(\tau c(\mathcal{H}))$, any measurement scheme $\mathcal{X}=[\mathcal{K},\sigma,\hat{U},\langle\hat{M}_1,\ldots,\hat{M}_n\rangle]$ which satisfies Eq. (2.1) is called a *realization* of $\mathbf{X}$. An operation-valued measure is called *realizable* if it has at least one realization. An importance of complete positivity for operation-valued measures is clear from the following.

Theorem 2.1. *An operation-valued measure $\mathbf{X}:\mathcal{B}(\mathbf{R}^n)\to\mathrm{P}(\tau c(\mathcal{H}))$ is realizable if and only if it is a CPOM. In particular, every CPOM $\mathbf{X}:\mathcal{B}(\mathbf{R}^n)\to\mathrm{P}(\tau c(\mathcal{H}))$ has a realization $\mathcal{X}=[\mathcal{K},\sigma,\hat{U},\langle\hat{M}_1,\ldots,\hat{M}_n\rangle]$ such that σ is a pure state and* $\dim(\mathcal{H})=\dim(\mathcal{K})$.

For a proof, see [16, Section 5]. A consequence from the above theorem is the following version of the Naimark extension of POM's.

Corollary 2.2. *For any POM $X:\mathcal{B}(\mathbf{R}^n)\to\mathcal{L}(\mathcal{H})$, there exists a measurement scheme $\mathcal{X}=[\mathcal{K},|\phi\rangle\langle\phi|,\hat{U},\langle\hat{M}_1,\ldots.\hat{M}_n\rangle]$ satisfying the relation*

$$X(\Delta_1\times\cdots\times\Delta_n)=V^{\dagger}[\hat{U}^{\dagger}(1\otimes M_1(\Delta_1)\cdots M_n(\Delta_n))\hat{U}]V, \tag{2.2}$$

for all $\Delta_1,\ldots,\Delta_n\in\mathcal{B}(\mathbf{R}^n)$, where V is the isometry from $\mathcal{H}$ to $\mathcal{H}\otimes\mathcal{K}$ such that $V\psi=\psi\otimes\phi$ for all $\psi\in\mathcal{H}$.

A measurement scheme $\mathcal{X}$ satisfying Eq. (2.2) is called an *interacting realization* of a POM X. A definition of the non-interacting version of realizations of POM's appears in [12, p. 68] and it should be noted that an interacting realization determines the state reduction but a non-interacting one does not.

The outcome variable X of a measurement scheme $\mathcal{X}$ is generally called a *quantum random variable* (q.r.v.). Thus any q.r.v. X has a CPOM $\mathbf{X}$ which determines the probability distributions of X. Let $\langle X_1, \dots, X_n \rangle$ be a finite sequence of q.r.v.s of measurement schemes $\mathcal{X}_1, \dots, \mathcal{X}_n$ and $\langle \mathbf{X}_1, \dots, \mathbf{X}_n \rangle$ the corresponding sequence of CPOM's. Then, from (DL1) and (DL2), the joint probability distribution $\Pr[X_1 \in \Delta_1, \dots, X_n \in \Delta_n \| \rho]$, $(\Delta_1, \dots, \Delta_n \in \mathcal{B}(\mathbf{R}^n))$, of $\langle X_1, \dots, X_n \rangle$ in a state ρ is given by the following *Davies-Lewis formula* [8]:

$$\Pr[X_1 \in \Delta_1, \dots, X_n \in \Delta_n \| \rho] = \mathrm{Tr}[\mathbf{X}_n(\Delta_n) \cdots \mathbf{X}_1(\Delta_1)\rho]. \tag{2.3}$$

Let $\hat{H}$ be the Hamiltonian of the system $\mathbf{S}$ and $\hat{U}_t = e^{-it\hat{H}/\hbar}$ the unitary operator of the time evolution. For the Heisenberg system state ρ, we shall write,

$$\rho(t) = \alpha(t)\rho = \hat{U}_t \rho \hat{U}_t^{\dagger},$$

for the time evolution of the states in the Schrödinger picture. Suppose that a finite sequence of measurements corresponding to a sequence $\langle \mathbf{X}_1, \dots, \mathbf{X}_n \rangle$ of CPOM is made successively at time $(0 <) t_1 < \cdots < t_n$, where it is supposed that $\tilde{\tau}_i \ll t_{i+1} - t_i$ for the durations $\tilde{\tau}_i$ of the coupling of measurement $\mathbf{X}_i$. We shall denote by $X_i(t_i)$ the outcome variable of the measurement $\mathbf{X}_i$ at time t_i. Then the joint probability distribution of the sequence $\langle X_1(t_1), \dots, X_n(t_n) \rangle$ of the outcomes in the state $\rho = \rho(0)$ is given by the following *Wigner-Davies-Lewis formula* [26,8]:

$$\begin{aligned} &\Pr[X_1(t_1) \in \Delta_1, X_2(t_2) \in \Delta_2, \dots, X_n(t_n) \in \Delta_n \| \rho] \\ &= \mathrm{Tr}[\mathbf{X}_n(\Delta_n)\alpha(t_n - t_{n-1}) \cdots \mathbf{X}_2(\Delta_2)\alpha(t_2 - t_1)\mathbf{X}_1(\Delta_1)\alpha(t_1)\rho], \\ &(\Delta_1, \Delta_2, \dots, \Delta_n \in \mathcal{B}(\mathbf{R})). \end{aligned}$$

Let $\mathbf{X} : \mathcal{B}(\mathbf{R}^n) \to \mathrm{P}(\tau c(\mathcal{H}))$ be a CPOM and ρ a density operator on $\mathcal{H}$. A family $\{\rho_x \mid x \in \mathbf{R}^n\}$ of density operators on $\mathcal{H}$ is called a family of *posterior states* for a *prior state* ρ and a CPOM $\mathbf{X}$ if it satisfies the following conditions (PS1)–(PS2):

(PS1) The function $x \mapsto \rho_x$ is strongly Borel measurable.

(PS2) For any $\Delta \in \mathcal{B}(\mathbf{R}^n)$,

$$\int_{\Delta} \rho_x \,\mathrm{Tr}[\mathbf{X}(dx)\rho] = \mathbf{X}(\Delta)\rho.$$

A family of posterior states always exists for any prior state ρ and it is unique in the following sense: If $\{\rho'_x \mid x \in \mathbf{R}\}$ is another family of posterior states for the prior state ρ, then $\rho'_x = \rho_x$ for $\mathrm{Tr}[\mathbf{X}(dx)\rho]$-almost everywhere [17]. Suppose that a measurement corresponding to $\mathbf{X}$ is made for the system $\mathbf{S}$ in a state ρ at the time of measurement and that the measurement gives the outcome $X = x$ $(x \in \mathbf{R}^n)$. Let $\{\rho_x \mid x \in \mathbf{R}^n\}$ be a family of posterior states for the prior state ρ. Then, with probability 1, ρ_x is the state of the system $\mathbf{S}$ at the time just after the measurement.

3. Noise of approximate measurement

Let $\mathcal{H}$ be a Hilbert space corresponding to a quantum system $\mathbf{S}$. Let A be an observable of $\mathbf{S}$. In this section, we consider a measuring instrument designed to measure the value of an observable A and discuss the noise contained in outcomes from the measuring instrument. Let X be a q.r.v. representing the outcome from the measuring instrument. Then the probability distribution $\Pr[X \in \Delta \| \rho]$ of X in a state ρ of the system $\mathbf{S}$ at the time of measurement is represented by a POM $\hat{\mathbf{X}}$ for some CPOM $\mathbf{X}$ satisfying

$$\Pr[X \in \Delta \| \rho] = \mathrm{Tr}[\hat{\mathbf{X}}(\Delta)\rho], \qquad (\Delta \in \mathcal{B}(\mathbf{R})).$$

For simplicity of notation, we shall write $X(\Delta) = \hat{\mathbf{X}}(\Delta)$, $(\Delta \in \mathcal{B}(\mathbf{R}))$. Let f be a real Borel function on $\mathbf{R}$. The *expectation* $\mathrm{Ex}[f(X)\|\rho]$ of the q.r.v. $f(X)$ in a state ρ is defined by

$$\mathrm{Ex}[f(X)\|\rho] = \int_{\mathbf{R}} f(x) \Pr[X \in dx\|\rho],$$

provided the integral is convergent. Denote by $\widehat{f(X)}$ the symmetric operator defined by

$$\begin{aligned} \langle\xi|\widehat{f(X)}|\xi\rangle &= \int_{\mathbf{R}} f(x)\langle\xi|X(dx)|\xi\rangle, \qquad (\xi \in \mathrm{dom}\left(\widehat{f(X)}\right)), \\ \mathrm{dom}\left(\widehat{f(X)}\right) &= \{\xi \in \mathcal{H} \mid \int_{\mathbf{R}} f(x)^2\langle\xi|X(dx)|\xi\rangle < \infty\}. \end{aligned} \tag{3.1}$$

Then we have

$$\mathrm{Ex}[f(X)\|\psi] = \langle\psi|\widehat{f(X)}|\psi\rangle,$$

for any vector state $\psi \in \mathrm{dom}(\widehat{f(X)})$. The *variance* $\mathrm{Var}[X\|\rho]$ and the *standard deviation* $\Delta X[\rho]$ of X in a state ρ are defined in the usual way, i.e.,

$$\begin{aligned} \mathrm{Var}[X\|\rho] &= \mathrm{Ex}[X^2\|\rho] - \mathrm{Ex}[X\|\rho]^2, \\ \Delta X[\rho] &= \mathrm{Var}[X\|\rho]^{1/2}. \end{aligned} \tag{3.2}$$

We say that a POM X has *finite second moment* in a state ρ if $\mathrm{Ex}[X^2\|\rho] < \infty$, or equivalently, if $\Delta X[\rho] < \infty$. Let X be a POM with finite second moment in ρ. Then we have $\mathrm{dom}(\hat{X}) \supset \mathrm{ran}(\sqrt{\rho})$ and that $\hat{X}\sqrt{\rho}$ is a Hilbert-Schmidt operator [12]. Thus, we shall write $\mathrm{Tr}[\hat{X}^2\rho] = \mathrm{Tr}[(\hat{X}\sqrt{\rho})^\dagger\hat{X}\sqrt{\rho}]$. For any POM's X, Y with $\Delta X[\rho], \Delta Y[\rho] < \infty$, the expression $\mathrm{Tr}[[\hat{X},\hat{Y}]\rho]$ is defined by

$$\mathrm{Tr}[[\hat{X},\hat{Y}]\rho] = \mathrm{Tr}[(\hat{X}\sqrt{\rho})^\dagger\hat{Y}\sqrt{\rho} - (\hat{Y}\sqrt{\rho})^\dagger\hat{X}\sqrt{\rho}].$$

By the Robertson uncertainty relation [24], for any state ρ and any pair of observables A, B with $\Delta A[\rho], \Delta B[\rho] < \infty$, we have

$$\Delta A[\rho]\Delta B[\rho] \geq \frac{1}{2}\left|\mathrm{Tr}[[\hat{A},\hat{B}]\rho]\right|. \tag{3.3}$$

The above relation is extended to any pair of POM's by Holevo [12, p.90], i.e., for any pair of POM's X, Y with $\Delta X[\rho], \Delta Y[\rho] < \infty$, we have

$$\Delta X[\rho]\Delta Y[\rho] \geq \frac{1}{2}\left|\mathrm{Tr}[[\hat{X},\hat{Y}]\rho]\right|. \tag{3.4}$$

We say that a POM $X : \mathcal{B}(\mathbf{R}) \to \mathcal{L}(\mathcal{H})$ is *compatible* with an observable A (or A-*compatible*, in short) if it satisfies the relation

$$[X(\Delta_1), A(\Delta_2)] = 0,$$

for all $\Delta_1, \Delta_2 \in \mathcal{B}(\mathbf{R})$. Let ρ be a state at the time of measurement. For an A-compatible POM X, the joint probability distribution $\Pr[X \in \Delta_1, A \in \Delta_2 \| \rho]$ of X and A in a state ρ is given by

$$\Pr[X \in \Delta_1, A \in \Delta_2 \| \rho] = \mathrm{Tr}[X(\Delta_1)A(\Delta_2)\rho], \qquad (\Delta_1, \Delta_2 \in \mathcal{B}(\mathbf{R})).$$

By a notational convention, for a Borel measure μ on $\mathcal{B}(\mathcal{R}^2)$, we shall write

$$\iint_{\mathbf{R}^2} f(x,y)\nu(dx,dy) = \int_{\mathbf{R}^2} f(x,y)\mu(d(x,y)),$$

where ν is the joint measure on $\mathcal{B}(\mathbf{R}) \times \mathcal{B}(\mathbf{R})$ defined by $\nu(\Delta_1, \Delta_2) = \mu(\Delta_1 \times \Delta_2)$, $(\Delta_1, \Delta_2 \in \mathcal{B}(\mathbf{R}))$, and we shall write $\mathrm{L}^p(\mathbf{R}^2, \nu(dx,dy)) = \mathrm{L}^p(\mathbf{R}^2, \mu)$ for the L^p-space of μ. We define the *root-mean-square error* (or *precision*, in short) $\epsilon[X|A,\rho]$ of X for measurement of an observable A in a state ρ by the relation

$$\epsilon[X\|A,\rho]^2 = \iint_{\mathbf{R}^2} (x-a)^2 \,\mathrm{Tr}[X(dx)A(da)\rho]. \tag{3.5}$$

Obviously, $\epsilon[X|A,\rho]$ represents the root-mean-square deviation of the outcome X of the measurement from the outcome A of the ideal measurement, when these two were made simultaneously in the state ρ.

Lemma 3.1. *Let μ be a finite Borel measure on $\mathbf{R}^2$. Then the relation*

$$\iint_{\mathbf{R}^2} (x-y)^2\,\mu(dx \times dy) = 0 \tag{3.6}$$

holds if and only if for any Δ_1, $\Delta_2 \in \mathcal{B}(\mathbf{R})$,

$$\mu(\Delta_1 \times \Delta_2) = \mu((\Delta_1 \cap \Delta_2) \times \mathbf{R}). \tag{3.7}$$

Proof. Suppose that Eq. (3.6) holds. Let $D = \{(x,y) \in \mathbf{R}^2 | x = y\}$. Then it follows from Eq. (3.6) that $\mu(\mathbf{R}^2 \setminus D) = 0$. Then we have

$$\begin{aligned} \mu(\Delta_1 \times \Delta_2) &= \mu((\Delta_1 \times \Delta_2) \cap D) = \mu((\Delta_1 \cap \Delta_2) \times \mathbf{R}) \cap D) \\ &= \mu((\Delta_1 \cap \Delta_2) \times \mathbf{R}). \end{aligned}$$

Conversely suppose that Eq. (3.7) holds. Then we have

$$\begin{aligned} \mu(\Delta_1 \times \Delta_2) &= \mu((\Delta_1 \cap \Delta_2) \times \mathbf{R}) \\ &= \int_{\Delta_1 \cap \Delta_2} \mu(dx \times \mathbf{R}) \\ &= \int_{\Delta_1} \delta_x(\Delta_2)\,\mu(dx \times \mathbf{R}) \\ &= \int_{\mathbf{R}} \mu(dx \times \mathbf{R}) \int_{\mathbf{R}} \chi_{\Delta_1 \times \Delta_2}(x,y)\,\delta_x(dy), \end{aligned}$$

where δ_x is the Dirac measure of $x \in \mathbf{R}$. Thus we obtain

$$\begin{aligned}\iint_{\mathbf{R}^2} (x-y)^2 \,\mu(dx \times dy) &= \int_{\mathbf{R}} \mu(dx \times \mathbf{R}) \int_{\mathbf{R}} (x-y)^2 \,\delta_x(dy) \\ &= 0.\end{aligned}$$

QED

Theorem 3.2. *An A-compatible POM X satisfies the relation*

$$\epsilon[X\|A,\rho] = 0, \tag{3.8}$$

for all density operator ρ on $\mathcal{H}$ if and only if $X = A$.

Proof. Obviously, if $X = A$ then Eq. (3.8) holds. Suppose that Eq. (3.8) holds. Let $\rho \in \mathcal{S}(\mathcal{H})$ and $\Delta \in \mathcal{B}(\mathbf{R})$. Let μ be the Borel measure on $\mathbf{R}^2$ such that $\mu(\Delta_1 \times \Delta_2) = \mathrm{Tr}[X(\Delta_1)A(\Delta_2)\rho]$, $(\Delta_1, \Delta_2 \in \mathcal{B}(\mathbf{R}))$. Then since $\epsilon[X\|A,\rho] = 0$, we have $\mu(\Delta \times \mathbf{R}) = \mu(\mathbf{R} \times \Delta)$ from Lemma 3.1, and hence

$$\mathrm{Tr}[A(\Delta)\rho] = \mathrm{Tr}[A(\Delta)X(\mathbf{R})\rho] = \mathrm{Tr}[A(\mathbf{R})X(\Delta)\rho] = \mathrm{Tr}[X(\Delta)\rho].$$

Since ρ and Δ are arbitrary, it is concluded that $A = X$. *QED*

Lemma 3.3. *For any A-compatible POM X, there exists a Hilbert space $\tilde{\mathcal{H}}$, an isometry $V : \mathcal{H} \to \tilde{\mathcal{H}}$ and self-adjoint operators $\tilde{X}$ and $\tilde{A}$ on $\tilde{\mathcal{H}}$ satisfying the following conditions:*

(1) $[\tilde{X}, \tilde{A}] = 0$ *and* $\tilde{A}V = V\hat{A}$.
(2) *For any* $\rho \in \mathcal{S}(\mathcal{H})$ *and* $f \in \mathrm{L}^1(\mathbf{R}^2, \mathrm{Tr}[X(dx)A(da)\rho])$,

$$\iint_{\mathbf{R}^2} f(x,a)\, \mathrm{Tr}[X(dx)A(da)\rho] = \mathrm{Tr}[f(\tilde{X},\tilde{A})V\rho V^\dagger].$$

Proof. Let $M : \mathcal{B}(\mathbf{R}^2) \to \mathcal{L}(\mathcal{H})$ be a POM such that $M(\Delta_1 \times \Delta_2) = X(\Delta_1)A(\Delta_2)$, for all $\Delta_1, \Delta_2 \in \mathcal{B}(\mathbf{R})$. Then, by the Naimark extension of M, there exist a Hilbert space $\tilde{\mathcal{H}}$, an isometry $V : \mathcal{H} \to \tilde{\mathcal{H}}$ and a projection valued measure $E : \mathcal{B}(\mathbf{R}^2) \to \mathcal{L}(\tilde{\mathcal{H}})$ such that $M(\Delta_1 \times \Delta_2) = V^\dagger E(\Delta_1 \times \Delta_2)V$ for all Δ_1, $\Delta_2 \in \mathcal{B}(\mathbf{R})$. Let $\tilde{X}$ and $\tilde{A}$ be self-adjoint operators on $\tilde{\mathcal{H}}$ defined by

$$\tilde{X} = \int_{\mathbf{R}} x\, E(dx \times \mathbf{R}),$$
$$\tilde{A} = \int_{\mathbf{R}} a\, E(\mathbf{R} \times da).$$

Then the assertion follows from a straightforward verification. *QED*

An A-compatible POM X is said to be *unbiased* if $\hat{A} = \hat{X}$; in this case, we have. $\mathrm{Ex}[A\|\rho] = \mathrm{Ex}[X\|\rho]$ for all state ρ with $\Delta X[\rho] < \infty$.

Theorem 3.4. *Let X be an unbiased A-compatible POM. Then for any state ρ with $\Delta X[\rho] < \infty$, we have*

$$\epsilon[X\|A,\rho]^2 = \Delta X[\rho]^2 - \Delta A[\rho]^2. \tag{3.9}$$

Proof. By Lemma 3.3, we have

$$\begin{aligned}\epsilon[X\|A,\rho]^2 &= \iint_{\mathbf{R}^2} (x-a)^2 \operatorname{Tr}[X(dx)A(da)\rho] \\ &= \operatorname{Tr}[(\tilde{X}-\tilde{A})^2 V\rho V^\dagger] \\ &= \operatorname{Tr}[\tilde{X}^2 V\rho V^\dagger] - \operatorname{Tr}[\tilde{A}^2 V\rho V^\dagger] \\ &= \Delta X[\rho]^2 - \Delta A[\rho]^2.\end{aligned}$$

QED

Remark. When X is A-compatible but $\hat{A} \neq \hat{X}$, we have

$$\epsilon[X|A,\rho]^2 = \Delta X[\rho]^2 - \Delta\hat{X}[\rho]^2 + \operatorname{Tr}[(\hat{A}-\hat{X})^2\rho],$$

where $\Delta\hat{X}[\rho]^2 = \operatorname{Tr}[\hat{X}^2\rho] - \operatorname{Tr}[\hat{X}\rho]^2$. It follows that $\epsilon[X|A,\rho]$ has a lower bound such that $\epsilon[X|A,\rho]^2 \geq \operatorname{Tr}[(\hat{A}-\hat{X})^2\rho]$.

4. Uncertainty principle for joint measurements

Consider a measuring instrument with two output variables X, Y designed to measure the values of observables A, B of a quantum system $\mathbf{S}$ described by a Hilbert space $\mathcal{H}$. Let $M : \mathcal{B}(\mathbf{R}^2) \to \mathcal{L}(\mathcal{H})$ be the joint POM of the pair $\langle X, Y\rangle$, and ρ be a state of $\mathbf{S}$ at the time of measurement. Then we have

$$\Pr[X \in \Delta_1, Y \in \Delta_2\|\rho] = \operatorname{Tr}[M(\Delta_1 \times \Delta_2)\rho],$$

for all Δ_1, $\Delta_2 \in \mathcal{B}(\mathbf{R})$. Let $\langle M_X, M_Y\rangle$ be the pair of marginal POM's of M, i.e., $M_X(\Delta) = M(\Delta \times \mathbf{R})$, $M_Y(\Delta) = M(\mathbf{R} \times \Delta)$, $(\Delta \in \mathcal{B}(\mathbf{R}))$. Then M_X and M_Y are the POM's of q.r.v.s X and Y, respectively, and hence it is natural to assume that M_X is an unbiased A-compatible POM and that M_Y is an unbiased B-compatible POM. In this case, it is known [2,11,27,1] that $\Delta X[\rho]$ and $\Delta Y[\rho]$ obeys a more stringent uncertainty relation than the Robertson-Holevo relation (3.4). A general proof of this fact is given below along with the ideas in [1].

A pair $\langle X, Y\rangle$ of POM's is called a *coexistent* pair if there is a POM $M : \mathcal{B}(\mathbf{R}^2) \to \mathcal{L}(\mathcal{H})$ such that $X(\Delta) = M(\Delta \times \mathbf{R})$, $Y(\Delta) = M(\mathbf{R} \times \Delta)$, for all $\Delta \in \mathcal{B}(\mathbf{R})$.

Theorem 4.1. *Let $\hat{A}$, $\hat{B}$ be self-adjoint operators on a Hilbert space $\mathcal{H}$. Let $\langle X, Y\rangle$ be a coexistent pair of POM's such that X is an unbiased A-compatible POM and Y is an unbiased B-compatible POM. Then, for any state ρ with $\Delta X[\rho], \Delta Y[\rho] < \infty$, we have*

(1) $\epsilon[X\|A,\rho]\epsilon[Y\|B,\rho] \geq \frac{1}{2}|\operatorname{Tr}[\,[\hat{A},\hat{B}]\rho]|$,

(2) $\Delta X[\rho]\Delta Y[\rho] \geq \left|\operatorname{Tr}[\,[\hat{A},\hat{B}]\rho]\right|$.

Proof. For simplicity, we shall prove the case where $\rho = |\psi\rangle\langle\psi|$. Let $M : \mathcal{B}(\mathbf{R}^2) \to \mathcal{L}(\mathcal{H})$ be a POM such that $M(\Delta \times \mathbf{R}) = X(\Delta)$ and $M(\mathbf{R} \times \Delta) = Y(\Delta)$. Let $\mathcal{X} = [\mathcal{K}, |\phi\rangle\langle\phi|, \hat{U}, \langle \hat{M}_1, \hat{M}_2\rangle]$ be an interacting realization of M. Set $X_1 = X$, $X_2 = Y$, $A_1 = A$, and $A_2 = B$. Define noise operators $\hat{N}_i$ $(i = 1, 2)$ by the relation

$$\hat{N}_i = \hat{U}^\dagger(1 \otimes \hat{M}_i)\hat{U} - \hat{A}_i \otimes 1.$$

Then we have

$$\begin{aligned} \langle\psi\otimes\phi|\hat{N}_i|\psi\otimes\phi\rangle &= \langle\psi\otimes\phi|\hat{U}^\dagger(1\otimes\hat{M}_i)\hat{U}|\psi\otimes\phi\rangle - \langle\psi|\hat{A}_i|\psi\rangle \\ &= \langle\psi|\hat{X}_i|\psi\rangle - \langle\psi|\hat{A}_i|\psi\rangle \\ &= 0, \end{aligned}$$

and hence

$$\begin{aligned} \Delta\hat{N}_i[\psi\otimes\phi]^2 &= \langle\psi\otimes\phi|\hat{N}_i^2|\psi\otimes\phi\rangle \\ &= \langle\psi\otimes\phi|(\hat{U}^\dagger(1\otimes M_i)\hat{U} - \hat{A}_i\otimes 1)^2|\psi\otimes\phi\rangle \\ &= \langle\psi|\hat{X}_i^2|\psi\rangle - \langle\psi|\hat{A}_i^2|\psi\rangle \\ &= \epsilon[X_i|A_i,\psi]^2. \end{aligned}$$

On the other hand, from the relations $(i, j = 1, 2)$

$$\begin{aligned} \langle\psi\otimes\phi|\hat{U}^\dagger(1\otimes M_i)\hat{U}(\hat{A}_j\otimes 1)|\psi\otimes\phi\rangle &= \langle\psi\otimes\phi|\hat{U}^\dagger(1\otimes M_i)\hat{U}|(\hat{A}_j\psi)\otimes\phi\rangle \\ &= \langle\psi|\hat{A}_i\hat{A}_j|\psi\rangle, \end{aligned}$$

we have

$$\langle\psi\otimes\phi|[\hat{N}_1,\hat{N}_2]|\psi\otimes\phi\rangle = \langle\psi|[\hat{A},\hat{B}]|\psi\rangle.$$

Thus by the Robertson uncertainty relation we have

$$\begin{aligned} \epsilon[X|A,\psi]\epsilon[Y|B,\psi] &= \Delta\hat{N}_1[\psi\otimes\phi]\Delta\hat{N}_2[\psi\otimes\phi] \\ &\geq \frac{1}{2}|\langle\psi\otimes\phi|[\hat{N}_1,\hat{N}_2]|\psi\otimes\phi\rangle| \\ &= \frac{1}{2}|\langle\psi|[\hat{A},\hat{B}]|\psi\rangle|. \end{aligned}$$

This concludes (1). From this relation, Theorem 3.4 and the Robertson uncertainty relation, we obtain

$$\begin{aligned} \Delta X[\rho]^2\Delta Y[\rho]^2 &= (\epsilon[X|A,\rho]^2 + \Delta A[\rho]^2)(\epsilon[Y|B,\rho]^2 + \Delta B[\rho]^2) \\ &\geq (\epsilon[X|A,\rho]\epsilon[Y|B,\rho] + \Delta A[\rho]\Delta B[\rho])^2 \\ &\geq |\mathrm{Tr}[[\hat{A},\hat{B}]\rho]|^2. \end{aligned}$$

This proves (2). *QED*

5. Standard quantum limit for repeated measurements

Let $\mathbf{X} : \mathcal{B}(\mathbf{R}) \to \mathrm{P}(\tau c(\mathcal{H}))$ be a CPOM and A an observable of a system $\mathbf{S}$ corresponding to $\mathcal{H}$. We define the *root-mean-square scatter* (or *resolution*, in short) $\sigma[\mathbf{X}\|A,\rho]$ of a CPOM $\mathbf{X}$ for measurement of an observable A in a state ρ by the relation

$$\sigma[\mathbf{X}\|A,\rho]^2 = \iint_{\mathbf{R}^2} (x-a)^2 \operatorname{Tr}[A(da)\mathbf{X}(dx)\rho]. \tag{5.1}$$

Let $\{\rho_x \mid x \in \mathbf{R}\}$ be a family of posterior state for a prior state ρ and $\mathbf{X}$. Then, we have

$$\sigma[\mathbf{X}\|A,\rho]^2 = \int_{\mathbf{R}} \operatorname{Tr}[\mathbf{X}(dx)\rho] \int_{\mathbf{R}} (x-a)^2 \operatorname{Tr}[A(da)\rho_x]. \tag{5.2}$$

Theorem 5.1. *Let $\hat{A}$ be a self-adjoint operator on a Hilbert space $\mathcal{H}$. Let $\mathbf{X} : \mathcal{B}(\mathbf{R}) \to \mathrm{P}(\tau c(\mathcal{H}))$ be a CPOM, ρ a density operator on $\mathcal{H}$ with $\Delta A[\mathbf{X}(\mathbf{R})\rho]$, $\Delta\hat{\mathbf{X}}[\rho] < \infty$ and $\{\rho_x \mid x \in \mathbf{R}\}$ a family of posterior states for ρ and $\mathbf{X}$. Then we have*

$$\sigma[\mathbf{X}\|A,\rho]^2 = \int_{\mathbf{R}} \Delta A[\rho_x]^2 \operatorname{Tr}[\mathbf{X}(dx)\rho] + \int_{\mathbf{R}} (\operatorname{Tr}[\hat{A}\rho_x] - x)^2 \operatorname{Tr}[\mathbf{X}(dx)\rho].$$

Proof. From $\Delta A[\mathbf{X}(\mathbf{R})\rho] < \infty$, we obtain

$$\int_{\mathbf{R}} \operatorname{Tr}[\hat{A}^2\rho_x] \operatorname{Tr}[\mathbf{X}(dx)\rho] = \operatorname{Tr}[\hat{A}^2\mathbf{X}(\mathbf{R})\rho] < \infty,$$

and hence $\operatorname{Tr}[\hat{A}^2\rho_x] < \infty$, $\operatorname{Tr}[\mathbf{X}(dx)\rho]$-almost everywhere. Thus the assertion follows from Eq. (5.2) and the relations

$$\begin{aligned}
\int_{\mathbf{R}} (a-x)^2 \operatorname{Tr}[A(da)\rho_x] &= \operatorname{Tr}[\hat{A}^2\rho_x] - 2x\operatorname{Tr}[\hat{A}\rho_x] + x^2 \\
&= \operatorname{Tr}[\hat{A}^2\rho_x] - \operatorname{Tr}[\hat{A}\rho_x]^2 + (\operatorname{Tr}[\hat{A}\rho_x] - x)^2 \\
&= \Delta A[\rho_x]^2 + (\operatorname{Tr}[\hat{A}\rho_x] - x)^2.
\end{aligned}$$

QED

Let $\mathbf{X}$ be a CPOM of a measuring instrument with one output variable X designed to make an unbiased measurement of an observable A of a system $\mathbf{S}$ corresponding to a Hilbert space $\mathcal{H}$. Suppose that the system $\mathbf{S}$ undergoes unitary evolution during the time τ between two identical measurements described by the CPOM $\mathbf{X}$. Let $\hat{U}_\tau$ be the unitary operator of the time evolution of the system $\mathbf{S}$, i.e., $\hat{U}_\tau = e^{-i\tau\hat{H}/\hbar}$, where $\hat{H}$ is the Hamiltonian of $\mathbf{S}$. Suppose that the system $\mathbf{S}$ is in a state ρ at the time of the first measurement. Then at the time just after the first measurement (say, $t = 0$) the system is in a posterior state ρ_x with the probability distribution $\Pr[X \in dx\|\rho] = \operatorname{Tr}[\mathbf{X}(dx)\rho]$. From this outcome $X = x$, the observer makes a prediction $X(\tau) = h(x)$ for the outcome of the second measurement at $t = \tau$. Then the squared uncertainty of this prediction is

$$\begin{aligned}
\Delta[\tau,\rho,x]^2 &= \int_{\mathbf{R}} (a - h(x))^2 \Pr[X \in da\|\rho_x(\tau)] \\
&= \int_{\mathbf{R}} (a - h(x))^2 \operatorname{Tr}[\mathbf{X}(da)\alpha(\tau)\rho_x].
\end{aligned} \tag{5.3}$$

As to determination of $h(x)$, the following mean-value-prediction strategy is naturally adopted:

$$h(x) = \mathrm{Tr}[\rho_x \hat{A}(\tau)], \tag{5.4}$$

where

$$\hat{A}(\tau) = \hat{U}_\tau^\dagger \hat{A}(0)\hat{U}_\tau. \tag{5.5}$$

The predictive uncertainty $\Delta[\tau,\rho]$ of this repeated measurement with the prior state ρ and the time duration τ is defined as the root-mean square of $\Delta[\tau,\rho,x]$ over all outcomes $X = x$ of the first measurement, i.e.,

$$\begin{aligned}\Delta[\tau,\rho]^2 &= \int \Delta[\tau,\rho,x]^2 \,\mathrm{Pr}[X \in dx\|\rho] \\ &= \iint_{\mathbf{R}^2} (a - h(x))^2 \,\mathrm{Tr}[\mathbf{X}(da)\alpha(\tau)\mathbf{X}(dx)\rho]. \end{aligned}\tag{5.6}$$

Theorem 5.2. *Let $\mathbf{X} : \mathcal{B}(\mathbf{R}) \to \mathrm{P}(\tau c(\mathcal{H}))$ be an unbiased A-compatible CPOM and ρ a density operator on $\mathcal{H}$ with $\Delta A(0)[\mathbf{X}(\mathbf{R})\rho]$, $\Delta A(\tau)[\mathbf{X}(\mathbf{R})\rho] < \infty$. If the relation*

$$\sigma[\mathbf{X}\|A,\rho] \le \epsilon[\hat{\mathbf{X}}\|A, \alpha(\tau)\mathbf{X}(\mathbf{R})\rho], \tag{5.7}$$

holds then we have

$$\Delta[\tau,\rho]^2 \ge \left|\mathrm{Tr}[\,[\hat{A}(0),\hat{A}(\tau)]\mathbf{X}(\mathbf{R})\rho]\right|. \tag{5.8}$$

Proof. From Eqs. (5.3)–(5.5) and Theorem 3.4,

$$\begin{aligned}\Delta[\tau,\rho,x]^2 &= \Delta X[\alpha(\tau)\rho_x]^2 \\ &= \epsilon[X\|\alpha(\tau)\rho_x]^2 + \Delta A(\tau)[\rho_x]^2,\end{aligned}$$

and hence by Theorem 5.1 and the Robertson uncertainty principle,

$$\begin{aligned}\Delta[\tau,\rho]^2 &= \int_{\mathbf{R}} \epsilon[\mathbf{X}\|\alpha(\tau)\rho_x]^2 + \Delta A(\tau)[\rho_x]^2 \,\mathrm{Tr}[\mathbf{X}(dx)\rho] \\ &= \epsilon[\hat{\mathbf{X}}\|A,\alpha(\tau)\mathbf{X}(\mathbf{R})\rho]^2 + \int_{\mathbf{R}} \Delta A(\tau)[\rho_x]^2 \,\mathrm{Tr}[\mathbf{X}(dx)\rho] \\ &\ge \sigma[\mathbf{X}\|A,\rho]^2 + \int_{\mathbf{R}} \Delta A(\tau)[\rho_x]^2 \,\mathrm{Tr}[\mathbf{X}(dx)\rho] \\ &\ge \int_{\mathbf{R}} \Delta A(0)[\rho_x]^2 + \Delta A(\tau)[\rho_x]^2 \,\mathrm{Tr}[\mathbf{X}(dx)\rho] \\ &\ge \int_{\mathbf{R}} 2\Delta A(0)[\rho_x]\Delta A(\tau)[\rho_x] \,\mathrm{Tr}[\mathbf{X}(dx)\rho] \\ &\ge \int_{\mathbf{R}} \left|\mathrm{Tr}[\,[\hat{A}(0),\hat{A}(\tau)]\rho_x]\right| \,\mathrm{Tr}[\mathbf{X}(dx)\rho] \\ &\ge \left|\mathrm{Tr}[\,[\hat{A}(0),\hat{A}(\tau)]\mathbf{X}(\mathbf{R})\rho]\right|.\end{aligned}$$

QED

The bound (5.8) is called the *standard quantum limit* (SQL) for repeated measurements with interval τ of an observable A. For the case where A is the position observable x of a free-mass m, relation (5.8) is reduced to the relation

$$\Delta[\tau,\rho]^2 \ge \frac{\hbar\tau}{m}, \tag{5.9}$$

which was posed in [4,6] and the validity of this standard quantum limit was the subject of a long controversy [28,5,14,21]. By the above theorem, any measuring instrument which beats the SQL must have the resolution lager than the precision. In [28], Yuen pointed out a flaw in the original derivation of the SQL (5.9) and proposed an idea of using contractive states to beat the SQL. A model which clears the above condition $\Omega$and beats the SQL was successfully constructed in our previous work [21,22] as a realization of Gordon-Louisell measurement $\{|\mu\nu a\omega\rangle\langle a|\}$ [9], where $|\mu\nu a\omega\rangle$ is a contractive state and $|a\rangle$ is a position eigenstate. Many linear-coupling models of position measurements which violates condition (5.7) are constructed in [23].

6. Concluding remarks

We have discussed quantum mechanical limitations on joint measurements and repeated measurements of a single object. It is shown that the uncertainty principle for joint measurements of noncommuting observables holds generally with a more stringent limit than the one usually supposed by the Robertson uncertainty relation. On the other hand, the SQL, which is also usually supposed from the Robertson uncertain relation, for repeated measurements of a single observable does not generally hold unless a certain additional condition is satisfied. The difference between these two problem is clear from the difference between those two uncertainties defined by Eq. (3.5) and Eq. (5.6) for which the optimizations are required. The crucial point is that in the latter problem we can use the result of the first measurement to predict the result of the second and hence the prediction can be based on posterior probability. However, in the problem of joint measurements, we are required to predict two outcomes only from prior probability given by the prior state. Thus we can circumvent the uncertainty principle in the problem of repeated measurements, when the measurement changes the prior state to the posterior state which has deterministic information about the *future value* of the observable to be measured. Of course, this future value must be significantly uncertain, if the prior state is of deterministic information about the *present value* and the measurement is *not* carried out. However, some measurement can give this present value precisely and further leaves the object in a state with deterministic information about the future value. Thus monitoring a mass in this way can give a precise information about *classical* force which drives the mass.

The author wishes to thank Professor Horace P. Yuen for hospitality at Northwestern University and Professor Roy J. Glauber for hospitality at Harvard University during his leave in 1988–1990. This work is supported in part by Hamamatsu Photonics K. K.

References

[1] E. Arthurs and M. S. Goodman. Quantum correlations — a generalized Heisenberg uncertainty relation. *Phys. Rev. Lett.*, **60**, 2447–2449, (1988).

[2] E. Arthurs and J. L. Kelly. On the simultaneous measurement of a pair of conjugate observables. *Bell. Syst. Tech. J.*, **44**, 725–729, (1965).

[3] L. E. Ballentine. The statistical interpretation of quantum mechanics. *Rev. Mod. Phys.*, **42**, 358–381, (1970).

[4] V. B. Braginsky and Yu. I. Vorontsov. Quantum-mechanical limitations in macroscopic experiments and contemporary experimental technique. *Uspehi Fiz. Nauk*, **114**, 41–53, (1974). [*Sov. Phys. Usp.* **17**, 644 (1975)].

[5] C. M. Caves. Defense of the standard quantum limit for free-mass position. *Phys. Rev. Lett.*, **54**, 2465–2468, (1985).

[6] C. M. Caves, K. S. Throne, R. W. P. Drever, V. D. Sandberg, and M. Zimmermann. On the measurement of a weak classical force coupled to a quantum-mechanical oscillator. I. Issues of principle. *Rev. Mod. Phys*, **52**, 341, (1980).

[7] E. B. Davies. *Quantum Theory of Open Systems.* Academic Press, London, 1976.

[8] E. B. Davies and J. T. Lewis. An operational approach to quantum probability. *Commun. Math. Phys.*, **17**, 239–260, (1970).

[9] J. P. Gordon and W. H. Louisell. Simultaneous measurement of noncommuting observables. In J. L. Kelly, B. Lax, and P. E. Tannenwald, editors, *Physics of Quantum Electronics*, pages 833–840, McGraw-Hill, New York, 1966.

[10] W. Heisenberg. The physical content of quantum kinematics and mechanics. In J. A. Wheeler and W. H. Zurek, editors, *Quantum Theory and Measurement*, pages 62–84, Princeton UP, Princeton, NJ, 1983. [translation into English by J. A. Wheeler and W. H. Zurek, 1981].

[11] A. S. Holevo. Statistical problems in quantum physics. In *Lecture Notes in Math.* **330**, pages 104–119, Springer-Verlag, Berlin, 1973.

[12] A. S. Holevo. *Probabilistic and Statistical Aspects of Quantum Theory.* North-Holland, Amsterdam, 1982.

[13] H. Martens and W. M. de Muynck. The inaccuracy principle. *Found. Phys.*, **20**, 357–380, (1990).

[14] W. -T. Ni. Quantum measurements and the standard quantum limit. *Phys. Rev. A*, **33**, 2225–2229, (1986).

[15] M. Ozawa. Conditional expectation and repeated measurement of continuous quantum observables. In *Lecture Notes in Math.* **1021**, pages 518–525, Springer-Verlag, Berlin, 1983.

[16] M. Ozawa. Quantum measuring processes of continuous observables. *J. Math. Phys.*, **25**, 79–87, (1984).

[17] M. Ozawa. Conditional probability and a posteriori states in quantum mechanics. *Publ. RIMS, Kyoto Univ.*, **21**, 279–295, (1985).

[18] M. Ozawa. Concepts of conditional expectations in quantum theory. *J. Math. Phys.*, **26**, 1948–1955, (1985).

[19] M. Ozawa. On information gain by quantum measurements of continuous observables. *J. Math. Phys.*, **27**, 759–763, (1986).

[20] M. Ozawa. Measuring processes and repeatability hypothesis. In *Lecture Notes in Math.* **1299**, pages 412–421, Springer-Verlag, Berlin, 1987.

[21] M. Ozawa. Measurement breaking the standard quantum limit for free-mass position. *Phys. Rev. Lett.*, **60**, 385–388, (1988).

[22] M. Ozawa. Realization of measurement and the standard quantum limit. In P. Tombesi and E. R. Pike, editors, *Squeezed and Nonclassical Light*, pages 263–286, Plenum, New York, 1989.

[23] M. Ozawa. Quantum mechanical models of position measurements. *Phys. Rev. A*, **41**, 1735–1737, (1990).

[24] H. P. Robertson. The uncertainty principle. *Phys. Rev.*, **34**(1), 163–164, (1929).

[25] J. von Neumann. *Mathematical Foundations of Quantum Mechanics.* Princeton UP, Princeton, NJ, 1955.

[26] E. P. Wigner. The problem of measurement. *Am. J. Phys.*, **31**, 6–15, (1963).

[27] H. P. Yuen. Generalized quantum measurements and approximate simultaneous measurements of noncommuting observables. *Phys. Lett. A*, **91**, 101–104, (1982).

[28] H. P. Yuen. Contractive states and the standard quantum limit for monitoring free-mass positions. *Phys. Rev. Lett*, **51**, 719–722, (1983).

A Quantum Experiment Uncertainty Principle: A General 2nd Moment Uncertainty Relation

E. Arthurs and M. S. Goodman
Bell Communications Research
Morristown, New Jersey 07960

Abstract

We present a new Quantum Experiment Uncertainty Relation which relates lower bounds on the fluctuations in an experimental measurement to the inherent fluctuations in the tracked observables. In particular, we extend a previous result the Generalized Heisenberg Uncertainty Relation, to obtain a more general quantum uncertainty relation which is useful for *n observables* and for *measurements at different times.* We show that all experiments that 'track' the same system operators, share a common lower uncertainty bound, independent of the technology, or the inventiveness of the experimenter. These results are useful for calculating universal lower bounds in different types of experiments, and examples are given for coherent homodyne detection and gravitational wave detection.

Lecture Notes in Physics, Vol. 378
C. Bendjaballah, O. Hirota, S. Reynaud (Eds.),
Quantum Aspects of Optical Communications

Introduction

Advances in quantum measurement techniques have required a better understanding of the fundamental limits with which quantum mechanical observables may be determined, based on the accuracy of the experiments. We consider an experiment in which an event results in a set of n experimental apparatus readings. We relate the fluctuations in the measurement (described by its covariance distribution) to the inherent fluctuations in the physical system observables. We refer to this general 2nd moment relation as a Quantum Experiment Uncertainty Relation which extends the results of our previous Generalized Heisenberg Uncertainty Relation[1], to permit any experiment involving n observables, and for experiments incorporating measurements at different times. We indicate two examples of experimental applications: coherent optical homodyne measurements, and gravity wave detection. As a consequence, it is shown that all experiments satisfying an equivalence criterion (experiments that 'track' the same physical observables) share a common lower uncertainty bound. That is, these universal lower bounds are independent of the inventiveness of the experimenter or the experimental implementation, or the technology used.

Quantum model of the experiment

We model the experiment theoretically, using the usual rules of quantum theory. In our development, $\hbar = 1$, and we indicate operators with the symbol caret above the operator symbol. A theoretical model of this experiment, considers a universe composed of two dynamic entities: a system and an apparatus. The function of the apparatus is to make n measurements of the system (not necessarily at the same time), producing the 'meter readings' $x_1, x_2, ...x_n$, for a given observation (event). The apparatus must be coupled to the system for an interval of time to perform the measurement function. We assume that the system and apparatus are prepared initially independently of one another, so that it is the deviation of the apparatus from its uncoupled behavior which constitutes the measurements. The quantum model of the experiment includes:

- a Hilbert space H_s for the system (in which the observables are represented by operators on that space);

- a Hilbert space H_m for the measuring apparatus, on which the 'meter' variables (also represented by operators on this space) are defined;

- a non-negative statistical operator of trace unity defined on the product space $\mathbf{H} = H_s \otimes H_m$, representing the initial state of the measurement $\hat{\rho}(0) = {}_s\hat{\rho}(0) \otimes {}_m\hat{\rho}(0)$ (where the prescript indicates the associated Hilbert space) and where the state of the apparatus ${}_m\hat{\rho}(0)$ must be appropriate to the calibration of the apparatus;

- the probability of observing the experimental values $(x_1, x_2, ...x_n)$, $P(x_1, x_2, ...x_n)$[1];

- The positive operator valued measure for computing the probability $P(x_1, x_2, ...x_n)$,

$$\hat{\Pi}(x_1, x_2, ...x_n), \qquad \hat{\Pi}(\cdot) \geq 0, \qquad \sum_{x_1,...,x_n} \hat{\Pi}(x_1, ...x_n) = \hat{I}$$

so that $P(x_1, ...x_n) = tr[\hat{\rho}(0)\hat{\Pi}(x_1, ...x_n)]$.

As an illustration of how $\hat{\Pi}$ is defined from the experimental arrangement, consider a time sequence of measurements. Suppose the total Hamiltonian is given by: $H = {}_s\hat{H} \otimes {}_m\hat{I} + {}_s\hat{I} \otimes {}_m\hat{H} + \hat{H}_{int}$ where $\hat{H}_{int}$ represents the interaction of the system and apparatus. The propagator $\hat{U}(t, t')$ will satisfy

$$\frac{i\ d\hat{U}(t,t')}{dt} = \hat{H}\hat{U}(t,t')$$

where $\hat{U}(t,t) = \hat{I}$. If ${}_m\hat{B}$ is an apparatus observable with spectral resolution ${}_m\hat{B} = \sum_x x_m \hat{P}_x$ and ${}_m\hat{B}$ is measured at times $(t_1, t_2, ..., t_n)$ where $0 = t_0 < t_1 < t_2 < ... < t_n$, then $\hat{\Pi}(x_1, ...x_n) = \hat{F}^+ \hat{F}$ where[2]

$$\hat{F} = ({}_s\hat{I} \otimes {}_m\hat{P}_{x_n})\hat{U}(t_n, t_{n-1})...({}_s\hat{I} \otimes {}_m\hat{P}_{x_2})\hat{U}(t_2, t_1)({}_s\hat{I} \otimes {}_m\hat{P}_{x_1})\hat{U}(t_1, t_o).$$

[1]The spectrum of the observables need not be discrete; they may be continuous or partly continuous. In such cases the formulas should be understood symbolically.

Analysis of A Quantum Experiment

The probability of observing a particular event, labelled by the readings of the apparatus (or meters), $(x_1, x_2, ..., x_n)$ is:

$$P(x_1, ...x_n) = tr[\hat{\rho}(0)\hat{\Pi}(x_1, ...x_n)]. \tag{1}$$

We define n classical random variables $X_1, ...X_n$, which take on values corresponding to the 'meter' readings $(x_1, ...x_n)$. We can characterize the spread of the measurements using the covariance matrix for the experiment,

$$C_x(j, k), \qquad j, k \,\epsilon[1...n],$$

which is given by:

$$C_x(j, k) = \sum_{x_1 ... x_n} (x_j - \overline{X}_j)(x_k - \overline{X}_k)P(x_1, ...x_n) \tag{2}$$

for $j, k \quad \epsilon[1...n]$ and where

$$\overline{X}_j = \sum_{x_1 ... x_n} x_j P(x_1 ... x_n). \tag{3}$$

The covariance matrix of the results is, of course, a real symmetric non-negative definite matrix, which we may denote by: $C_x \geq 0$. Now we can rewrite (1) as:

$$P(x_1, ...x_n) = tr_s[{}_s\hat{\rho}(0) {}_s\hat{\tau}(x_1, ...x_n)] \tag{4}$$

where:

$${}_s\hat{\tau}(x_1, ...x_n) = tr_m[({}_s\hat{I} \otimes {}_m\hat{\rho}(0))\hat{\Pi}(x_1, ...x_n)] \tag{5}$$

and

$${}_s\hat{\tau}(\cdot) \geq 0, \qquad \sum_{x_1, ..., x_n} {}_s\hat{\tau}(x_1, ...x_n) = {}_s\hat{I}.$$

Using (3) and (4), we can express $\overline{X}_j$ as:

$$\overline{X}_j = tr_s[{}_s\hat{\rho}(0) {}_s\hat{A}_j] \tag{6}$$

where

$${}_s\hat{A}_j = \sum_{x_1, ..., x_n} x_j \; {}_s\hat{\tau}(x_1 ... x_n); \quad j = 1...n. \tag{7}$$

Since ${}_s\hat{\tau}(x_1, ...x_n)$ does not depend on ${}_s\hat{\rho}(0)$, we say that the 'meter' reading X_j <u>tracks</u> the Hermitian operator ${}_s\hat{A}_j$, i.e., they both have the same average value for <u>all</u> ${}_s\hat{\rho}(0)$. *In fact, we can relate the meter readings X_j, to what is measured in the system by characterizing the entire experiment as 'tracking' the set of system operators:* $({}_s\hat{A}_1, ..., {}_s\hat{A}_n)$. If we define the quantum covariance matrix[3] of the ${}_s\hat{A}$ operators as: $QC_A(j,k)$, for $j,k \;\; \epsilon[1...n]$, then this covariance is:

$$QC_A(j,k) = tr_s[{}_s\hat{\rho}(0)({}_s\hat{A}_j - < {}_s\hat{A}_j >)({}_s\hat{A}_k - < {}_s\hat{A}_k >)] \tag{8}$$

where

$$< {}_s\hat{A}_j > = tr_s[{}_s\hat{\rho}(0){}_s\hat{A}_j]. \tag{9}$$

QC_A is an Hermitian, non-negative matrix with:

$$Re[QC_A] = \frac{1}{2} tr_s[{}_s\hat{\rho}(0)[{}_s\hat{A}_j - < {}_s\hat{A}_j >, {}_s\hat{A}_k - < {}_s\hat{A}_k >]_+] \tag{10}$$

and

$$Im[QC_A] = \frac{1}{2} tr_s[{}_s\hat{\rho}(0)[{}_s\hat{A}_j, {}_s\hat{A}_k]_-] \tag{11}$$

where Re QC_A is a non-negative real symmetric matrix. Using these definitions, and by direct calculation (see Appendix):

$$C_x \geq QC_A \tag{12}$$

which is the statement of the Quantum Experiment Uncertainty Relation. That is,

$$C_x = QC_A + QC_N \tag{13}$$

where QC_N is a Hermitian non-negative definite 'noise' matrix with

$$ImQC_N = -ImQC_A. \tag{14}$$

This Quantum Uncertainty Relation shows that any measurement which tracks $({}_s\hat{A}_1, ..., {}_s\hat{A}_n)$ *must* have at least the fluctuations inherent in the tracked system observables. That is, experiments which track the same system observables, share a common lower uncertainty bound, independent of how the measurement is performed. Of course, different experiments may exhibit *different* amounts of noise depending on experimental details, but none will have less than QC_A. Moreover (12) says that the $QC_A(j,k)$ where $j \neq k$ contains an imaginary part equal to $\frac{1}{2}$ the expected value of the commutator of the

${}_s\hat{A}_j$ and ${}_s\hat{A}_k$. Since a diagonal element of a non-negative definite matrix can vanish only if all other elements of the same row and column also vanish, we have:

$$QC_N(j,j) > 0, \qquad QC_N(k,k) > 0 \tag{15}$$

and

$$Var(X_j) = Var({}_s\hat{A}_j) + QC_N(j,j) \tag{16}$$
$$Var(X_k) = Var({}_s\hat{A}_k) + QC_N(k,k) \tag{17}$$

so there must be 'extra' noise when tracking non-commuting observables, beyond that already present in the ${}_s\hat{A}$'s.

Examples

An example of a measurement of a pair of non-commuting observables occurs in the balanced homodyne detection of an optical signal[1,4]. This type of detection is used in the measurement of squeezed states of light and in coherent communications. The observables tracked, $\hat{A}_1$ and $\hat{A}_2$ are the in-phase and quadrature amplitudes of the electromagnetic field, with:

$$[{}_s\hat{A}_1, {}_s\hat{A}_2]_- = i.$$

QC_A will then be of the form:

$$QC_A(j,k) = ReQC_A(j,k) + \frac{i(-1)^{j+1}}{2}(1-\delta_{j,k}), \quad j,k \,\epsilon[1,2]. \tag{18}$$

Since $QC_N \geq 0$,

$$QC_N(1,1)\; QC_N(2,2) \geq \frac{1}{4} \tag{19}$$

and

$$Var(X_1) = Var(\hat{A}_1) + QC_N(1,1) \tag{20}$$
$$Var(X_2) = Var(\hat{A}_2) + QC_N(2,2). \tag{21}$$

Equations (19), (20), and (21) represent an instance of the Generalized Heisenberg Uncertainty Relation[1], which is a special case of the Quantum Experiment Uncertainty Relation (equation 12).

As a second example, consider an experiment to detect gravitational radiation[5]. A gravity wave detector is usually modelled by a quantum harmonic oscillator, while the gravitational radiation is represented by a (very weak) classical force perturbing the oscillator. The Hamiltonian for the system to be measured, ${}_s\hat{H}$, is given by:

$$ {}_s\hat{H} = \frac{1}{2m}{}_s\hat{P}^2 + \frac{m\omega^2 {}_s\hat{Q}^2}{2} - {}_s\hat{Q}G(t) \tag{22} $$

where G(t) represents the force induced by a gravity wave. Let ${}_s\hat{Q}(t)$ be the response of the oscillator in the Heisenberg representation, i.e.,

$$ {}_s\dot{\hat{Q}}(t) = i[{}_s\hat{Q}, {}_s\hat{H}]. $$

From Ehrenfest's theorem, we have:

$$ m < {}_s\ddot{\hat{Q}} > + m\omega^2 < {}_s\hat{Q}(t) >= G(t). $$

Therefore a possible experiment would be to measure X(t), where X(t) continuously tracks $\hat{Q}(t)$. The Quantum Covariance of ${}_s\hat{Q}(\cdot)$, is given by the non-negative function $QC_Q(t,t')$:

$$ QC(t,t') = tr({}_s\hat{\rho}(0)({}_s\hat{Q}(t) - < {}_s\hat{Q}(t) >)({}_s\hat{Q}(t') - < {}_s\hat{Q}(t') >) \tag{23} $$

and the covariance of the $X(\cdot)$ process, $C_x(t,t')$ is given by:

$$ C_x(t,t') = [(X(t) - \bar{X}(t))(X(t') - \bar{X}(t'))] \tag{24} $$

and the Quantum Uncertainty Relation states $C_x \geq QC_A$, or equivalently

$$ C_x(t,t') = QC_A(t,t') + QC_N(t,t') \tag{25} $$

where $ImQC_N(t,t') = -ImQC_A(t,t')$, and for all finite sequences $z_1, z_2, ... z_N$ and $t_1 < t_2 < ... < t_n$ then:

$$ \sum_{j,k} \bar{z}_j QC_N(t_j, t_k) z_k \geq 0 \tag{26} $$

where $\bar{z}$ is the complex conjugate of z.

Conclusions

To conclude, we have presented a new, general, 2nd moment quantum experiment uncertainty relation which relates the measured covariance distribution, C_x, to the inherent fluctuations in the tracked observables, described by their quantum covariance QC_A; $C_x \geq QC_A$. This relation extends previous work[1] to the case of $n > 2$ observables and to measurements made at different times. This result is useful in obtaining fundamental lower bounds, and in determining when inherent 'extra' noise is required, beyond that already present in the 'tracked' observables. Finally, we have shown that experiments which 'track' the same physical observables, share a common technology independent lower uncertainty bound, which cannot be circumvented.

Appendix

In this section we prove equation (12) above. Consider the non-negative operator ${}_s\hat{D}$, where:

$${}_s\hat{D} = \sum_x \sum_{j,k} {}_s\hat{\tau}(x_1, x_2, ...x_n)\bar{z}_j(x_j\ {}_s\hat{I} - {}_s\hat{A}_j)(x_k\ {}_s\hat{I} - {}_s\hat{A}_k)z_k \tag{27}$$

and $z = (z_1, z_2, ...z_n)$ is an arbitrary complex vector. If we expand the right hand side of this equation, we obtain:

$${}_s\hat{D} = \sum_{j,k} \bar{z}_j(x_j x_k\ {}_s\hat{\tau}(x) - {}_s\hat{A}_j {}_s\hat{A}_k)z_k \tag{28}$$

or:

$$tr_s[{}_s\hat{\rho}(0) {}_s\hat{D}] = \sum_{j,k} \bar{z}_j[C_x(j,k) - QC_A(j,k)]z_k \geq 0 \tag{29}$$

and (12) is proved. QED.

References:

1. E. Arthurs and M. S. Goodman, Phys. Rev. Lett. 60, 2447, 1988.
2. E. P. Wigner, 'Interpretation of Quantum Mechanics', p 287, in *Quantum Theory and Measurement*, Ed. J. A. Wheeler and W. H. Zurek, Princeton Univ. Press, Princeton, 1983).
3. A. Golebiewski and M. Witko, Phys. Rev. A, 38, 2715, 1988.
4. E. Arthurs and J.L. Kelly, Bell System Tech. J. 44, 725, 1965.
5. *Quantum Optics, Experimental Gravity, and Measurement Theory*, Ed. P. Meystre and M. O. Scully, NATO ASI series, V. 94, New York, 1983.

Quantum mechanical description of light propagation

Izo I. Abram

Centre National d'Études des Télécommunications
196, Avenue Henri Ravera, 92220 Bagneux, France

Within the conventional approach to Quantum Optics, the problem of the propagation of a short pulse of light through a transparent linear or nonlinear medium is addressed through the technique of modal analysis [1]. That is, the pulse is expressed as a linear superposition of the modes of the cavity of quantization and thus moves in space as the relative phases of the modes evolve in time. In a nonlinear medium, the field Hamiltonian contains also an interaction term among the modes and, thus, new modes are excited as the pulse evolves in time under the nonlinear interaction term.

In principle, this modal approach should be able to account for all optical phenomena, since the modes constitute a complete basis set that can describe all possible configurations of the electromagnetic field. In practice, however, the application of the modal approach to nonlinear propagative phenomena presents many difficulties. For example, consider a single-mode Gaussian beam propagating through a medium with a nonlinear refractive index, and undergoing catastrophic self-focusing down to a point. A description of this process through the modal approach requires the use of an increasingly larger number of modes, as the beam propagates, all interacting with each other. In addition, the structure of each mode will have to depend on the degree of focusing of the beam, since the refractive index of the medium is a local function of the beam intensity. Soon, then, the problem becomes inextricable.

The reason that such problems arise in the modal analysis of propagative nonlinear optical phenomena can be traced to the fact that the modal decomposition of the pulse is equivalent to a Fourier transformation of the spatial distribution of the field. Fourier techniques are very successful in solving **linear** differential equations, since they convert these equations into algebraic equations, which are much easier to solve. However, the use of Fourier techniques to nonlinear differential equations, most often, does not simplify the solution of the original differential equations.

Lecture Notes in Physics, Vol. 378
C. Bendjaballah, O. Hirota, S. Reynaud (Eds.),
Quantum Aspects of Optical Communications

One way of circumventing the problems of the modal approach of Quantum Optics is by the development of an alternate approach, formulated in the direct–space representation. Such a direct–space version of Quantum Optics would use local quantities for the electromagnetic field, rather than delocalized modes, a view-point similar to that used in the classical theory of nonlinear optics [2] . This latter theory is very successful in describing all propagative nonlinear optical phenomena except, of course, those that involve the quantum statistics of light.

In order to develop a direct–space approach to Quantum Optics, we have to go back to the canonical quantization procedure for the electromagnetic field inside an effective nonlinear medium [3]. The Lagrangian density that gives equations of motion for the field that are identical to the macroscopic (phenomenological) Maxwell equations is,

$$\mathcal{L} = \frac{E^2 - B^2}{2} + \frac{1}{2}\chi^{(1)}E^2 + \frac{1}{3}\chi^{(2)}E^3 + \frac{1}{4}\chi^{(3)}E^4 + \dots \tag{1}$$

where E and B are the electric and magnetic fields. The medium is introduced through its linear and nonlinear optical susceptibilities $\chi^{(n)}$, as in the classical theory of nonlinear optics. In this simplified treatment, the dispersion and the tensorial properties of $\chi^{(n)}$ are neglected ; E and B are treated as scalars, and propagation is considered in the z-direction. The Lagrangian density (1) permits us to obtain the two canonical conjugate variables of the field : they are the vector potential A and the displacement field D. The displacement field is given by $D = E + P$, and thus includes the response of the medium in the form of the induced (linear and nonlinear) polarization P.

The field can be quantized, by postulating that the two canonical variables A and D are non–commuting operators whose equal–time commutator is given by

$$[D(r,t), A(r',t)] = i\,\delta(r - r') \tag{2}$$

From this point on, field is usually expanded into modes and photon (modal) creation and annihilation operators are defined as linear combinations of the modal components of A and D, such that they follow Bose commutation relations. However, these last few steps **are not necessary** in the quantization procedure. Quantization is already complete, as soon as the equal–time commutator (2) is postulated, which, it should be emphasized, **is postulated in direct–space**. Thus, starting with the direct–space commutator (2), it is possible to develop a direct–space approach to Quantum Optics, without using the modal decomposition of the field.

The question may arise, however, how can propagation be described within the Hamiltonian formulation of Quantum Optics, without the use of modes which, in the conventional theory, take care of the spatial aspect of propagation, while the Hamiltonian accounts for the temporal aspect. Within the Lagrangian formalism, this problem does not arise since the Lagrangian density addresses the spatial and temporal coordinates of the field on the same footing. Within the Hamiltonian formalism, all the temporal and spatial information for the field is contained in the electromagnetic energy–momentum tensor, which can be calculated directly from the Lagrangian density [4] . One of the components of the energy–momentum tensor is the canonical Hamiltonian

$$H = \int_V \frac{E^2 + B^2}{2} + \frac{1}{2}\chi^{(1)}E^2 + \frac{2}{3}\chi^{(2)}E^3 + \frac{3}{4}\chi^{(3)}E^4 + \ldots \tag{3}$$

that is used also in the conventional theory of Quantum Optics. As in the conventional theory, the integration in eq. (3) is performed over the volume of the (infinite) cavity of quantization under periodic boundary conditions. For any field operator Q, the Hamiltonian generates infinitesimal displacements along the temporal coordinate through the Heisenberg equation :

$$\frac{dQ}{dt} = i[H, Q] \tag{4}$$

The Heisenberg equation (4) can be solved to give the temporal variation of Q as

$$Q(t) = \exp[iH(t - t_0)]Q(t_0)\exp[-iH(t - t_0)] \tag{5}$$

in terms of the exponential time–evolution operator $\exp(iHt)$. Another component of the energy–momentum tensor that is of interest here is the canonical momentum operator

$$G = \int_V BD = \int_V BE + \chi^{(1)}BE + \chi^{(2)}BE^2 + \chi^{(3)}BE^3 + \ldots \tag{6}$$

that corresponds to the Minkowski definition for the electromagnetic momentum [5] . This canonical momentum operator permits us to generate infinitesimal displacements along the spatial coordinate of the field through a Heisenberg–like equation

$$\frac{dQ}{dz} = -i[G, Q] \tag{7}$$

which can be solved, in a way analogous to eq. (4), to give the spatial dependence of any field operator as

$$Q(z) = \exp[-iG(z - z_0)]Q(z_0)\exp[iG(z - z_0)] \tag{8}$$

in terms of the exponential spatial progression operator $\exp(-iGz)$.

In order to describe the propagation of a short pulse of light within this direct–space formulation of Quantum Optics, we have to establish a relationship between the Heisenberg equations for the Hamiltonian (4) and momentum operators (7), such that it can permit us to convert the exponential time–evolution operator (5) into a function of the exponential spatial progression operator (8). This transformation will permit us to express the time evolution of a short pulse in terms of its motion through space.

It is relatively easy to obtain the relationship between the Hamiltonian and momentum operators, simply by using their definitions (3) and (6) and the basic commutator (2). A little algebra, then, yields this relationship in the form of a set of two commutatorial equations

$$[G, E] = \frac{1}{c}\,[H, B] \tag{9a}$$

and

$$[G, B] = \frac{1}{c}\,[H, D] \tag{9b}$$

Using the Heisenberg equations (4) and (7) we can see that these commutatorial equations are nothing else but the Maxwell equations for our simplified propagation geometry,

$$\frac{dE}{dz} = -\frac{1}{c}\frac{dB}{dt} \tag{10a}$$

and

$$\frac{dB}{dz} = -\frac{1}{c}\frac{dD}{dt} \tag{10b}$$

expressed in operatorial form. Thus, the two commutatorial equations (9) contain all the information on the propagative dynamics of the field that is contained in the Maxwell equations (10) and can, therefore, be used to describe the propagation of a short pulse of light, much in the same way that the classical Maxwell equations are used in classical linear or nonlinear optics. However, in addition to this, the operatorial nature of eqs. (9) implies that these equations can treat at the same time the quantum statistics of light, since the operator algebra that they involve can keep track of the commutation and ordering of the field operators in the course of propagation.

We shall examine here the use of the commutatorial Maxwell equations (9) in the description of the propagation of a short pulse of light through a linear dielectric medium. We consider a short electromagnetic pulse which consists of the simultaneous presence of the electric and magnetic fields, such that at time t, it consists of a field distribution peaked around the point z :

$$W(z, t) = E(z, t) + v\; B(z, t) \tag{11}$$

where $v = \frac{c}{\sqrt{\epsilon}}$ is the speed of light in the dielectric medium, with $\epsilon = 1 + \chi^{(1)}$ being the dielectric function. Following eq.(5), the time evolution of the electromagnetic wave W can be written as

$$\begin{aligned} W(z, t) &= \exp(iHt) W(z, 0) \exp(-iHt) \\ &= W(z, 0) + (it)\,[H, W(z, 0)] + \frac{(it)^2}{2!}\,[H, [H, W(z, 0)]\,] + \ldots \end{aligned} \tag{12}$$

We note that for the electromagnetic wave (11), the two operatoratorial Maxwell equations (9) can be combined into a single operatorial wave equation

$$[G, W] = \frac{1}{v}\,[H, W] \tag{13}$$

which provides a simple relationship between the Hamiltonian and momentum operators that permits us to substitute the multiple commutators of the Hamiltonian in eq.(12) by multiple commutators of the momentum operator. Use of eq.(13) thus permits us to re–write eq.(12) as

$$W(z,t) = W(z,0) + (ivt)\,[G, W(z,0)] + \frac{(ivt)^2}{2!}\,[G,[G,W(z,0)]\,] + \ldots$$
$$= \exp(iGvt)W(z,0)\exp(-iGvt) = W(z-vt,0) \qquad (14)$$

Thus, eq.(14) indicates that the time–evolution of a short electromagnetic pulse in a linear medium consists of the translation of the pulse in space, at a steady speed v, as expected from classical optics. In a subsequent publication [6] we examine the application of this direct–space formulation of the theory of Quantum Optics explicitly to the case of a nonlinear medium.

In conclusion, in this paper we have presented a direct–space formulation of the theory of Quantum Optics, that is based on the canonical quantization procedure for the electromagnetic field in a effective nonlinear medium. To describe light propagation, this formalism does not use the conventional modal decomposition of the electromagnetic field, but relies on the simultaneous use of the Hamiltonian and momentum operators. These two operators contain all the information on the propagative dynamics of the field inside the effective medium, while at the same time their operatorial algebra permits us to account for the quantum statistics of the field in the course of its propagation. In this paper, only the case of the linear medium was examined. However, the formalism developed here is more general, and permits the description of nonclassical effects that may occur in the course of the propagation of a short light pulse in a transparent medium.

References

1. J. Tucker, D.F. Walls : Phys. Rev. **178**, 2036 (1969).
2. Y. R. Shen : *The Principles of Nonlinear Optics* (Wiley–Interscience, New York 1984).
3. M. Hillery, L.D. Mlodinow: Phys. Rev. **A 30**, 1860 (1984).
4. C. Itzykson, J. B. Zuber : *Quantum Field Theory* (McGraw–Hill, New York 1980).
5. R. Peierls : in *Highlights of Condensed–Matter Physics*, Ed. by F. Bassani, F. Fumi, and M.P. Tosi, (North-Holland, Amsterdam 1985) pp. 237–255.
6. I. Abram, E. Cohen : Phys. Rev. A (to be published).

QUANTUM PROPAGATION IN OPTICAL FIBRES

K. J. Blow, R. Loudon† and S. J. D. Phoenix

British Telecom Research Laboratories, Martlesham Heath, Ipswich. IP5 7RE. England.

† *Physics Department, Essex University, Colchester, CO4 3SQ. England.*

I. INTRODUCTION

Calculations in quantum optics have traditionally made use of a form of quantum electrodynamics in which the field is assumed to be confined within an optical cavity. However, many optical experiments take place in an open system geometry with no easily identifiable cavity in which the electromagnetic field should be quantized. This is the case, for example, in pulse propagation along an optical fibre where the cross section is well-defined but the total length plays little or no role in governing the natures of the excitations and their propagation characteristics. Without the notional cavity the modes of the system are continuous in both the time and frequency domain and this enables us to work directly with continuum operators rather than their discrete mode counterparts. We shall, in the present work, restrict ourselves to the physically important case where the field can be adequately described in one spatial dimension along the direction of propagation. This will be the case for propagation in an optical fibre.

Optical fibres have been very successful media for the demonstration of nonlinear propagation effects such as self-phase modulation (SPM), [1] solitons, [2] all-optical switching, [3] and squeezed light generation. [4] The generation of squeezed states used CW light where four wave mixing is the relevant interaction. However, CW light suffers from guided acoustic wave Brillouin scattering (GAWBS). [5] It is expected that if pulsed light of sufficiently short duration is used the Brillouin process can be dephased. Pulsed squeezed light generation has so far been limited to degenerate parametric downconversion [6] but SPM can also induce squeezing. [7] In this paper we study the

Lecture Notes in Physics, Vol. 378
C. Bendjaballah, O. Hirota, S. Reynaud (Eds.),
Quantum Aspects of Optical Communications

effects of SPM in optical fibres following a continuous mode approach from the outset. The equation of motion for the field can be solved exactly for an arbitrary coherent state input. We shall illustrate the theory with examples of the squeezing obtained for a single pulse in a balanced homodyne detector with a CW local oscillator.

II. CONTINUOUS-MODE FORMALISM

The discrete-mode quantized electric field operator can be readily converted to continuous-mode form by allowing the cavity length to tend to infinity. The mode spacing, $\Delta\omega$, then tends to zero, the discrete operators convert into continuum operators and the Kronecker delta becomes a Dirac delta function according to

$$\hat{a}_i \rightarrow (\Delta\omega)^{1/2}\,\hat{a}(\omega) \qquad \text{and} \qquad \delta_{ij} \rightarrow \Delta\omega\,\delta(\omega - \omega') \,. \tag{1}$$

The field operator commutation relation becomes in the continuum limit

$$[\hat{a}(\omega), \hat{a}^{\dagger}(\omega')] = \delta(\omega - \omega') \,. \tag{2}$$

Any discrete-mode summations can now be transformed into an integral in the usual way. The positive frequency part of the electric field operator is given by

$$\hat{E}^{+}(z,t) = i\int_0^{\infty} d\omega \left(\frac{\hbar\omega}{4\pi\varepsilon_0 cA}\right)^{1/2} \hat{a}(\omega)\,e^{-i\omega(t - z/c)} \,, \tag{3}$$

where A is the cross-sectional area. The magnetic field operator is defined in a similar fashion.

The integration in (3) strictly extends over positive frequencies only. However, in many cases of practical interest the mode excitation is restricted to a narrow bandwidth, B, centred on a frequency, ω_0, such that $B \ll \omega_0$. In such cases the lower limit on the integral can be extended to $-\infty$ to a very good approximation. In this narrow bandwidth approximation it is convenient to define Fourier transformed operators

$$\hat{a}(t) = \frac{1}{\sqrt{2\pi}}\int d\omega\,\hat{a}(\omega)\,e^{-i\omega t} \tag{4}$$

so that from (2) we find that the operators in the time domain have the commutator $[\hat{a}(t), \hat{a}^{\dagger}(t')] = \delta(t - t')$. The normal-order Poynting vector, $\hat{S}(z,t)$, also simplifies in this limit to give

$$\hat{S}(z,t) = \left(\frac{\hbar\omega}{A}\right) \hat{a}^{\dagger}(t - z/c)\hat{a}(t - z/c) \tag{5}$$

so that the flux of a light beam, in units of photons per unit time, is represented by the operator $\hat{f}(t) = \hat{a}^{\dagger}(t)\hat{a}(t)$.

A continuous-mode coherent state is defined by a generalization of the usual single-mode definition,

$$\begin{aligned} |\{\alpha\}\rangle &= \exp\left\{\int d\omega[\alpha(\omega)\hat{a}^{\dagger}(\omega) - \alpha^{*}(\omega)\hat{a}(\omega)]\right\}|\{0\}\rangle \\ &= \exp\left\{\int dt\,[\alpha(t)\hat{a}^{\dagger}(t) - \alpha^{*}(t)\hat{a}(t)]\right\}|\{0\}\rangle \end{aligned} \tag{6}$$

where $\alpha(t)$ is the Fourier transform of $\alpha(\omega)$ and $|\{0\}\rangle$ is the continuous-mode vacuum state. The coherent state defined in this fashion has the following properties

$$\begin{aligned} \hat{a}(\omega)\,|\{\alpha\}\rangle &= \alpha(\omega)\,|\{\alpha\}\rangle \\ \hat{a}(t)\,|\{\alpha\}\rangle &= \alpha(t)\,|\{\alpha\}\rangle\ . \end{aligned} \tag{7}$$

The mean flux of the coherent state is therefore given by $\langle\hat{f}(t)\rangle = |\alpha(t)|^2$ and this function can have any arbitrary time-dependence. Continuum number states and squeezed states can also be defined [8] but the coherent states defined above are sufficient for our present purposes.

III. NORMAL-ORDERING THEOREM

It is convenient to express the continuous-mode operators $\hat{a}(t)$ in terms of a linear superposition of operators associated with an arbitrary noncontinuous set of basis functions which are not necessarily modes of the system. Thus let $\phi_j(t)$ be a complete orthonormal set of functions on t,

$$\int dt\, \phi_i(t)\phi_j^*(t) = \delta_{ij}$$
$$\sum_j \phi_j^*(t)\phi_j(t') = \delta(t - t') \tag{8}$$

where i and j label the members of a denumerably infinite set. Each function in the set is assigned a destruction operator according to the relations

$$\hat{c}_j = \int dt\, \phi_j^*(t)\, \hat{a}(t)$$
$$\hat{a}(t) = \sum_j \phi_j(t)\hat{c}_j \tag{9}$$

with commutator $[\hat{c}_i, \hat{c}_j^\dagger] = \delta_{ij}$. The operators defined by (9) therefore represent a set of independent bosons. The above analysis can also be performed in frequency space using the Fourier transformed orthonormal functions $\phi_i(\omega)$, and the relation (4). A coherent state $|\gamma_j>$ can be defined for each noncontinuous mode. Any continuum coherent state can be expressed in terms of these states by the relations

$$|\{\alpha\}> = \prod_j |\gamma_j> = |\{\gamma_j\}>$$
$$\gamma_j = \int dt\, \alpha(t)\, \phi_j^*(t)\ . \tag{10}$$

The noncontinuous-mode states $|\gamma_j>$ have all the usual properties of Glauber coherent states with amplitudes γ_j.

Consider the operator $\exp(\hat{O})$ where $\hat{O}$ is

$$\hat{O} = \int dt\, g(t)\, \hat{a}^\dagger(t)\, \hat{a}(t) = \sum_i \sum_j g_{ij}\, \hat{c}_i^\dagger \hat{c}_j \tag{11}$$

and the g_{ij} are given by

$$g_{ij} = \int dt\, g(t)\, \phi_i^*(t)\phi_j(t)\ . \tag{12}$$

Let U be the unitary transform that converts $\underline{g}$ to a diagonal matrix $\underline{G}$ so that $\underline{\hat{d}} = U\underline{\hat{c}}$ and $|\delta_j>$ are eigenstates of the operator $\hat{d}_j$. The coherent state matrix element of $\exp(\hat{O})$ then becomes

$$\prod_j <\delta_j| \exp\left[G_{ii}\hat{d}_j^\dagger \hat{d}_j\right] |\delta_j> = \prod_j <\delta_j| :\exp\left[\left(e^{G_{jj}} - 1\right)\hat{d}_j^\dagger \hat{d}_j\right]: |\delta_j> \tag{13}$$

where the colons denote normal ordering and we have used the standard discrete-mode normal-ordering theorem. Use of the inverse unitary transformation now converts (13) to the coherent state $|\{\alpha\}>$ matrix element. The normal-ordering theorem for continuous-mode operators is thus established:

$$\exp\left[\int dt\, g(t)\hat{a}^\dagger(t)\hat{a}(t)\right] = :\exp\left[\int dt \left(e^{g(t)} - 1\right)\hat{a}^\dagger(t)\hat{a}(t)\right]: \ . \tag{14}$$

IV. SELF-PHASE MODULATION

The Heisenberg equation of motion which describes the evolution of a field in a non-dispersive $\chi^{(3)}$ medium is given by a generalization of the classical evolution equation [9] [10]

$$i\frac{\partial \hat{a}(t)}{\partial x} = \kappa \int d\tau\, g(\tau)\hat{a}^\dagger(t-\tau)\hat{a}(t-\tau)\hat{a}(t) \ . \tag{15}$$

Here we have included a response function $g(\tau)$ and written t as the time in the frame moving with the group velocity, x and κ are normalized variables given by

$$x = \frac{n_0\omega_0}{c} z$$

$$\kappa = \frac{n_2 \hbar\, \omega_0}{2n_0^2 \varepsilon_0 c A} \tag{16}$$

where n_0 is the linear refractive index of the fibre mode, z is the propagation distance, n_2 is the Kerr coefficient, ω_0 is the central frequency and A is the nonlinear effective area of the fibre mode. Note that x is simply the system length in units of wavenumbers. The response function is normalized and in silica the typical response time τ_0 for the Kerr effect is about 10^{-15} s. Thus for picosecond timescales the precise details of $g(\tau)$ are not important and we may choose a rectangular response function without loss of generality. The dimensionless quantity $\kappa x/\tau_0$ is of the order of 10^{-6} for a fibre length of 1km and a wavelength of 1.5μm. The instantaneous Kerr limit is recovered by setting $g(\tau) = \delta(\tau)$. When the response is not instantaneous the susceptibility has both a real and imaginary part which leads to a lack of conservation of the field commutator. Equation (15) is then no longer valid and noise operators must be included to preserve the commutator. However, in the limit where any input pulse has no structure on the timescale of the response these extra noise operators can be neglected. [10] Equation (15) is easily integrated and with the use of the above normal-ordering theorem we find that

$$<\hat{a}(x,t)> = \exp\left(\int_0^\infty d\tau \left[e^{-i\kappa x g(\tau)} - 1\right] |\alpha(0,t-\tau)|^2\right) \alpha(0,t) . \qquad (17)$$

It is now clear why it was necessary to include the effects of a finite medium response time in the equation of motion (15). If the response function were to be made equal to a delta function, thus recovering the instantaneous response limit, the amplitude expectation value (16) would then contain the term $e^{-i\kappa x\delta(\tau)}$ which is a non-integrable singularity. Assuming a simple step response and considering a fibre of length L we find that the exponent $\kappa x g(\tau)$ becomes

$$\kappa x g(\tau) = \frac{n_2 L \hbar \omega_0^2}{2 n_0 \varepsilon_0 c^2 A \tau_0} \equiv \frac{\omega_0}{n_0 c} n_2 L |E_0|^2 \qquad (18)$$

where $|E_0|^2 = \dfrac{\hbar\omega_0}{2\varepsilon_0 V}$ is the mean square vacuum field confined to a volume V. In our case the volume is given by $cA\tau_0$ whereas the usual value implicit in the quantized field approach is AL. The exponent $\kappa x g(\tau)$ is the classical nonlinear phase shift of the RMS vacuum field E_0 confined to a volume $cA\tau_0$ propagating in a Kerr medium. When the reponse is instantaneous this phase shift becomes infinite as the RMS field fluctuations diverge. Equivalently we can interpret this as due to the vacuum field fluctuation over a bandwidth $2\pi/\tau_0$ centred on ω_0. These singularities do not appear in approximate treatments such as linearising about a strong classical wave [11] or using a time-dependent Hartree-Fock wavefunction. [12]

The squeezing spectrum for a field as it propagates through a fibre is determined by

$$S(\omega,t_m) = \frac{1}{T}\int_{t_m - T/2}^{t_m + T/2} dt \int_{t_m - T/2}^{t_m + T/2} dt' V(t,t') e^{i\omega(t - t')} \quad ; \tag{19}$$

the two-time correlation, $V(t,t')$, of the quadrature operator $\hat{X}(t) = \hat{a}(t)e^{-i\phi} + \hat{a}^\dagger(t)e^{i\phi}$, is given by

$$V(t,t') = \langle\hat{X}(t)\hat{X}(t')\rangle - \langle\hat{X}(t)\rangle\langle\hat{X}(t')\rangle \tag{20}$$

where ϕ is the local oscillator phase, t_m is the time at which the spectrum is measured, T is the integration time of the spectrum analyzer and the normalization has been chosen so that $S(\omega,t_m) = 1$ in the shot noise limit. Equation (18) is only strictly valid with a CW local oscillator, the signal pulse is, of course, arbitrary. The two-time correlation function is 'nearly' a delta function in the sense that $V(t,t')$ is only non zero for $|t - t'| < \tau_0$. Thus for frequencies which satisfy the inequality $\omega < 2\pi/\tau_0$ the spectrum is independent of ω. In the physical limits applicable to propagation in a fibre we need only calculate the spectrum at $\omega = 0$.

In the figure below we plot the logarithm of the squeezing spectrum as a function of the local oscillator phase at a fixed length $x = 10^{10}$. Curve (a) is the spectrum for a square pulse and (b) is for a sech pulse. In both cases the measurement time t_m corresponds to the peak of the pulse and the integration time T is equal to the width of the square pulse which is chosen to be 100ps. The two input pulses are chosen so that they have the same peak power and the same total photon number. The input pulse peak power was chosen to be 1W so that the classical phase shift is about π at a distance of 1km. The square pulse shows a high degree of squeezing whereas the sech pulse shows excess noise at all values of the phase. The optimum squeezing obtained for the above conditions is about 15dB. The difference in squeezing between the square and sech pulses can be understood by noting that as the squeezing is 'instantaneous' the sech pulse acquires a noise ellipse whose angle varies along the pulse and, since the local oscillator has a constant phase, the spectrum contains some contributions with reduced noise and some with enhanced noise. The net result is that the oscillation in the evolution of the noise is washed out and the squeezing is destroyed at higher intensities. This pulse-shape mechanism is extremely well-known in all-optical switching where it leads to a degradation in the CW contrast ratio. [3]

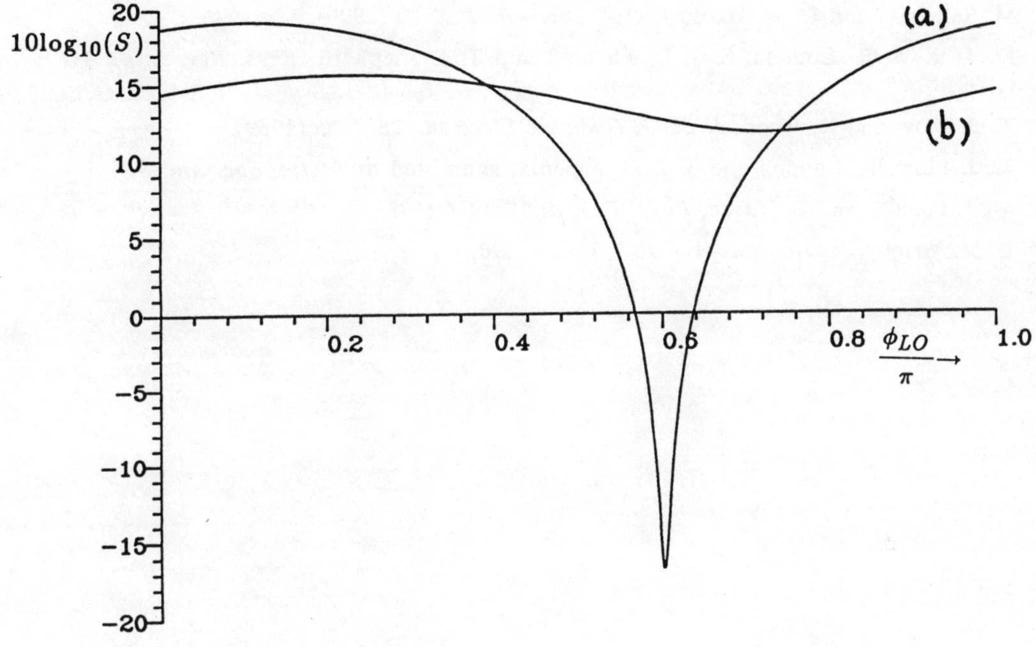

Figure 1 : Plot of the logarithm of the squeezing spectrum for (a) a square pulse and (b) a sech pulse as a function of local oscillator phase.

REFERENCES

1. P. N. Kean, K. Smith and W. Sibbett, *IEE Proceedings*, **J 134** 163 (1987).
2. L. F. Mollenauer, R. H. Stolen and J. P. Gordon, *Phys. Rev. Lett.*, **45** 1095 (1980).
3. K. J. Blow, N. J. Doran and B. K. Nayar, *Opt. Lett.*, **14** 754 (1989).
4. R. M. Shelby, M. D. Levenson, S. H. Perlmutter, R. G. DeVoe and D. F. Walls, *Phys. Rev. Lett.*, **57** 691 (1986).
5. R. M. Shelby, M. D. Levenson and P. W. Bayer, *Phys. Rev. B*, **31** 5244 (1985).
6. R. E. Slusher, P. Grangier, A. LaPorta, B. Yurke and M. J. Potasek, *Phys. Rev. Lett.*, **59** 2566 (1987).
7. M. Shirasaki and H. A. Haus, *J. Opt. Soc. Am. B*, **7** 30 (1990).
8. K. J. Blow, R. Loudon, S. J. D. Phoenix and T. J. Shepherd, *Phys. Rev. A*, **42** 4102 (1990).
9. K. J. Blow and D. Wood, *IEEE J. Quantum Electron.*, **25** 2665 (1989).
10. K. J. Blow, R. Loudon and S. J. D. Phoenix, submitted to *J. Opt. Soc. Am. B*.
11. M. J. Potasek and B. Yurke, *Phys. Rev. A*, **35** 3974 (1987).
12. E. M. Wright, *J. Opt. Soc. Am. B*, **7** 1142 (1990).

The Uncertainty Principle in a QND Measurement of Photon Number

Hans Martens & Willem M. de Muynck
Department of Theoretical Physics, Eindhoven University of Technology,
PO Box 513, 5600 MB Eindhoven, The Netherlands.

Abstract

We study a simple setup for non–destructive detection of photon number. We in particular compare the device's inaccuracy and its phase disturbance to the uncertainty principle. As the traditional uncertainty principle involves neither inaccuracy nor disturbance, a new type of uncertainty relation is employed. Self–phase modulation is taken into account, losses are neglected.

Introduction

Quantum optics makes a number of fundamental quantum measurements feasible. Heisenberg's γ–microscope(1) is such an experiment. Heisenberg intended to show how a position measurement in quantum mechanics must disturb momentum. The size of the disturbance is inversely proportional to the inaccuracy of the position meter. A non–destructive measurement of photon number via the optical Kerr–effect may be seen as an analog of Heisenberg's γ–microscope. Such a Kerr–device would then be expected to disturb phase, the variable conjugate to photon number.

The setup in its simplest form is sketched in fig. 1. A probe beam P, initially in a coherent state $|\alpha\rangle_P$, and a signal beam S (frequency ω_S) are passed through a medium

Lecture Notes in Physics, Vol. 378
C. Bendjaballah, O. Hirota, S. Reynaud (Eds.),
Quantum Aspects of Optical Communications

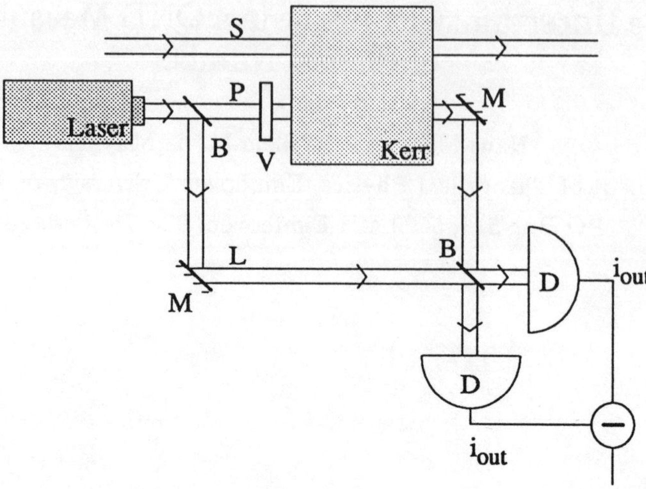

fig. 1 *Kerr QND measurement of photon number. A signal beam* S *is mixed with a probe beam* P *into a non-linear medium. The outgoing probe beam is fed into a homodyne detector. [*V *is a delay,* B*'s are beamsplitters,* M*'s are mirrors,* D*'s are detectors, and* L *indicates the homodyne local oscillator beam.]*

with a third order non–linearity. During a time $\tau = \ell/c$ the two modes interact via a Hamiltonian

$$H_I = 4\chi_{SP} N_S N_P + \chi_{PP} {N_P}^2 + \chi_{SS} {N_S}^2 \quad ; \tag{1}$$

with ($\hbar = 1$)

$$\chi_{SP} := \frac{9}{8V\epsilon^2}\, \omega_S \omega_P\, \chi^{(3)}(\omega_P;\, \omega_P, -\omega_S, \omega_S) \quad ; \quad N_S := a_S^\dagger\, a_S \quad .$$

The operator a_S is the annihilation operator for the S mode. N_P, a_P, χ_{SS} and χ_{PP} are defined analogously. After the interaction the probe beam P is detected in a homodyne detector. This may be seen[2] as a measurement of

$$Q := \tfrac{1}{2}\sqrt{2}\, a_P \exp(-i\varphi) + \text{h.a.} \quad . \tag{2}$$

The phase φ can be controlled *via* the delay V.

Photon Number

The setup of fig. 1 is intended to measure photon number "inaccurately" [3]. But in what sense? We first look at the device without self–phase modulation, i.e., we neglect the χ_{PP} and χ_{SS} terms in (1). Then, tracing over the P Hilbert space,

$$\langle Q(\tau)\rangle = \sqrt{2}\,|\alpha|\,\cos(4\tau\chi_{SP}N_S + \varphi) \quad . \tag{3}$$

More generally, the q outcome distribution can be written as

$$P(\mathrm{d}q) = \sum_n f_n(\mathrm{d}q)\, P_n \quad ; \tag{4}$$

with $f_n(\Delta q) \geq 0$ and $\int f_n(\mathrm{d}q) = 1$. Here

$$f_n(\mathrm{d}q) = \frac{\mathrm{d}q}{\sqrt{2\pi}\,\sigma} \exp\left[-\frac{[q - \sqrt{2}|\alpha|\cos(4\tau\chi_{SP}n + \varphi)]^2}{2\sigma^2}\right] \quad ; \tag{5}$$

$$P_n = {}_S\langle n|\rho_S|n\rangle_S \quad ; \quad \sigma^2 = \tfrac{1}{2} \quad .$$

The function $f_n(\mathrm{d}q)$ may be thought of as a conditional probability distribution, giving the probability of outcome q given the accurate value n. The q outcome distribution is the N_S distribution smeared by means of $f_n(\mathrm{d}q)$ [4]. This is a consequence of the form of the interaction Hamiltonian (1): H_I is a function of the operator N_S, which is conserved [5]. Thus (4) is a relation between *probability distributions*, and not only between *expectation values*. The measurement is 'inaccurate' in the sense of (4).

If we assume that S contains relatively few photons, i.e.

$$4\tau\chi_{SP}n \ll 1 \quad , \tag{6}$$

we can linearize (5), so that it becomes a convolution Gaussian. Then f is a function of $q-n$ alone: the measurement is *covariant*. Moreover, (4) can be inverted (deconvolution): the *whole* N_S distribution can be estimated using the Kerr–device outcomes.

Looking at (4), we see that the amount of inaccuracy of the device is characterized by the n–width of the likelihood-type function $f_n(\mathrm{d}q)$ for fixed q. Equivalently, we can characterize the amount of inaccuracy by the distinguishability of the distributions $f_n(\mathrm{d}q)$ and $f_{n+1}(\mathrm{d}q)$. Given the form (5) of the smearing function f, the width or distinguishability can be characterized by the noise–gain ratio:

$$\delta_N := \langle \Delta^2 Q(\tau)\rangle^{\frac{1}{2}}/G \quad ; \tag{7}$$

with

$$\langle \Delta^2 Q(\tau)\rangle := \langle (Q(\tau) - \langle Q(\tau)\rangle)^2\rangle = \tfrac{1}{2} \quad ; \tag{8}$$

$$G := \left| \frac{\partial \langle Q(\tau)\rangle}{\partial N_S} \right| = 4\sqrt{2}\,\tau\chi_{SP}\,|\alpha \sin(\varphi + 4\tau\chi_{SP}N_S)| \quad . \tag{9}$$

In the low photon number regime the inaccuracy is therefore given by

$$\delta_N \simeq \frac{1}{8\tau\chi_{SP}\,|\alpha\,\sin(\varphi)|} \quad . \tag{10}$$

The delay phase φ is to be chosen such that accuracy is optimal[6,7]. In this simplest case the optimal choice is $\varphi_{opt} = \pi/2$.

If we include self-phase modulation, (4) is still valid. The smearing function f does not have the form (5) anymore, however. Nevertheless the approximate covariance of (5) can be expected to remain intact for low non–linearities[6]. Thus the inaccuracy characterization via noise/gain ratio remains adequate. We get ($\tau\chi_{PP} \ll 1$)

$$\langle Q(\tau)\rangle \simeq \sqrt{2}\,\gamma|\alpha|\cos(4\tau\chi_{SP}N_S + \varphi') \quad ; \tag{3'}$$

$$\langle \Delta^2 Q(\tau)\rangle \simeq \tfrac{1}{2} + |\alpha|^2(1-\gamma^2) + $$

$$+ |\alpha|^2\gamma^2\left[\gamma^2\cos(8\tau\chi_{SP}N_S + 2\varphi' + \varphi_0) - \cos(8\tau\chi_{SP}N_S + 2\varphi')\right] \quad ; \tag{8'}$$

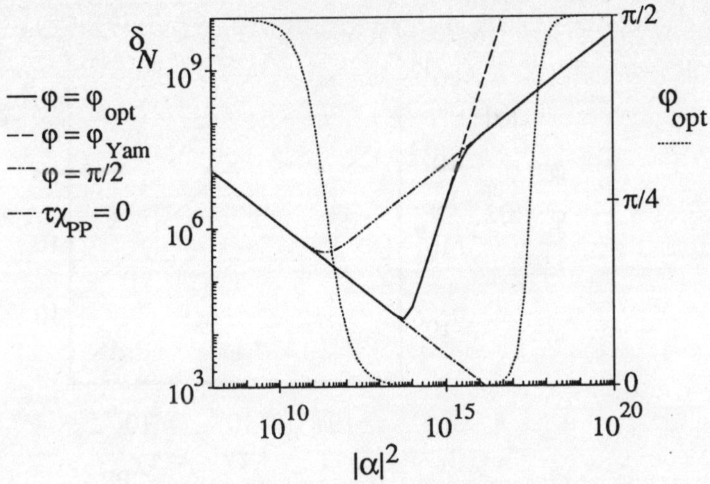

fig. 2 *N–inaccuracy [left y-axis] vs. laser power for different choices of phase φ [φ is controllable by V in fig. 1]. Note that, even for the optimal choice $\varphi_{\rm opt}$ [dotted, right y–axis] for the phase, there is a lower bound to the accuracy achievable. [*$\cot(\varphi_{\rm Yam}) = 4\tau\chi_{\rm PP}|\alpha|^2$ *(ref. 6)*; $\tau\chi_{\rm PP} = \tau\chi_{\rm SP} = 10^{-12}$*]*

with

$$\gamma = \exp(-2(|\alpha|\tau\chi_{\rm PP})^2) \quad ;$$

$$\varphi' = \varphi - \gamma\tau\chi_{\rm PP} - \gamma|\alpha|^2 \sin(2\tau\chi_{\rm PP}) \quad ;$$

$$\varphi_0 = 2\tau\chi_{\rm PP} - 8|\alpha|^2 (\tau\chi_{\rm PP})^3 \quad .$$

Assuming the low photon number regime, noise is minimal if $\varphi' \simeq \pi/4$ [ref.6, fig. 26]. Gain is maximal, however, if $\varphi' = \pi/2$. Thus optimal accuracy requires a trade-off between noise and gain. But even for optimal phase $\varphi_{\rm opt}$ (determined numerically; fig.2, dotted line), performance is limited. This deterioration of measurement quality is a consequence of the effect of self–phase modulation on the P–state. Initially coherent, it becomes extended in the phase direction ("crescent squeezing" (6)). Minimal inaccuracy $\delta_{N,\rm opt} \sim (\tau\chi_{\rm PP})^{3/5}(\tau\chi_{\rm SP})^{-1}$ is achieved for probe strength $|\alpha|_{\rm opt} \sim (\tau\chi_{\rm PP})^{-3/5}$ (fig.2, fig.3). Low inaccuracies require high non-linearities.

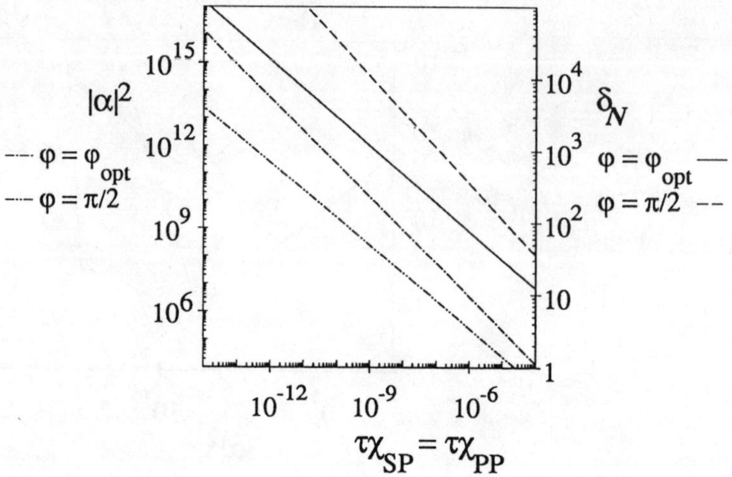

fig. 3 *N–inaccuracy [right y-axis] at optimal laser power* $|\alpha|^2$ *[left y–axis], for* $\varphi = \pi/2$ *and* $\varphi = \varphi_{\rm opt}$ *, vs. non-linearity* $\tau\chi_{\rm PP} = \tau\chi_{\rm SP}$. *Large coefficients of non-linearity are necessary for low inaccuracies.*

Phase

As stated in the introduction to the paper, we want to investigate the device's performance compared to the uncertainty principle, analogously to Heisenberg's γ–microscope. Thus we study phase in the Kerr–device. To describe phase we choose[8] the "states"

$$|\phi\rangle \;:=\; (2\pi)^{-\frac{1}{2}} \sum_{n\in\mathbb{N}} \exp(i\phi n)\,|n\rangle \quad . \tag{11}$$

They are not orthogonal, but do satisfy

$$\int_{-\pi}^{\pi} |\phi\rangle\langle\phi|\,d\phi \;=\; 1 \quad . \tag{12}$$

Eq. (12) implies that $\{|\phi\rangle\langle\phi|\,d\phi\}$ is a *positive operator–valued measure*[9] (POVM). The phase states are eigenstates of the operator

$$e := \sum_{n\in\mathbb{N}} |n\rangle\langle n+1| \quad . \tag{13}$$

We prefer this phase representation (due to Lévy–Leblond) over the more well–known Carruthers & Nieto phase description[10], because of the clumsiness of the latter. Here we have a nice Fourier relation (viz. (11)) between phase and photon number, as well as a Weyl–type commutation relation

$$e^{a} \exp(ibN) = \exp(ibN)\, e^{a} \exp(iab) \quad (a\in\mathbb{Z},\ b\in\mathbb{R}) \quad . \tag{14}$$

We look again first at the case without self–phase modulation. If we would want to know the signal's phase as it was before the interaction with the probe beam, we must try to compensate for this interaction as much as possible. This compensation cannot be complete, however. We can measure on the S beam emerging from the Kerr device $\{|\phi+\phi_0\rangle\langle\phi+\phi_0|\,d\phi\}$. This POVM becomes (Heisenberg picture)

$$|\phi+\phi_0\rangle_S\langle\phi+\phi_0|\,d\phi \longrightarrow \int_{-\pi}^{\pi} d\phi'\ |\phi'\rangle_S\langle\phi'|\ \mu(d\phi,\phi') \quad ; \tag{15}$$

with phase bias given by

$$\phi_0 = 4|\alpha|^2\tau\chi_{SP} - \tau\omega_S \quad .$$

Here, again, $\int \mu(d\phi,\phi') = 1$ and $\mu(\Delta\phi,\phi') \geq 0$ (cf.(4)). Thus measurement of $\{|\phi+\phi_0\rangle\langle\phi+\phi_0|\,d\phi\}$ on the outgoing signal mode would be interpretable as a smeared measurement of initial phase. An accurate measurement of initial phase has become impossible. Phase is *disturbed*[5]. The amount of disturbance is given by the lowest inaccuracy with which initial phase can conceivably be determined. Here that is the width of μ in (15). In the low photon number regime,

$$\mu(d\phi,\phi') \simeq \vartheta_3\,[\tfrac{\phi-\phi'}{2};\ \exp(-8|\alpha\tau\chi_{SP}|^2)]\ d\phi \quad . \tag{16}$$

ϑ_3 denotes the third of Jacobi's theta functions. Given (16)'s covariance, a suitable measure for phase disturbance is[9]

$$V_\phi := -1 + \left|\int_{-\pi}^{\pi} \mu(d\phi,0)\exp(i\phi)\right|^{-2} \quad . \tag{17}$$

Substituting (16) in (17), we get

$$V_\phi = -1 + \exp(2|\alpha|^2 [1 - \cos(4\tau\chi_{SP})]) \quad . \tag{18}$$

If we include self–phase modulation, (15) turns into

$$|\phi+\phi_0'; -2\tau\chi_{SS}\rangle_S\langle\phi+\phi_0'; -2\tau\chi_{SS}|\mathrm{d}\phi \longrightarrow \int_{-\pi}^{\pi} \mathrm{d}\phi' \; |\phi'\rangle_S\langle\phi'| \; \mu(\mathrm{d}\phi,\phi') \quad ; \tag{15'}$$

with μ as in (15), and

$$\phi_0' = \phi_0 + \tau\chi_{SS} \quad ;$$

$$|\phi;\nu\rangle := (2\pi)^{-\frac{1}{2}} \sum_{n\in\mathbb{N}} \exp(\mathrm{i}\phi n + \tfrac{1}{2}\mathrm{i}\nu n(n+1)) \, |n\rangle \quad . \tag{11'}$$

Self–phase modulation *distorts* phase into a slightly different degree of freedom. But, like the phase bias ϕ_0, this distortion can be compensated for in principle. By measuring after the Kerr–device the POVM corresponding to (11′), we obtain information precisely about phase (11). Thus phase *disturbance,* which cannot be compensated for, is unaffected by self–phase modulation.

Uncertainty principle

Intuitively, it is clear that phase disturbance and N measurement inaccuracy are complementary[1]. But the traditional uncertainty principle (either in its Heisenberg or in its Robertson form), has nothing to say about the subject. It bounds statistical *scatter* in *independent, accurate* measurements[11]. Inaccuracy is not involved. The measurement transformation, essential for the 'disturbance' concept, is not at all taken into account. As indicated above, the phase is disturbed in the sense that it has become impossible to measure accurately the pre–measurement phase. The device consisting of the non–destructive N meter followed by a measurement of initial phase, would be a joint phase–number meter. Therefore an uncertainty relation limiting the accuracy achievable in joint measurements of incompatible observables, also limits the phase-disturbance in a non–destructive photon number measurement[5]. Such uncertainty relations have become available only relatively recently, however[12,13,14]. For the

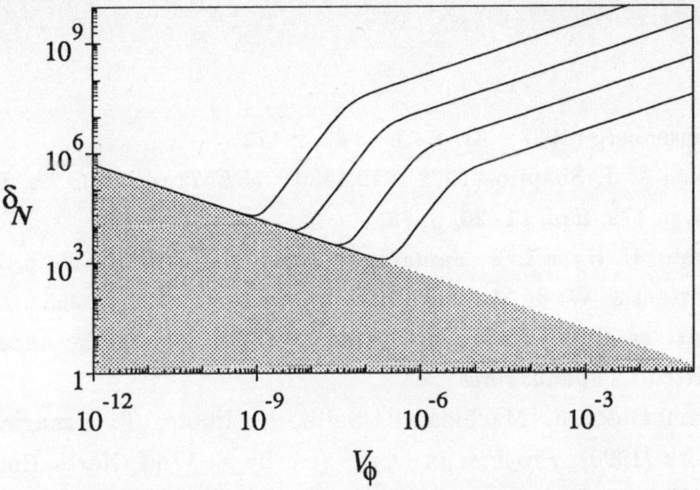

fig. 4 *Possible combinations of N–inaccuracy and phase disturbance for the Kerr-device with self-phase modulation (the lines correspond to* $\tau\chi_{PP} = \tau\chi_{SP} = 10^{-12}, 10^{-11}, 10^{-10}, 10^{-9}$). *The combinations forbidden by relation (19) are in the shaded area.*

phase-number case an *inaccuracy inequality* can be derived for covariant measurements[15] (cf. ref. 8). For the inaccuracy-disturbance complementarity we study here, it yields the theoretical limit

$$\delta_N^2 \, V_\phi \geq \tfrac{1}{4} \ . \tag{19}$$

Since the measurement here can be considered approximately covariant [see (16) and (5)], the bound (19) is suitable to compare the Kerr–device to. In the simple case that we have no self–phase modulation, the inaccuracy and disturbance can be seen to satisfy ($\tau\chi_{SP} \ll 1$)

$$\delta_N^2 \log(1 + V_\phi) = \tfrac{1}{4} \ , \tag{20}$$

which is slightly above (19). If we include self–phase modulation the performance deteriorates, as was to be expected from (8′). This is indicated in fig. 4.

References

1. W. Heisenberg (1927): *Zs. f. Phys.* **43**, p.172
2. H. Yuen & J. Shapiro (1978,1979,1980): *IEE Trans. Inf. Th.* **IT–24**, p. 657; *ibid.* **IT–25**, p. 179; *ibid.* **IT–26**, p. 78
3. N. Imoto, H. Haus & Y. Yamamoto (1985). *Phys. Rev. A* **32**, p. 2287
4. H. Martens & W. de Muynck (1990): *Found. Phys.* **20**, p. 255
5. H. Martens & W. de Muynck (1990): "Disentangling the uncertainty principle", submitted to *Found. Phys.*
6. Y. Yamamoto, S. Machida, S. Saito, N. Imoto, T. Yanagawa, M. Kitagawa, G. Björk (1990): *Progress in Optics* (ed. by E. Wolf, North Holland, Amsterdam) **28**, p. 87
7. R. Shelby, M. Levenson, S. Perlmutter, R. DeVoe & D. Walls (1986): *Phys. Rev. Lett.* **57**, p. 2409
8. J.–M. Lévy–Leblond (1976); *Ann. of Phys.* **101**, p. 319
9. A. Holevo (1982): *Probabilistic & Statistical Aspects of Quantum Theory* (North Holland, Amsterdam)
10. P. Carruthers & M. Nieto (1968): *Rev.Mod.Phys.* **40**, p. 411
11. L. Ballentine (1970): *Rev. Mod. Phys.* **42**, p. 358
12. H. Martens & W. de Muynck (1990): *Found. Phys.* **20**, p. 357
13. P. Busch (1987): *Found. Phys.* **17**, p. 905
14. S. Ali (1985): *Riv. Nuovo Cim.* **8**, p. 1
15. H. Martens & W. de Muynck (1990): "Non-destructive measurements of photon number", in preparation.

Some remarks on the phase operator in quantum optics

H. le Pas de Sécheval and C. Bendjaballah

Laboratoire des Signaux et Systèmes du CNRS
École Supérieure d'Électricité, Plateau de Moulon,
91192 Gif sur Yvette , France

1 Introduction

The problem of a phase operator, satisfying the laws of quantum mechanics, has inspired many physicists since Dirac's assumption of the existence of a Hermitian phase operator conjugate to the number operator [1]. Several studies have been devoted to this subject [2,3] and recently, a new approach [4]-[7] has renewed the interest in the phase operator.
In order to apply techniques such as heterodyning, use squeezed light or the PSK modulation in optical space communications, it is important to compare the different models and to determine which one is the most adapted to quantum optical communications.

2 The different models

Three main models are available for the description of the phase. We have compared them from a probabilistic point of vue plotting the probability density for a coherent state

$$| \alpha >= \exp(i\theta) \tag{1}$$

For optical-frequency heterodyning with the image band in the vacuum state, Shapiro and Wagner [10] established that, for given θ :

$$\begin{aligned} p_S(\phi) = \tfrac{\exp -|\alpha|^2}{2\pi} \left(1 + 2 \mid \alpha \mid \sqrt{2} \cos(\phi - \theta)\right) \times \\ Q\left(-\sqrt{2} \mid \alpha \mid \cos(\phi - \theta)\right) \exp\left(\mid \alpha \mid^2 \cos(\phi - \theta)\right) \end{aligned} \tag{2}$$

Lecture Notes in Physics, Vol. 378
C. Bendjaballah, O. Hirota, S. Reynaud (Eds.),
Quantum Aspects of Optical Communications

where Q(x) is the error function:

$$\int_x^\infty dt \, \exp\left(\frac{-t^2}{2}\right) \tag{3}$$

In their approach, Barnett and Pegg introduce the phase operator from a finite-dimensional space by considering a limiting procedure. However this procedure involves serious difficulties of convergence and must be analyzed from a mathematical point of vue.
In a finite dimensional Hilbert space, the trace of all commutators is zero. This is a real problem because of the procedure to apply in the limiting process when dealing with finite-dimensional spaces. The state-space dimension d is namely allowed to tend to infinity only after physical results, such as expectation values, are calculated. Following the limiting procedure, all commutators in the usual (i. e. infinite-dimensional) space should therefore have a zero trace, which is in contradiction with the uncertainty principle and Heisenberg inequalities whose demonstrations are exactly based on the non-nullity of the commutator trace of a couple of non commuting observables.
Indeed, Barnett's construction of the phase operator raises another theoretical problem. It is, namely, constructed from a set of well-defined phase states:

$$\mid \phi >= \lim_{d\to\infty} \frac{1}{\sqrt{d+1}} \sum_{n=0}^{d} \exp(in\phi) \mid n > \tag{4}$$

These states are "well defined" in a very particular sense [8].
Because the cosine and sine operators do not commute, it is not possible to find a complete set of eigenvectors of C and S simultaneously. "Well defined" then means, that for such states, $\Delta \cos\phi = \Delta \sin\phi = 0$. But the C and S operators are defined from the phase operator of Susskind and Glogower. Consequently, these states are dependent on another description of the phase while the purpose was to construct a different operator from that of these authors. It would have been a much more difficult but more logical mathematical problem to look for a phase operator whose eigenvectors are "well-defined" phase states for the cosine and sine operators defined from the phase operator to be found.
Nevertheless, the above-mentioned approach is very interesting because of the construction procedure of the phase operator from phase states which are known *a priori*.
From the finite-dimensional approach of D. T. Pegg and S. M. Barnett one can derive the following expression:

$$p_P(\phi) = \lim_{d\to\infty} \left| \frac{\exp(-|\alpha|^2/2)}{\sqrt{d+1}} \sum_{n=0}^{d} \frac{|\alpha|^n \exp\left(in\left(\frac{m2\pi}{d+1} - \phi\right)\right)}{\sqrt{n!}} \right|^2, (m = 0, 1, \ldots) \tag{5}$$

In radio electricity, where phase problems also arise, a simple detection model is currently used yielding the following probability density [12] :

$$p_C(\phi) = \frac{1}{2\pi} \frac{\exp\left((1 + |\alpha|) \cos \varphi\right)}{I_0(1 + |\alpha|)} \tag{6}$$

where $I_0(x)$ is the zeroth-order Bessel function.

3 Results

All the distributions have been computed and plotted (Fig. 1-Fig. 6).

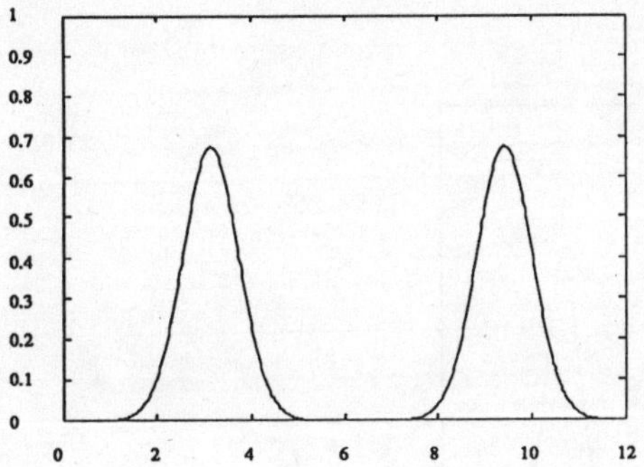

Fig. 1. $p_S(\phi)$, $|\alpha|^2 = 2$

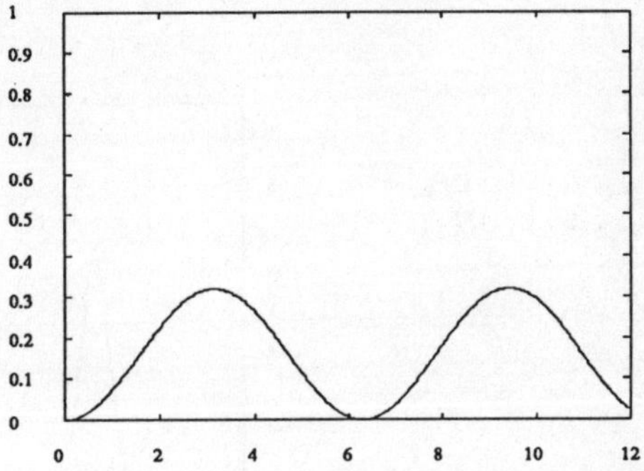

Fig. 2. $p_P(\phi)$, $|\alpha|^2 = 2$

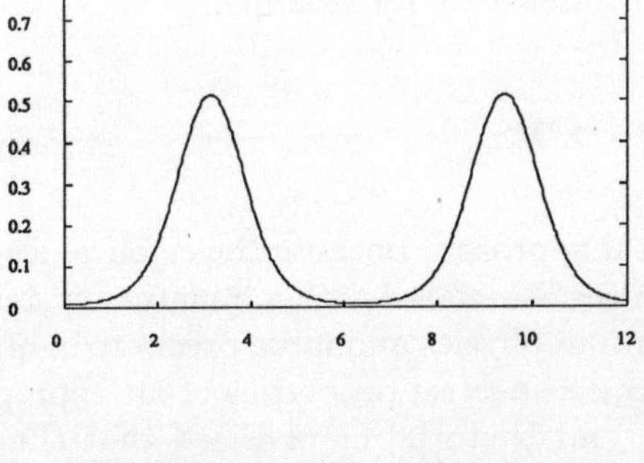

Fig. 3. $p_C(\phi)$, $|\alpha|^2 = 2$

They all tend to a Dirac distribution when the mean number of photons $|\alpha|^2$ tend to infinity but when $|\alpha|^2$ is small they are all different and there is no case for which the first two models give the latter, i. e. $p_S = p_C$ or $p_P = p_C$.

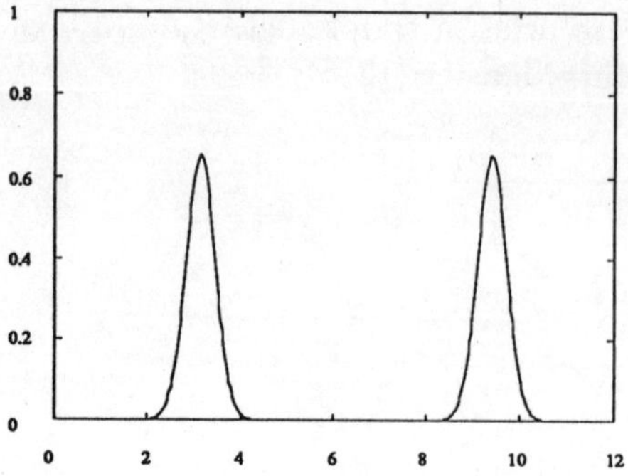

Fig. 4. $p_S(\phi)$, $|\alpha|^2 = 10$

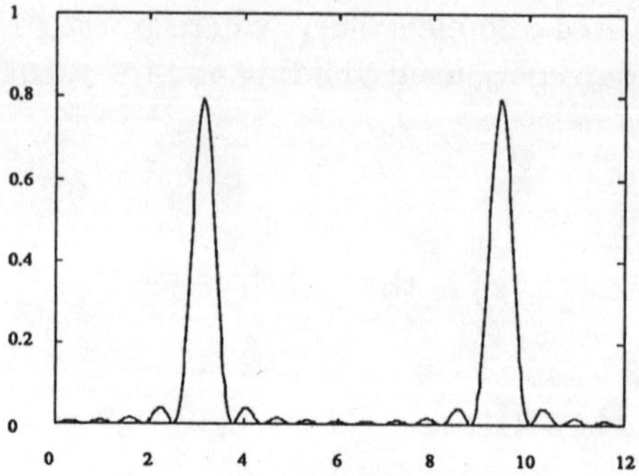

Fig. 5. $p_P(\phi)$, $|\alpha|^2 = 10$

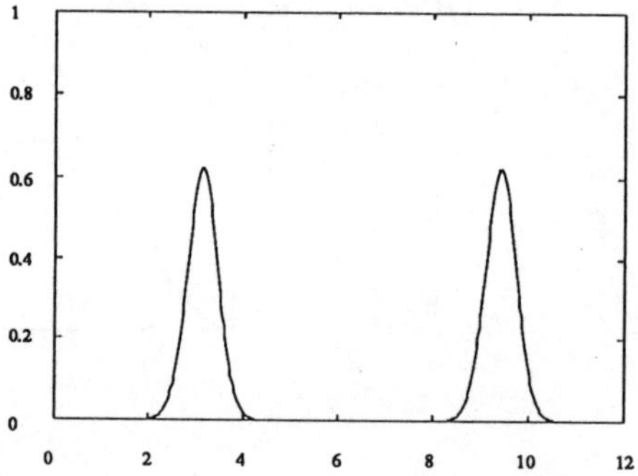

Fig. 6. $p_C(\phi)$, $|\alpha|^2 = 10$

No classical limit may be obtained from them as it is the case for the photon number:$|< \alpha|n >|^2$ is a Poisson probability density with a classical analogue (a light beam with constant intensity), for example.

4 Spectral point of vue

Our approach is a spectral approach. Because the eigenvalues represent the admissible values of the properties associated with a quantum system, a great number of problems, arising in quantum physics and more precisely in quantum measurement theory, can be related to the spectral properties of an appropriate operator.
Particularly, Carruthers and Nieto [2] emphasized that the absence of a proper phase operator results from the boundedness of the spectrum of the number operator. In the same way, Misra, Prigogine and Courbage [9] pointed out the incompatibility between the existence of an entropy operator and the semiboundedness of the Hamiltonian.

From a heuristical point of vue, it is, therefore, of interest to compare the spectra of the operators which have to represent the same property.
Let us recall some of the basic properties of the shift operator:

Definition

The shift operator is defined as follows :

$$U|n>=|n-1> \tag{7}$$

Its hermitian conjugate verifies:

$$U^{\dagger}|n>=|n+1> \tag{8}$$

Spectral properties

Theorem 1 $\forall\lambda, |\lambda|<1$, *$\lambda$ is an eigenvalue of U associated with the eigenvector:*

$$|\lambda>=\sqrt{1-|\lambda|^2}\sum_{n=0}^{\infty}\lambda^n|n>=\Lambda\sum_{n=0}^{\infty}\lambda^n|n> \tag{9}$$

Proof.

$$U|\lambda>=\Lambda\sum_{n=1}^{\infty}\lambda^n|n-1>=\lambda\Lambda\sum_{n=1}^{\infty}\lambda^{n-1}|n-1>=\lambda|\lambda>$$

Corollary $Sp(U)=D(0,1)$

Theorem 2 $\forall\lambda, |\lambda|=1$, *$\lambda$ is an eigenvalue of U associated with the eigenvector of infinite norm:*

$$|\lambda>=\sum_{n=0}^{\infty}\lambda^n|n> \tag{10}$$

The spectrum of the downwards shift operator is then the closed unity disc of the complex plane $\mathbb{C}$. All the eigenvalues on the unit circle are associated with unbounded eigenvectors. This remark is of importance since the notion of the angle between two vectors is only defined for vectors of finite norm. Therefore it is no longer possible to consider an orthonormal basis of eigenvectors when dealing with unbounded eigenvectors; Furthermore, it can be shown that the normalization of such vectors is not unique.

By construction, in the finite-dimensional case, the eigenvalues of Barnett's hermitian phase operator are evenly spread over the unit circle of $\mathbb{C}$. So, as the dimension of the space tends to infinity, the spectrum of the phase operator tends to be dense in the unit circle of $\mathbb{C}$.
Thus, the spectrum of both operators are very different.
Nevertheless, a "well-defined" phase may be interpreted as a particular normalization of the unbounded vector $E_\phi = \sum_{n=0}^{\infty} \exp(in\phi)|n>$. Another normalization corresponds to the coherent state with a mean number of photons equal to 1,

$$|\exp(i\phi)>= \exp(-1/2)\sum_{n=0}^{\infty}\frac{\exp(in\phi)}{n!}|n> \tag{11}$$

But the inconvenient of this normalization is that it gives a weighting role to the photon number.
It is then possible to use the construction process to define other phase operators based on different normalizations.
In Barnett's procedure, the "well-defined" phase states involve a limiting procedure on the states $|n>$, that is, on the energy or more precisely on the state space dimension. The infinite-dimensional Hilbert space of quantum optics is then considered as the limit of a pre-Hilbert space of finite dimension when its dimension tends to infinity.

5 The new normalization

The eigenvalues of an operator being the admissible values of the associated properties, the idea is to define the phase states in relation to their physical properties. These states are then normed and represented by a complex number on the unity circle. The normalization is realized via a limiting procedure on the eigenstates of the shift operator. The interest of this procedure is that it does not involve the dimensionality of the state space. It is a limiting procedure inside the usual Hilbert space.
Based upon the fact that physical states may only be represented by vectors of finite norm we now define new vectors from the coherent phase states introduced by Shapiro [13] by means of the following limiting procedure:

$$|\exp(i\phi)>= \lim_{\lambda\to 1, \lambda<1}\sqrt{1-|\lambda|^2}\sum_{n=0}^{\infty}\lambda^n \exp(in\phi)|n> \tag{12}$$

Many of the properties of these phase states can be derived via a limiting procedure from the calculations of Shapiro [13]. In particular:

$$<\exp(i\phi)|N|\exp(i\phi)>= \lim_{\lambda\to 1, \lambda<1}\frac{|\lambda|^2}{1-|\lambda|^2} = \infty \tag{13}$$

$$| < \exp(i\phi)|n > |^2 = 0 \tag{14}$$

$$< \exp(i\phi)|N^2|\exp(i\phi) >= \infty \tag{15}$$

$$< \Delta N^2 >= \infty \tag{16}$$

$$| < E_\theta|\exp(i\phi) > |^2 = \delta(\theta - \phi) \tag{17}$$

These new phase states also have the following properties:

Theorem *The* $|\exp(i\phi) >$ *are simultaneous eigenvalues of* U *and* $U^\dagger$ *and consequently of* C *and* S.

Proof.

$$\begin{aligned} U|\exp(i\phi) >= & \lim_{\lambda\to 1, \lambda<1} \sqrt{1-|\lambda|^2} \sum_{n=0}^{\infty} \lambda^n \exp(in\phi) U|n > \\ = & \lim_{\lambda\to 1, \lambda<1} \sqrt{1-|\lambda|^2} \sum_{n=0}^{\infty} \lambda^n \exp(in\phi)|n-1 > \\ = & \lim_{\lambda\to 1, \lambda<1} \sqrt{1-|\lambda|^2} \lambda \exp(i\phi) \sum_{n=0}^{\infty} \lambda^n \exp(in\phi)|n > \\ = & \exp(i\phi)|\exp(i\phi) > \end{aligned} \tag{18}$$

$$\begin{aligned} U^\dagger|\exp(i\phi) >= & \lim_{\lambda\to 1, \lambda<1} \sqrt{1-|\lambda|^2} \sum_{n=0}^{\infty} \lambda^n \exp(in\phi) U^\dagger|n > \\ = & \lim_{\lambda\to 1, \lambda<1} \sqrt{1-|\lambda|^2} \sum_{n=0}^{\infty} \lambda^n \exp(in\phi)|n+1 > \\ = & \lim_{\lambda\to 1, \lambda<1} \frac{\sqrt{1-|\lambda|^2}}{\lambda} \exp(-i\phi) \left(\sum_{n=0}^{\infty} \lambda^n \exp(in\phi)|n > -|0 > \right) \\ = & \exp(-i\phi)|\exp(i\phi) > \end{aligned} \tag{19}$$

C and S being defined by $C = \frac{U+U^\dagger}{2}$ and $S = \frac{U-U^\dagger}{2i}$, the proof of the second part of the theorem follows immediately.

Corollary *Consequently all classical moments and trigonometrical formulas are conserved and in particular the states* $|\exp(i\theta) >$ *do verify* $\Delta C = 0$ *and* $\Delta S = 0$ *and are "well-defined" states in the sense of Loudon.*

Theorem *These phase states build an orthonormal set of the usual Hilbert space of quantum optics and*

$$J = \int_0^{2\pi} d\phi |\exp(i\phi) >< \exp(i\phi)| \tag{20}$$

is a trace operator.

We then obtain from the phase states another expression of the shift operator:

$$U = \int_0^{2\pi} d\phi \exp(i\phi) |\exp(i\phi) >< \exp(i\phi)| \tag{21}$$

$$U^\dagger = \int_0^{2\pi} d\phi \exp(-i\phi) |\exp(i\phi) >< \exp(i\phi)| \tag{22}$$

6 Conclusions

We have shown that, from a probabilistic point of vue, the phase operators of Shapiro on one hand and Pegg and Barnett on the other hand are not equivalent. Following the construction procedure suggested by these authors, we have built new phase states and proposed another expression of the phase operator.

References

1. P. A. M. Dirac : Proc. Roy. Soc. **A 114**, 243 (1927)
2. P. Carruthers, M. M. Nieto : Rev. of Mod. Phys., Vol **40**, n 2, 411 (1968)
3. J. M. Lévy-Leblond : Ann. of Phys. **101**, 319 (1976)
4. S. M. Barnett, D. T. Pegg : J. Phys. A: Math. Gen., **19**, 3849-3862 (1986)
5. S. M. Barnett, D. T. Pegg : J. of Mod. Opt., Vol. **36**, n 1, 7-19 (1989)
6. D. T. Pegg, S. M. Barnett : Phys. Rev. A, Vol. **39**, n 4, 1665 (1989)
7. J. A. Vaccaro, D. T. Pegg : Opt. Comm., Vol. **70**, n 6, 529 (1989)
8. R. Loudon : *The quantum theory of light* (Clarendon Press, Oxford 1973) p. 143.
9. B. Misra, I. Prigogine, M. Courbage : Proc. of the Nat. Acad. of Sci. of the United States of America, **76** 4768-4772 (1979)
10. J. H. Shapiro, S. S. Wagner : IEEE Journ. of Quant. Electr., Vol. **QE-20**, n 7 (1984)
11. H. le Pas de Sécheval, C. Bendjaballah : *in preparation*
12. H. L. van Trees : *Detection, Estimation and Modulation Theory* (Jonn Wiley & Sons , New York 1968) Part I, p. 338.
13. J. H. Shapiro, S. R. Shepard, N. C. Wong: in *Coherence and Quantum Optics VI*, ed. by J. H. Eberly et al., Plenum Press, New York 1990, pp. 1177-1081.

Quantum State Control and Nonlinear Schrődinger Equation

Yoshitomi MORIKAWA*, Yoshié Otake TAKAHARA** and Osamu HIROTA***

*Institute for Nuclear Study, University of Tokyo, Tanashi, Tokyo 188, Japan[1]
**Dept. of Electronic and Computer Engineering
Ibaraki National College of Technology, Katsuta, Ibaraki 312, Japan
***Faculty of Engineering, Tamagawa University, Machida, Tokyo 194, Japan

1. Introduction

In the communication theory, the most important aim is always as how to maximize the S/N ratio. Recently the realization of the squeezed state of light reveals us the possibility to break the standard quantum limit of the noise which is given by the theory of the coherent states. Nowadays, the analysis and discussions on the squeezed states from the quantum mechanical point of view becomes more important in the communication theory. On the other hand, the transformation of the noise through the communication channel such as the transmitter, transfer process and receiver, also plays important roles. We simplify these processes to those where an input light is transformed through some transformation to an output port (see Figure 1) and we introduce a new approach to the transformation process.

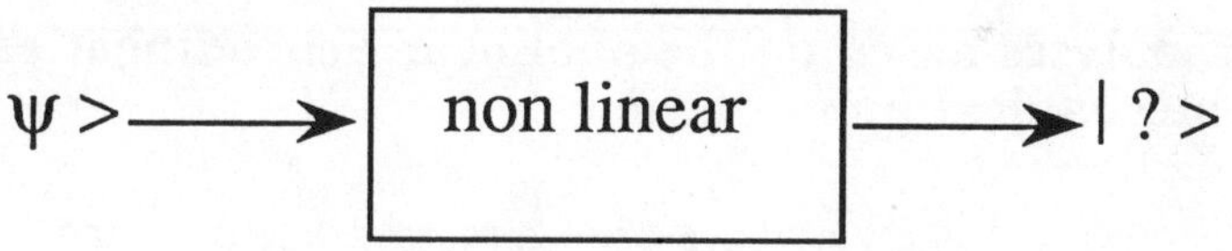

Fig.1. The simplified communication channel

Our ultimate aim is to control the quantum states such as squeezed states including their noise which can be obtained as an output state through some transformation process and to obtain the new quantum states what we want. Since usual processes of the transformation are unitary transformation, we cannot increase the S/N ratio. On the contrary, though the non-unitary transformation has not yet realized experimentally, if such a transformation can be achieved, it will be epoch-making transformation which increases the S/N ratio [1]. The total system is obeyed by unitary evolution, we do not see the total system but some parts of it. The parts can be obeyed by non-unitary evolution, though the total system are obeyed unitary evolution. Many theories and analyses have been discussed based on the nonlinear

[1] Present address : Research Institute for Polymers and Textiles, 1-1-4 Higashi, Tsukuba, Ibaraki 305, Japan

Lecture Notes in Physics, Vol. 378
C. Bendjaballah, O. Hirota, S. Reynaud (Eds.),
Quantum Aspects of Optical Communications

Schrődinger equation in various fields [2,6], not only from the classical point of view but also from the quantum mechanical point of view. Some types of the nonlinear Schrődinger equations are widely used on many fields in classical theory: optical fiber, plasma, chaos, cosmology and so on.

In this paper, we consider the quantum state, we then introduce a nonlinear term into the Schrődinger equation so we introduce the nonlinear Schrődinger equation into the transformation process and we also analyze the non-unitary transformation. There are many theories and analyses based on the nonlinear Schrődinger equation from the quantum mechanical point of view. We pick up the simple type of equation in $1 + 1$ dimension in order to investigate the effects caused by the nonlinearity :

$$i\frac{\partial u}{\partial t} + \frac{\partial^2 u}{\partial x^2} + |u|^2 u = 0 \tag{1}$$

and treat this equation from both the quantum mechanical point of view and the quantum field theoretical point of view. In the section 2, we pick up some theories and analyses which have been discussed based on the nonlinear Schrődinger equation from the quantum mechanical point of view and briefly review them. We also develop the classical treatment of the above type of nonlinear Schrődinger equation to quantum version. We take into account not only the absolute value, but also the phase of the wave function u. In the section 3, we briefly review the analysis based on the nonlinear Schrődinger equation from the quantum field theoretical point of view [7]. We must investigate the effect of noise before we take into account the effect of nonlinearity. We have a method to investigate such an interaction generally, which was applied to the case of neutron interferometry previously [8].

In the section 4, we briefly survey this work. Finally, we outline the plan to treat not only the effect of the noise but also the effect of the nonlinearity.

2. Some analyses based on the nonlinear Schrődinger equation from the quantum mechanical point of view

Many theories and analyses have been discussed based on the nonlinear Schrődinger equation in various fields [2,6], not only from the classical point of view but also from the quantum mechanical point of view. We pick up some theories and analyses based on the nonlinear Schrődinger equation from the quantum mechanical point of view and briefly review them. We consider the quantum mechanics for which there is a non-unitary evolution, for instance, Schrődinger equation with non-linear terms. This possibility was pointed out by many authors: Bialynicki-Birula and Mycielski, Kamesberger and Zeilinger, and S.Weinberg. Many experiments, including simulations, are also carried out by this standpoint. From now on, the results reveal that the strength of the nonlinearity is very weak, even if it exists [5]. In the section 2.1, we briefly review the theory of the nonlinear Schrődinger equation by Weinberg and the experiment [4,5]. In the section 2.2, we briefly review the numerical analysis by Kamesberger and Zeilinger which is based on the nonlinear Schrődinger equation proposed by Bialynicki-Birula and Mycielski [6]. We also develop the classical treatment of one type of nonlinear Schrődinger equation to quantum version in section 2.3 and consider the role of the phase of the wave function.

2.1. Theory by Weinberg

Weinberg has recently proposed the following equation from the standpoint testing the linearity of quantum mechanics, which is the general nonlinear Schrődinger equation keeping the homogeneity condition and without keeping the superposition condition [4].

$$i\frac{d\psi_k}{dt} = \frac{\partial h(\psi, \psi^*)}{\partial \psi_k^*} \tag{2}$$

In this system, if ψ_1 and ψ_2 are solutions, then $\lambda\psi_1$ is also solution for complex number λ, but $\psi_1 + \psi_2$ is not. This equation reduces to the ordinary linear Schrődinger equation when the real function $h(\psi, \psi^*)$ has a bilinear form $h = \psi_k^* H_{kl} \psi_l$. Some groups carried out the tests of the linearity of quantum mechanics experimentally by means of Ramsey's resonance methods or hydrogen maser [5]. The results reveal us a very small upper bound for the strength of the nonlinearity.

2.2. Simulation by Kamesberger and Zeilinger

Kamesberger and Zeilinger investigated the effect of nonlinearity on Fresnel pattern by numerical simulation [6]. They took the nonlinear Schrődinger equation followed by Bialynicki-Birula and Mycielski [...].

$$i\hbar\frac{\partial}{\partial t}\psi(\vec{r}, t) = \left[-\frac{\hbar^2}{2m}\nabla^2 + V(\vec{r}) - b\ln(|\,\psi\,|^2)\right]\psi(\vec{r}, t) \tag{3}$$

This equation does not keep the homogeneity condition, either. Upper limit for the strength of the nonlinearity b was investigated by experiments by Lamb shift, neutron interferometry and cold neutron Fresnel diffraction pattern and the upper limit for b is very small. They studied numerically how large the strength b is if the effects of nonlinearity appear on the Fresnel diffraction pattern. For simplicity they put the absorbing potential along to x-axis

$$V(x, y, z) = V(y, z) \tag{4}$$

$$\psi(\vec{r}, t) = \chi(x)\phi(y, z, t) \tag{5}$$

$$i\hbar\frac{\partial}{\partial t}\phi(y, z, t) = \left[-\frac{\hbar^2}{2m}\left(\frac{\partial^2}{\partial y^2} + \frac{\partial^2}{\partial z^2}\right) + V(y, z) - b\ln(|\,\phi\,|^2)\right]\phi(y, z, t) \tag{6}$$

2.3. Phase, probability conservation

In this section, we pick up a solvable nonlinear Schrődinger equation as a simple example in order to investigate the effects of nonlinearity on the transformation process

$$i\frac{\partial u}{\partial t} + \frac{\partial^2 u}{\partial x^2} + |u|^2 u = 0 \tag{7}$$

This equation does not keep the homogeneity condition, either. This type of equation has soliton-like solution and is widely used phenomenologically in classical theory, for instance, to describe the envelop curve of wave packet in the transfer of the light in optical fiber though the roles of time t and coordinate x are interchanged [2]. Only the absolute value has been considered, because the value describes an envelope curve in the optical fiber case. However we consider this type of equation as the equation of motion of wave function from quantum mechanical point of view in unit of $\hbar = 1$. Namely we take into account not only the absolute value, but also the phase. In the classical theory, the competition between second term and third term causes the stable soliton solution. On the other hand in quantum mechanics, we can control the quantum state u by adjusting the κ and ω. Hereafter, only for simplicity, we take $\omega = 2$ in equation (7).

In $\kappa > 0$ case, the stationary solution exists, which is described as

$$\phi(\xi,\tau) = \sqrt{\rho(\xi)} \exp i\sigma(\xi,\tau) \tag{8}$$

Then, the current conservation law is broken in this system

$$\frac{\partial \rho}{\partial t} + \mathrm{div}\vec{j} = i\kappa|\phi|^2(\phi - \phi^*) \tag{9}$$

$$= -2\kappa\rho^{\frac{3}{2}} \sin\sigma \tag{10}$$

where

$$\rho \equiv \phi^*\phi, \tag{11}$$

$$\vec{j} \equiv \phi^* \frac{\partial \phi}{\partial \xi} - \phi \frac{\partial \phi^*}{\partial \xi} \tag{12}$$

Here, ρ and $\vec{j}$ are defined by standard manner and we assumed σ is real in deriving the equation (10) from the equation (9). This phase σ plays very important roles. The phase σ depends on time τ and the phase is oscillating with τ. This phase is important concerning to the frequency modulation. As expected, the current conservation law recovers as the strength of nonlinearity $\kappa \to 0$. Physical meaning of the breaking of this current conservation law should be considered.

3. Analysis based on the nonlinear Schrődinger equation from the quantum field theoretical point of view

We briefly review the analysis on the nonlinear Schrődinger equation from the quantum field theoretical point of view by Sogo and Wadati [7]. For the sake of simplicity, we consider 1 + 1 dimensional nonlinear Schrődinger equation

$$i\frac{\partial \phi}{\partial t} + \frac{\partial^2 \phi}{\partial x^2} - 2\kappa|\phi|^2\phi = 0 \tag{13}$$

This equation has the same form denoted in section 2.3. However we treat this equation as not quantum, but field theoretical one : we consider ϕ as boson field following Sogo and Wadati. We postulate the same time commutation relation of boson fields

$$[\phi(x,t),\phi^\dagger(y,t)] = \delta(x-y) \tag{14}$$

$$[\phi(x,t),\phi(y,t)] = [\phi^\dagger(x,t),\phi^\dagger(y,t)] = 0 \tag{15}$$

The equation (13) is derived from Heisenberg equation $i\phi_t = [\phi, H]$ with the Hamiltonian

$$H = \int dx(\phi_x^\dagger \phi_x + \kappa\phi^\dagger\phi^\dagger\phi\phi). \tag{16}$$

The solution under the boundary condition $\phi \to 0 \ \ as \ |\xi| \to 0$ is solved as following by quantum inverse scattering method. Let's consider the following scattering problem

$$\begin{cases} \psi_{1x} + i\lambda/2\ \psi_1 = i\alpha\phi^\dagger\psi_2 \\ \psi_{2x} - i\lambda/2\ \psi_2 = -i\epsilon\alpha\psi_1\phi \end{cases}$$

$$\begin{aligned} \psi(x,\lambda) &= \begin{pmatrix}1\\0\end{pmatrix}\exp(-i\lambda x/2) & x \to -\infty \\ &= \begin{pmatrix}A(\lambda)\exp(-i\lambda x/2)\\ B(\lambda)\exp(i\lambda x/2)\end{pmatrix} & x \to \infty \end{aligned} \tag{17}$$

where $\psi(x,\lambda)$ is Jost function and $A(\lambda)$, $B(\lambda)$ are scattering data and the solution $\phi(x)$ is written, using $A(\lambda)$, $B(\lambda)$, in the form

$$\phi(x) = \sum_{n=0}^{\infty} \phi^{(n)}(x) \tag{18}$$

with

$$\phi^{(n)}(x) = (-c)^n \int \left(\prod_1^n \frac{dp_i}{2\pi}\right)\left(\prod_0^n \frac{dk_i}{2\pi}\right) \frac{\exp\left[i\left(\sum_0^n k_i - \sum_1^n p_i\right)x\right]}{\prod_{i=1}^n\left[(p_i - k_{i-1} - i\epsilon)(p_i - k_i - i\epsilon)\right]}$$

$$\times R^*(p_1)\cdots R^*(p_n)R(k_n)\cdots R(k_0)$$

where R is defined by scattering data $A(\lambda)$, $B(\lambda)$

$$R^\dagger(k) = -\frac{i\epsilon}{\tilde{\alpha}}B^\dagger\left(A^\dagger\right)^{-1} \tag{19}$$

The commutation relations between the scattering data and Hamiltonian are

$$[H, A(\lambda)] = 0, \qquad [H, B(\lambda)] = -\lambda^2 B(\lambda) \tag{20}$$

The commutation relations between these scattering data are

$$[A(\lambda), A(\mu)] = \left[A(\lambda), A^\dagger(\mu)\right] = [B(\lambda), B(\mu)] = 0,$$

$$A(\lambda)B^\dagger(\mu) = \left(1 + \frac{i\kappa}{\lambda - \mu}\right)B^\dagger(\mu)A(\lambda),$$

$$A^\dagger(\lambda)B^\dagger(\mu) = \left(1 - \frac{i\kappa}{\lambda-\mu}\right) B^\dagger(\mu)A^\dagger(\lambda),$$

$$B(\lambda)B^\dagger(\mu) = \left(1 + \frac{\kappa^2}{(\lambda-\mu)^2}\right) B^\dagger(\mu)B(\lambda)$$

$$+2\pi\alpha^2 A(\lambda)A^\dagger(\mu)\delta(\lambda-\mu) \tag{21}$$

Namely, $A(\lambda)$ is a conserved quantity. We define the vacuum state $|0>$ so that $\phi(x)|0>= 0$, then we can construct the eigenstate of H

$$|k_1, k_2, \cdots k_n> \equiv B^\dagger(k_1)B^\dagger(k_2)\cdots B^\dagger(k_n)|0> \tag{22}$$

$$H|k_1, k_2, \cdots k_n> = (k_1^2 + k_2^2 + \cdots + k_n^2)|k_1, k_2, \cdots k_n> \tag{23}$$

In the attractive case $\kappa < 0$,

$$\lambda_j = \lambda + i\frac{\kappa}{2}(n+1-2j) \qquad j = 1, 2, \cdots, n \tag{24}$$

the energy has the characteristic form of n^3

$$E_n = \sum_{j=1}^{n} \left\{\lambda + i\frac{\kappa}{2}(n+1-2j)\right\}^2$$

$$= n\lambda^2 - \frac{\kappa^2}{12}(n^3 - n) \tag{25}$$

This eigenstate is described by ϕ the boson field

$$|k_1, k_2, \cdots k_n> = i\epsilon\tilde{\alpha}^n \int \prod_{j=1}^{n} dx_j \exp(ik_j x_j)$$

$$\times \left[\prod_{1\le i<j\le n} \left(1 - \frac{i\kappa}{k_i - k_j}\epsilon(x_i - x_j)\right)\right]$$

$$\times \phi^\dagger(x_1)\phi^\dagger(x_2)\cdots\phi^\dagger(x_n)|0> \tag{26}$$

We would like to apply this established analysis to the transformation process from the field theoretical point of view. For this purpose, we have to take into account the noise. We have carried out the analysis though it is a linear case from the quantum mechanical point of view. In the next secion, we briefly survey the method.

4. Analysis of the effects of noise

For the analysis of the transfer of the signal (or noise), we must precisely take into account the interaction between the wave packet of the input light and some media on the transformation process. The wave packet of the light has the vaccum fluctuation and this fluctuation corresponds to the noise in quantum optical communication theory. At first, we have to investigate such an interaction generally, before we take into account the effect of the nonlinearity. The coherence pattern can be measured when, in the experiment of the interferometry, each of the wave function after divided into two paths, interacts with the different macroscopic systems. We have analyzed the effects of the fluctuation of the wave length of the incident neutron $\delta\lambda$ on the damping of the coherence pattern in the neutron interferometry [8]. Let's survey the method briefly. In the interferometer a neutron interacts with the macroscopic system like a device, and the effect of the interaction emerges as the function of the wave number $F(k)$ in the phase of the state of the wave packet. This formulation is a sort of generalized forms of the Glauber approximation or the optical approximation [9]. The wave function after passing through the device is described as follows :

$$\Psi_{k_0}^{out}(x,t) = \frac{1}{2(\pi^3(\delta k)^2)^{1/4}} \int dk \exp[-\frac{(k-k_0)^2}{2(\delta k)^2} + iF(k) + i(kx - \omega t)] \tag{27}$$

However, the forms of $F(k)$ differ from each other according to their interactions. The generalized form of $F(k)$ is described by the polynomial function of k, and the order of k corresponds to the magnitude of the fluctuation of the wave number of incident wave packet. When this fluctuation is not large, it is sufficient to consider up to the second order of k

$$F(k') = \alpha + \frac{\beta}{k_0} k' + \frac{\gamma}{{k_0}^2} k'^2 \tag{28}$$

Here, $k = k_0 + k'$, $-\delta k \lesssim k' \lesssim \delta k$.

The observed intensity becomes

$$I = 1 + (1+b^2)^{-1/4} \exp[-\frac{1}{4}(\frac{\delta k}{k_0})^2 \beta^2 \frac{1}{1+b^2}]$$
$$\cos[\chi - \alpha + \theta/2 - \frac{1}{4}(\frac{\delta k}{k_0})^2 \beta^2 \frac{b}{1+b^2}] \tag{29}$$

where, $b \equiv -(\delta k/k_0)^2 \gamma$ and $\theta = \tan^{-1} b$.

The coefficient of the second order of k is included in the damping term. The damping term which appeared because of the path difference in passing the device emerges in the equation (29). The longitudinal coherence length of neutron can be measured in terms of this damping caused by the path difference. In the case of thermal neutron [10], since it is impossible to vary its path length of Si interferometer, the measurement of the longitudinal coherence length should be made by means of varying the path difference β/k_0 arose from inserting a device, Al crystal in most cases. As it was mentioned, the damping effects which are caused by the fluctuations of inserted crystal emerge by this manner of measuring. We have already investigated these effects of the fluctuation, such as the macroscopic characteristics of the devices, on the damping terms theoretically. For simplicity, we take the potential series as a model of the device and use the Glauber approximation. $F(k)$ is described as follows

$$F(k) = -\frac{mVNa}{\hbar^2 k} \tag{30}$$

We expand $F(k)$, using the following coefficients

$$\alpha = -\beta = \gamma = -\frac{mVNa}{\hbar^2 k_0} \equiv -gV \tag{31}$$

Then, $F(k)$ is described in terms of k' as follows

$$F(k') = -gV\{1 - \frac{k'}{k_0} + (\frac{k'}{k_0})^2\} \tag{32}$$

This potential series contain some fluctuations such as the strength of the potentials δV, the interval of the potentials δa , the number of the potentials δN. For simplicity, we consider that fluctuations of potential strength exist only among the fluctuations above and obey a Gaussian distribution with the central value V_0 and the deviation δV

$$\mathcal{V}(V_0, \delta V) = \frac{1}{\sqrt{2\pi(\delta V)^2}} \int dV \exp[-\frac{(V - V_0)^2}{2(\delta V)^2}] \tag{33}$$

Then the intensity of the superposed state with the compensation of the path difference is as follows

$$\begin{aligned} I = \frac{1}{2} + \frac{1}{2(2d-1)^{1/2}} \exp[-(gV_0)^2(\frac{\delta V}{V_0})^2 + \frac{4c}{2d-1}] \\ +[d^2 + b^2]^{-1/4} \exp[-\frac{(gV_0)^2}{2}(\frac{\delta V}{V_0})^2 + \frac{dc}{d^2+b^2}] \\ \cos(\chi + gV_0 + \frac{\theta}{2} + \frac{bc}{d^2+b^2}) \end{aligned} \tag{34}$$

where

$b \equiv gV_0(\delta k/k_0)^2, \qquad c \equiv \frac{1}{4}(gV_0)^4(\delta V/V_0)^4(\delta k/k_0)^2,$

$d \equiv 1 + \frac{3}{2}(gV_0)^2(\delta V/V_0)^2(\delta k/k_0)^2, \qquad$ and $\theta \equiv \tan^{-1}(b/d)$

The resultant damping effect which originates from the fluctuation as the macroscopic characteristics becomes an important factor for the study of the wave packet reduction [11].

Our next study is the analysis of the effects of both noise and nonlinearity. We would like to apply our method, which was briefly surveyed in section 4, to the analysis of the transformation process based on the nonlinear Schrődinger equation and investigate the effects of nonlinearity on the transformation process.

Acknowledgments

We deeply thank Dr. K. Sogo for fruitful discussions. Two of us (Y.M.) and (Y.O.T.) would like to acknowledge the financial supports by Masamichi Igarashi Foundation and the Telecommunications Advancement Foundation (TAF) respectively.

References

[1] O. Hirota: The Trans. IEICE, **E72**, 5, 460 (1989).

[2] A. Hasegawa : in *Optical Soliton in Fiber* (in japanese), Y. Otsuki, Ed., Kyoritsu Publishers,1987, p. 69

[3] B. M. Herbst and M. J. Ablowitz: Phys. Rev. Lett., **62**, 2065 (1989). preprint, March 1989.

[4] S. Weinberg: Phys. Rev. Lett., **62**, 485 (1989); S. Weinberg: Ann. Phys. (N. Y.), **194**, 336 (1989).

[5] J. J. Bollinger, D. J. Heinzen, W. M. Itano, S. L. Gilbert, and D. J. Wineland: Phys. Rev. Lett., **63**, 1031 (1989); T. E. Chupp and R. J. Hoare: Phys. Rev. Lett., **64**, 2261 (1990); R. L. Walsworth, I. F. Silvera, E. M. Mattison and R. F. C. Vessot: Phys. Rev. Lett., **64**, 2599 (1990).

[6] A. G. Klein: Physica B **151**, 44 (1988); J. Kamesberger and A. Zeilinger: Physica B **151**, 193 (1988).

[7] H. B. Thacker: Preprint FermiLab-Conf-81/47-THY (1981); K. Sogo and M. Wadati: Prog. Theor. Phys., **68**, 85 (1982); K. Sogo and M. Wadati: Prog. Theor. Phys., **69**, 431 (1983); M. Wadati: Proceedings of KSI'84 (Springer, 1985).

[8] Y. Morikawa and Y. Otake: Il Nuovo Cimento **105** B 507 (1990).

[9] R. J. Glauber: in *Lectures in Theoretical Physics*, W. E. Brittin and L. G. Dunham Eds. (Interscience Publishers, Inc., New York, 1959), p.135.

[10] H. Kaiser, S. A. Werner and E. A. George: Phys. Rev. Lett., **51**, 1106 (1983).

[11] M. Namiki, Y. Otake and H. Soshi: Prog. Theor. Phys., **77**, 508 (1987).

Session II

– Quantum Measurements and Information –

Chairman : I. Ojima

QUANTUM ENTROPY AND QUANTUM MEASUREMENTS

Göran Lindblad

Department of Theoretical Physics
Royal Institute of Technology
S-100 44 Stockholm, Sweden

The object of this talk is to review some aspects of the theory of quantum measurements and the transmission of information between quantum systems. The model used involves the interaction of finite quantum systems, one observed system and one which represents the measuring apparatus. We refer to them by indices 1 and 2, respectively. The Hilbert space of the combined system is thus $K = K_1 \otimes K_2$. We make no allowance for the fact that a real measuring apparatus is macroscopic in some sense. The dynamics is given by unitary or isometric operators on K, the states by density operators ρ on K, and we denote the partial states by e.g. $\rho_1 = \mathrm{Tr}_2\rho$. If an outside observer can look at system 2, how much can he say about the quantum state of 1? We will look at this problem from a rather restricted point of view, where the main tool is provided by a collection of inequalities for the entropy and relative entropy functionals. On the other hand the idea of a measurement will be given a very wide interpretation which involves general processes of information transfer.

Entropy inequalities

We introduce some standard notations. Let $\{p_k\}$ be a probability distribution and ρ,μ density operators on K, then

$$H\{p_k\} = - \sum_k p_k \ln p_k$$

$$S(\rho) = - \mathrm{Tr}(\rho \ln \rho) \geq 0$$

$$S(\rho|\mu) = \mathrm{Tr}(\mu \ln \mu - \mu \ln \rho) \geq 0$$

The last two quantities are the entropy and the relative entropy, respectively. They are both invariant under isometric transformations of K. Furthermore $S(\rho) = 0$ if and only if (iff) ρ is pure (a vector state), $S(\rho|\mu) = 0$ iff $\rho = \mu$. We now give a very short list of the most important properties of these functionals. We have for any set of states $\{\rho_k\}$,

$$\sum_k p_k S(\rho_k) \leq S(\sum_k p_k \rho_k) \leq \sum_k p_k S(\rho_k) + H\{p_k\}. \qquad (1)$$

where the first equality holds iff all $\rho_k = \rho$, the second equality holds iff all the ρ_k are orthogonal. For any state of the system 1+2 we can define the logarithmic correlation

Lecture Notes in Physics, Vol. 378
C. Bendjaballah, O. Hirota, S. Reynaud (Eds.),
Quantum Aspects of Optical Communications

$$C_{12}(\rho) = S(\rho_1 \otimes \rho_2 \mid \rho) = S(\rho_1) + S(\rho_2) - S(\rho) \quad (2)$$

which vanishes iff ρ is of tensor product form. The other extreme case is that ρ is pure, in which case ρ_1 and ρ_2 are isometric (they have the same spectrum with multiplicities, except, possibly, for the multiplicity of the zero eigenvalue) and consequently they have the same entropy. Conversely, if ρ is any state on K we can construct a pure state on $K \otimes K$ such that the partial states are ρ and a state isometric to it. Let $\{|k)\}$ be any complete orthonormal eigenbasis for ρ and introduce the notation $E_{kl} = |k)(l|$. Then ρ is of the form $\rho = \sum_k p_k E_{kk}$ and we can choose the pure state to be

$$\omega = \sum_{k,l} (p_k\, p_l)^{\frac{1}{2}}\, E_{kl} \otimes E_{kl} \quad (3)$$

but this choice is far from unique. Using this we can prove the triangle inequalities

$$|S(\rho_1) - S(\rho_2)| \leq S(\rho) \leq S(\rho_1) + S(\rho_2) \quad (4)$$

plus permutations of ρ , ρ_1 and ρ_2 . We just do the construction (3) in a space $K = K_1 \otimes K_2 \otimes K_3$ and use (2) with index 2 replace by 3, noting that $S(\rho_3) = S(\rho_{12})$, $S(\rho_{13}) = S(\rho_2)$. The classical relation $S(\rho) \geq \max_i S(\rho_i)$ no longer holds.

Let T be a linear map of B(K) (the space of bounded linear operators on K) into itself which satisfies a continuity condition, which we leave out, and

$$T(X)^+T(X) \leq T(X^+X) \text{ , all } X \in B(K) \quad \text{(strong positivity)} \quad (5)$$

$$T(1) = 1$$

Then T maps the set of density operators on K into itself through $\rho \to \rho \cdot T$. It then holds that the relative entropy is non-increasing under T : for all ρ , μ

$$S(\rho \cdot T \mid \mu \cdot T) \leq S(\rho | \mu) \quad (6)$$

This means that T makes the states less easy to distinguish from each other, it represents an irreversible loss of information. Note that the change in the entropy may be of either sign. There is a subclass of maps which do not decrease the entropy. In Hilbert spaces of finite dimension these are the maps which leave the tracial state (the state of maximal entropy) invariant. We note some examples of maps satisfying (5).

1) The conditional expectation associated with an ideal (von Neumann) quantum measurement. Let $\{P_\alpha\}$ be orthogonal projections, $\sum_\alpha P_\alpha = 1$, and define

$$T(X) = \sum_\alpha P_\alpha X P_\alpha \quad (7)$$

In this case $T \cdot T = T$, $T(XT(Y)) = T(X)T(Y)$ and T does not decrease the entropy:

$$S(\rho \cdot T \mid \rho) = S(\rho \cdot T) - S(\rho) \geq 0$$

The entropy increase vanishes iff $\rho \cdot T = \rho$, which means that the state is a mixture (incoherent superposition) of states in the subspaces $P_\alpha K$. The measurement is complete or maximal if the P_α are 1-dimensional projections.

2) The map which describes the irreversible evolution of each initial state to a fixed final state μ : $\rho \cdot T = \mu$, all ρ , or $T(X) = \mu(X)\, 1$.

3) The map which gives the partial trace: $\rho \cdot T = \rho_1$ or $T(X) = X \otimes 1$, all $X \in B(K_1)$. This maps $B(K_1)$ into $B(K)$, but it is still of the type (5).

For proofs of the properties of the entropy and relative entropy functionals see the review by Wehrl[1] and for later developments (especially for infinite systems) the work of Kosaki[2] and the review by Araki[3].

CP dynamical maps

The examples of maps given above have an even stronger positivity property than (5), they are completely positive (CP). This means that they satisfy: For all $\{X_i \in B(K)\}_1^n$ and all n, $T(X_i^+ X_j) \geq 0$ as a matrix with B(K)-valued matrix elements, i.e. as an operator in $M_n(B(K))$. With suitable continuity conditions (for dim $K = \infty$) such a map, if it also preserves the unit, must be of the form

$$T(X) = \textstyle\sum_\alpha V_\alpha^+ X V_\alpha \qquad (8)$$

$$V_\alpha \in B(K), \qquad \textstyle\sum V_\alpha^+ V_\alpha = 1,$$

However, this decomposition is not unique in general. There is an obvious generalization to maps between different spaces of operators. The space of such unit-preserving CP maps is closed under the following operations.

1) Convex combination: $\{T_i\} \rightarrow \sum_i p_i T_i$

2) Composition: $(T_1,T_2) \rightarrow T_1 \cdot T_2$

3) Tensor product: $(T_1,T_2) \rightarrow T_1 \otimes T_2$

4) Partial trace: If T is CP on $K_1 \otimes K_2$, μ a density operator on K_2, then the following map is CP in K_1

$$T_1(X) = Tr_2[\mu\, T(X \otimes 1)] \qquad (9)$$

The last point means that if we have a unitary evolution in a closed system and if we can define a subdynamics on an (open) subsystem, then this subdynamics is given by CP maps. Essentially the only case where such a subdynamics exists in the quantum case is where the initial state of the whole system is of a tensor product form. Introductions to the subject of CP maps can be found in refs. 4-6.

The CP maps have a useful representation constructed as follows. Let T be CP in K and ρ a faithful state in K (all eigenvalues strictly positive). Construct ω according to the recipe (3). Then the state

$$\tau(T,\rho) = \omega\cdot(T \otimes id) \tag{10}$$

is an affine function on the convex set of CP maps. Conversely, τ defines T uniquely

$$T(X) = \sum_{i,j} (p_i\, p_j)^{-\frac{1}{2}}\, \tau(X \otimes E_{ij})\, E_{ij}$$

Here τ must satisfy $\tau_2 = \rho$ when $T(1) = 1$, i.e. $\tau(1 \otimes E_{ij}) = \delta_{ij}\, p_i$. We see that τ is pure iff T is an isometry. If we extend the affine bijection $T \leftrightarrow \tau$ to CP maps which do not preserve the unit, then there is a decomposition of τ into a convex combination of pure states which corresponds to the decomposition in (8)

$$\tau(T,\rho) = \sum_\alpha p_\alpha\, \tau_\alpha \tag{11}$$

$$p_\alpha\, \tau_\alpha = (V_\alpha \otimes 1)\, \omega\, (V_\alpha \otimes 1)^+, \qquad p_\alpha = \rho(V_\alpha{}^+ V_\alpha)$$

For a given choice of ω we can make the decomposition (11) unique, and hence that of (8) as well, by choosing the pure states to be orthogonal.

We now introduce an entropy associated with the pair (T,ρ) (see ref. 7). It is defined as the entropy of the state (10)

$$S(T,\rho) = S(\tau(T,\rho)) \tag{12}$$

A priori this definition could depend on the choice (3), but it does not as can be seen as follows. It is easily checked that the expression

$$(V_\alpha \otimes 1)\, \omega\, (V_\beta \otimes 1)^+$$

defines a pure state on the space of operators $M(B(K \otimes K))$, where $n = \mathrm{card}\{\alpha\}$. By tracing over $\alpha = \beta$ we obtain the partial state (11), while the complementary partial state is given by the matrix

$$\sigma_{\alpha\beta} = \rho(V_\alpha{}^+ V_\beta) \tag{13}$$

We then know that (11) and (13) have the same entropy. (11) is independent of the arbitrariness in the decomposition (8), while (13) is independent of the arbitrariness in the construction (3). The common value (12) of the entropy is thus uniquely defined by T and ρ. From (1) and (11) follows that

$$S(T,\rho) \leq H\{p_\alpha\} \tag{14}$$

with equality iff the decomposition (11) is orthogonal. We propose to use the entropy (12) as a measure of the irreversibility inherent in T, and now turn to some inequalities satisfied by this functional. If ρ'_α is defined by

$$p_\alpha\, \rho'_\alpha = V_\alpha\, \rho\, V_\alpha{}^+ \tag{15}$$

then

$$\sum_\alpha p_\alpha S(\rho'_\alpha) \leq S(\rho\cdot T) \leq \sum_\alpha p_\alpha S(\rho'_\alpha) + S(T,\rho) \tag{16}$$

and the same relation holds with $S(\rho\cdot T)$ replaced by $S(\rho)$. The proof follows from

$$\sum_\alpha p_\alpha S(\tau \mid \tau_\alpha) = S(\tau) = S(T,\rho)$$

and (6) applied to the two partial traces if we note that there is for each α an isometry between (15) and the operator

$$p_\alpha \rho_\alpha = \rho^{\frac{1}{2}} V_\alpha^+ V_\alpha \rho^{\frac{1}{2}} \tag{17}$$

giving $S(\rho'_\alpha) = S(\rho_\alpha)$. There are also the triangle inequalities

$$| S(\rho) - S(\rho\cdot T)| \leq S(T,\rho) \leq S(\rho) + S(\rho\cdot T) \tag{18}$$

(plus permutations) which are derived from (4) by noting that $V_\alpha \rho V_\beta^+$ is a density matrix isometric to ρ, hence of the same entropy and with partial states (13) and $\rho\cdot T$.

In the case of a von Neumann measurement (7) the entropy is easily calculated to be $S(T,\rho) = H\{p_\alpha\}$, which is the conventionally accepted measure for the average information gain in such an ideal measurement. We will now turn to the role of the entropy functionals for more general measurement processes.

The measurement process

It is one of the fundamental restrictions in quantum theory that there is no linear process which can be used to make a copy of a general quantum state. In classical probability theory the copying of information has no theoretical limits, but the nearest quantum counterpart to a copying process is the construction (3) which is non-linear and cannot result from a linear dynamical map. A single quantum measurement can only capture part of the information in a general state, e.g. that associated with a commuting set of observables. We will deal with measurements in a very general sense and in general there is no associated observable. These processes include e.g. photon counting, but we will stay with finite quantum systems for simplicity and avoid field theory.

We take a general unit-preserving CP map acting in K_1 and show that there is an isometric dilation to a space $K_1 \otimes K_2$ such that T is a subdynamics defined according to the scheme (9). Furthermore the outcome (α) can be read off from the final state of the measuring instrument (system 2), and the partial state of system 1 after such an observation is ρ'_α defined by (15). K_2 is spanned by a complete orthonormal basis $\{|e_\alpha)\}$ and we again use $E_{\alpha\beta} = |e_\alpha)(e_\beta|$. There is also a particular initial state of the measuring instrument called $|e_M)$. Define an isometry in K by

$$W = \sum_\alpha V_\alpha \otimes E_{\alpha M}$$

It is then found that

$$W^+(X \otimes 1)\, W = \sum_\alpha V_\alpha{}^+ X\, V_\alpha \otimes E_{MM}$$

The partial trace over K_2 gives $T(X)$. This highly simplified structure is sufficient for a discussion of the information transfer in the measurement process. If the initial state of the composite system (1+2) is of the product form $\rho = \rho_1 \otimes E_{MM}$ then the final state is

$$\rho' = W \rho W^+ = \sum_{\alpha\,\beta}\ V_\alpha \rho_1 V_\beta{}^+ \otimes E_{\alpha\beta}$$

Projecting out the state of system 2 which corresponds to the outcome (α) we find

$$(1 \otimes E_{\alpha\alpha})\, \rho' \,(1 \otimes E_{\alpha\alpha}) = p_\alpha\, \rho'_{1\alpha} \otimes E_{\alpha\alpha} \tag{19}$$

which shows the effect of an outside observer reading the outcome on the instrument. The effect of the interaction is to create a correlation between the subsystems in the final state. The final partial states are $\rho'_1 = \rho_1 \cdot T$ and

$$\rho'_2 = \sum_{\alpha,\beta} \rho_1(V_\beta{}^+ V_\alpha)\, E_{\alpha\beta} = \sigma = (13)$$

From $S(\rho') = S(\rho_1)$ we find the correlation (2) defined by quantities relating to subsystem 1:

$$C_{12}(\rho') = S(T,\rho_1) + S(\rho_1 \cdot T) - S(\rho_1) \tag{20}$$

In order to investigate the interpretation of this expression we consider the standard case (7). If the initial state is an incoherent mixture ($\rho_1 = \rho_1 \cdot T$) then the correlation is equal to the information gain

$$C_{12} = S(T,\rho_1) = H\{p_\alpha\}.$$

If ρ_1 is pure then $S(\rho_1 \cdot T) = H\{p_\alpha\}$ and

$$C_{12} = 2\, H\{p_\alpha\} \tag{21}$$

There is then an additional correlation between the subsystems which does not carry information about the observable defined by the projectors $\{P_\alpha\}$, but which reflects the existence of the quantum coherence in the initial state. This is an example of EPR correlations in a pure quantum state.

We can define an 'observable' correlation I_{12} in any state ρ as the maximal mutual information (in the classical sense) between the outcomes when we perform any two von Neumann measurements on systems 1 and 2, respectively. It is clearly enough to consider only maximal measurements and we find that

$$I_{12}(\rho) = \sup_E \{C_{12}(\rho \cdot E)\} \leq C_{12}(\rho)$$

$$E(X) = \sum_{k,l} (P_k \otimes Q_l)\, X\, (P_k \otimes Q_l)$$

where 'sup' is over all choices of complete orthogonal 1D projections $\{P_k, Q_l\}$ in K_1 and K_2 , respectively, and where the inequality comes from (6). It is possible to show that we have equality iff ρ and $\rho_1 \otimes \rho_2$ have a common spectral resolution given by such a set of projections. This can be deduced from the results of Petz [8,9]. On the other hand it can be shown that for pure states we have the situation of (21):

$$I_{12}(\rho) = \tfrac{1}{2} C_{12}(\rho) \quad \text{if } \rho \text{ is pure.} \tag{22}$$

The remaining half of C_{12} can then be taken as a measure of the EPR correlations. It seems natural to conjecture that for general states

$$I_{12}(\rho) \geq \tfrac{1}{2} C_{12}(\rho) \quad (?)$$

We now return to the measurement process. Some of the EPR correlation in the final state ρ' can be removed by the projection of (19), i.e. we choose for system 2 the observable defined by $\{E_{\alpha\alpha}\}$. The final states of 1+2 and 2 are then

$$\rho'' = \sum_\alpha p_\alpha \rho'_{1\alpha} \otimes E_{\alpha\alpha} \tag{23}$$

$$\rho''_2 = \sum_\alpha p_\alpha E_{\alpha\alpha}$$

while the state of 1 is unchanged. Then a simple calculation and (16) gives

$$C_{12}(\rho'') = S(\rho_1 \cdot T) - \sum_\alpha p_\alpha S(\rho'_{1\alpha}) \leq S(T, \rho_1) \tag{24}$$

In the case (7) this is equal to $I_{12}(\rho'') = H\{p_\alpha\}$, but this is not generally true. If we again use the results of Petz it is found that $I_{12}(\rho'') = C_{12}(\rho'')$ iff all the density operators $\{\rho'_{1\alpha}\}$ commute.

A detailed discussion of the relation between the correlations and the amount of information is given in ref. 10 for a more restricted class of measurements. We will now see that in the general case treated here the correlations C_{12} and I_{12} do not tell us everything.

<u>The information transfer</u>

We observe that there is a nontrivial process with $C_{12}(\rho') = 0$, namely

$$\rho_1 \to \rho_1 \cdot T = \mu , \quad \mu \text{ is pure.} \tag{25}$$

System 2 absorbs the information in the initial state of 1 and gives nothing back. This is cleary not a 'measurement' in a conventional sense but there is still a kind of information transfer which is not seen in the final state correlations. In order to deal with this and similar examples we have to extend the formalism. Introduce an auxilary space K_0 of the same dimension as K_1 and a pure state ω_{01} for 0+1 constructed from a faithful state ρ_1 using (3). The interaction acts in $K = K_0 \otimes K_1 \otimes K_2$ as $1 \otimes W$, i.e. it leaves the system 0 alone. The pure initial state for the total system is $\omega = \omega_{01} \otimes E_{MM}$, and we find the following correlation between 0+1 and 2 in the final state

$$C_{(01)2}(\omega') = 2\, S(T,\rho_1).$$

By (18) this is not smaller than (20) and in general it is strictly larger. It is zero iff T is isometric and for (25) we find that $S(T,\rho_1) = S(\rho_1)$. From (22):

$$I_{(01)2}(\omega') = S(T,\rho_1) \tag{26}$$

In case (7) this is still $H\{p_\alpha\}$, but by (24) we will in general obtain a larger mutual information through this seemingly trivial extension of the system by adding a non-interacting part. This behaviour is not possible in a classical system and the 'explanation' lies in the EPR correlations in the initial state. We can perform a reduction to obtain a state ω'' analogous to (23), but in general with a larger correlation

$$C_{(01)2}(\omega'') = S(T,\rho_1)$$

If the decomposition in (8) is that corresponding to an orthogonal decomposition in (11) then we find that this correlation is observable and equal to $H\{p_\alpha\}$, otherwise the set of observables giving the information (26) will not include the $\{E_{\alpha\alpha}\}$.

The pure state ω_{01} contains a perfect correlation between systems 0 and 1 in the initial state. The partial state of 0 is unchanged by the interaction, hence it is independent of the initial state of 2. Thus we can consider the 0-2 correlation in the final state of the composite system to be a measure of the information transfer directed from the initial state of 1 to the final state of 2. We find after some calculations that $S(\omega'_{02}) = S(\rho_1 \cdot T)$ and that the correlation is again expressed in quantities relating to system 1

$$C_{02}(\omega') = S(T,\rho_1) + S(\rho_1) - S(\rho_1 \cdot T).$$

If we use instead the final state ω'', then, with the notation (17) and inequality (16)

$$C_{02}(\omega'') = S(\rho_1) - \textstyle\sum_\alpha p_\alpha\, S(\rho_{1\alpha}) \le S(T,\rho_1) \tag{27}$$

Noting again that $S(\rho_{1\alpha}) = S(\rho'_{1\alpha})$ and that $\sum_\alpha p_\alpha\, S(\rho'_{1\alpha})$ is the average entropy of system 1 after the observation of the outcome (α), (27) says that the entropy of system 1 can be decreased on the average by an amount equal to $C_{02}(\omega'')$ (compare refs. 10-12). We also note that equality holds in $I_{02}(\omega'') \le C_{02}(\omega'')$ iff the $\{\rho_{1\alpha}\}$ are commuting density operators.

When talking of 'measurements' we usually have an ideal of 'minimal disturbance': the interaction should copy some of the information in the initial state of system 1 without destroying it, such that it exists in the final states of both systems 1 and 2. In the present scheme this wll lead to correlations in the system pairs 1-2 and 0-2 as we have seen above, but also in the pair 0-1 to indicate that the relevant information is not removed from system 1 (unlike what happens in the case (25)). An example is provided by a complete von Neumann measurement for which $S(T,\rho_1) = S(\rho_1 \cdot T)$ and the correlations in the final state ω" are

$$C_{01} = C_{02} = S(\rho_1), \qquad C_{12} = S(\rho_1 \cdot T).$$

Note that ρ_1 should be faithful here, the most interesting case is $S(\rho_1) \simeq S(\rho_1 \cdot T)$. For a general map T we find in the final state ω' or ω"

$$C_{01} = S(\rho_1) + S(\rho_1 \cdot T) - S(T,\rho_1)$$

which shows that it does not pay to make $S(T,\rho_1)$ too large. Instead the optimal case is when C_{01}, C_{02} and C_{12} are of the same magnitude in order to have a measurement giving maximal information. If we consider 'weak' interactions in the sense that $S(T,\rho_1) \ll S(\rho_1) \simeq S(\rho_1 \cdot T)$, then C_{12} and C_{02} will be the limiting quantities and the information transfer is bounded by $S(T,\rho_1)$

The trick of introducing the system 0 and the pure initial state of 0+1 was used for a slightly different purpose by Vaidman et al[13].

Repeated measurements

The present general formalism can rather easily be exteded to cope with the situation where system 1 is repeatedly probed by different instruments. This generalization is straightforward if we can make the simplification of allowing the interaction with each instrument to take place instantaneously, while 1 evolves as a closed system (by unitary maps) between interactions. If the action of the instruments are described by CP maps $T_1,\ldots,T_N$ acting at instants $t_1 < t_2 < \ldots < t_N$ and the unitary dynamics in the interval (t_{k-1}, t_k) is given by U_k , then there is a composite CP map

$$T = T'_1 \cdot T'_2 \cdot \quad \cdot T'_N$$

$$T'_k(X) = U_k{}^+ T_k(X) U_k ,$$

describing the whole process including intrinsic dynamics and the action of the instruments in the time interval $(t_0,t_N]$. We can then discuss the map T using the methods introduced above.

Consider the example of 1 being a spin-½ system without proper dynamics ($U_k = 1$, all k) and T_k being von Neumann spin projection measurements in alternating x- and y-directions . We know that the sequence of outcomes will be as random as coin tosses (i.e. a Bernoulli process), hence the total entropy of the set of outcomes is $N \log 2$. The information transfer as defined by C_{02}, where 2 now represents N spin measuring devices, is still no more than $\log 2$. This correlation refers to the information coming from the initial state of 1. The randomness in the set of N outcomes is instead matched by the complex quantum correlations in the composite system which includes the N instruments. In this way it is possible to make a division of the randomness in the set of outcomes into that coming from the initial state of the observed system and that due to the noncommuting nature of the quantum measurements. This distinction seems to be essential in the discussion of 'quantum chaos'. In that context we also expect the intrinsic dynamics of the observed system to be essential (see refs. 7, 14-16).

Acknowledgement. This work was supported by the Swedish Natural Science Research Council.

References

1. A. Wehrl: Rev. Mod. Phys. 50, 221-260 (1978)
2. H. Kosaki: J. Operator Theory 16, 335-348 (1986)
3. H. Araki, in: VIIIth International Congress on Mathematical Physics. Eds. M. Mebkhout, R. Seneor, World Scientific 1987, p 354-365.
4. V.I. Paulsen: Completely bounded maps and dilations. Longmans 1986.
5. D.E. Evans, J.T. Lewis: Comm. Dubl. Inst. Adv. Stud. Ser.A 24 (1977)
6. K. Kraus: States, Effects and Operations. Lecture Notes in Physics 190, Springer 1983.
7. G. Lindblad: Commun. Math. Phys. 65, 281-294 (1979)
8. D. Petz, in: Quantum Probability and Applications II. Eds. L. Accardi, W. von Waldenfels, Lecture Notes in Mathematics 1136, Springer 1985, p 428-441.
9. D. Petz: Commun. Math. Phys. 105, 123-131 (1986)
10. G. Lindblad: Commun. Math. Phys. 33, 305-322 (1973)
11. G. Lindblad: ibid 28, 245-249 (1972)
12. R. Balian: Eur. J. Phys. 10, 208-213 (1989)
13. L. Vaidman, Y. Aharonov, D.Z. Albert: Phys. Rev. Lett. 58, 1385-1387 (1987)
14. G. Lindblad, in: Fundamental Aspects of Quantum Theory. NATO ASI B144. Eds. V. Gorini, A. Frigerio, Plenum Press 1986, p 199-208.
15. G. Lindblad, in: Quantum Probability and Applications II. Eds. L. Accardi, W. von Waldenfels, Lecture Notes in Mathematics 1136, Springer 1985, p 348-360.
16. G. Lindblad, in: Quantum Probability and Applications III. Eds. L. Accardi, W. von Waldenfels, Lecture Notes in Mathematics 1303, Springer 1988, p 183-191.

Information Dynamics and Its Applications to Optical Communication Processes

Masanori OHYA

Department of Information Sciences
Science University of Tokyo
Noda City, Chiba 278, Japan

Abstract

Various physical or nonphysical systems can be described by states, so that the dynamics of a system is described by the state change. One of essential characters of a state is expressed by its complexity. Complexity such as entropy is a key concept in Information theory. We call the study of the state change together with such complexities "Information dynamics", which is a kind of synthesis of dynamics of state change and information theory. Here we explain what information dynamics is and indicate how it can be used in optical communication. Some of concrete applications of information dynamics are discussed in the papers [9,10] of this volume.

§1. What is Information Dynamics ?

Let an input dynamical system and an output dynamical system be described by $(\mathcal{A}, \mathfrak{S}, \alpha)$ and $(\overline{\mathcal{A}}, \overline{\mathfrak{S}}, \overline{\alpha})$, respectively. Here $\mathcal{A}$ is a set of all objects to be observed and $\mathfrak{S}$ is a set of all means getting the observed value for each element A in $\mathcal{A}$, and α describes an inner evolution of the input system. We call $\mathfrak{S}$ a "state space" here. Same for the output system $(\overline{\mathcal{A}}, \overline{\mathfrak{S}}, \overline{\alpha})$. Thus we may say

[Giving a mathematical structure to input and output triples $\equiv$ Having a theory].

A map providing a bridge between two systems is called a "channel" if it sends a state of the input system to that of the output system ; $\Lambda^* : \mathfrak{S} \to \overline{\mathfrak{S}}$. There exist several channels, whose properties specify the character of two systems.

Mathematical structure of almost all systems can be expressed by the following charts.

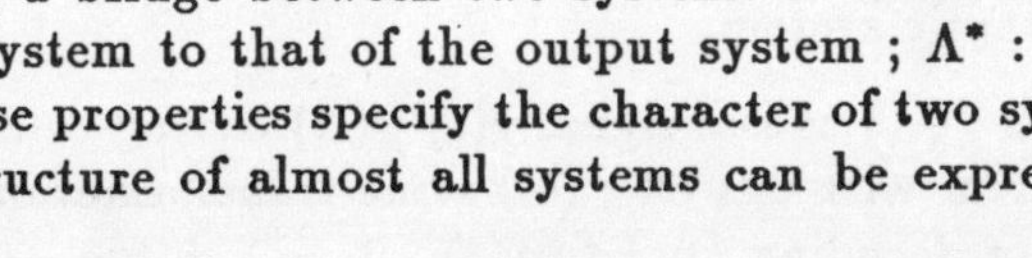

(chart 1)

Lecture Notes in Physics, Vol. 378
C. Bendjaballah, O. Hirota, S. Reynaud (Eds.),
Quantum Aspects of Optical Communications

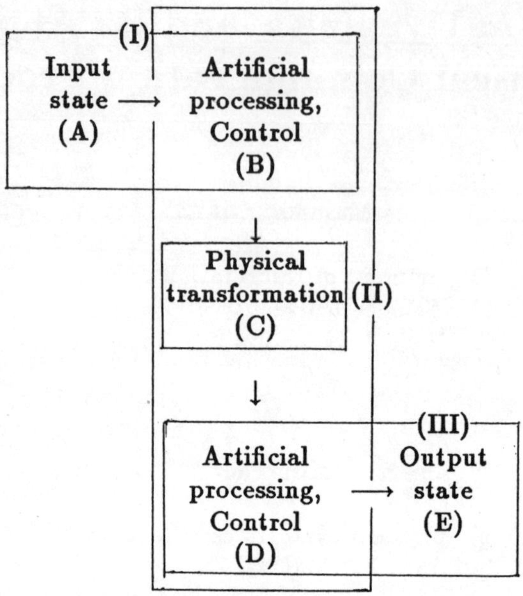

(Chart 2)

In the above chart 2, (I) (= (A) + (B)) corresponds to an input system, (II) (= (B) + (C) + (D)) does to a transformation (channel) system (III) (= (D) + (E)) to an output system. More complex systems are constructed from this first structure.

A "naked" state (A) is artificially processed or controlled to a "dressed" state, and it is suffered to change by a physical (natural) transformation, and it is again artificially processed and controlled. The fundamental part of the process for this state change is obviously "A→ C→ E".

Let us give some examples of an input system (I), a transformation system (channel) (II) and an output system (III).

(1) Causal System : (A) $x \in R^n$; (B) $\dot{x} = f(x)$; (C) $x(t) = \Phi_t(x)$, where Φ_t is an evolution (semi)group generated by f.

Remark: If $x(t)$ can not be obtained directly, then examines the properties of f and find some proper approximation to obtain Φ_t.
$\Longrightarrow$ We may have a chaotic system.

(2) Signal Transmission : (A) causal signal $x(t)$; (B) coding causal signal $x(t)$ to $\{x_n\}$ by e.g., sampling theorem with cut off ; (C) some transformation $y_n = f(x_n)$ for any n ; (III) interpolating or decoding $\{y_n\}$ to $y(t)$.

(3) Discrete systems : (A) input probability distribution $p = \{p_1, \ldots, p_n\}$ of events $X = \{x_1, \ldots, x_n\}$; (C) transition probability $(p(i|j))$; (E) output probability distribution $q = (q_i); q_i = \sum_j p(i|j)p_j$.

(4) Continuous systems : (A) probability measure μ on measurable space $(\Omega, \mathfrak{F})$; (C) Markov kernel $\lambda : \Omega \times \overline{\mathfrak{F}} \to [0,1]$ s.t. $\lambda(x, \cdot) \in P(\overline{\Omega})$ and $\lambda(\cdot, A) \in M(\Omega)$; (E) $\overline{\mu} \equiv \int_\Omega \lambda(x, \cdot) d\mu$.

(5) Quantum system 1 : (A) $x \in \mathcal{H}$ (Hilbert space) ; (C) unitary operator U or isometric operator V ; (E) $y = Ux$ or $Vx \in \mathcal{H}$.

(6) Quantum system 2 : (A) density operator $\rho \in T(\mathcal{H})_{+,1}$; (C) $\Lambda^* \equiv \mathrm{Ad}U$ or $\mathrm{Ad}V$. (E) $\overline{\rho} = \Lambda^* \rho$.

(7) C*-system containing all above: (A) $(\mathcal{A}, \mathfrak{S}, \alpha(G))$ C*-triple; (C) $\Lambda^* : \mathfrak{S}(\mathcal{A}) \to \mathfrak{S}(\overline{\mathcal{A}})$ a dual map of a completely positive map $\Lambda : \overline{\mathcal{A}} \to \mathcal{A}$, where $(\overline{\mathcal{A}}, \overline{\mathfrak{S}}, \alpha(\overline{G}))$ C*-triple; (E) $\overline{\varphi} = \Lambda^* \varphi$.

Once input and output systems are mathematically fixed and a transformation rule (channel) is given, we next consider some complexities of the state associated with the systems, which are a corner stone of information dynamics. The first complexity is one for a state itself : For a state φ, the complexity seen from a reference system $\mathcal{S}$, a certain subset of $\mathfrak{S}$, is dented by $C^{\mathcal{S}}(\varphi)$. The second complexity is determined by both input and output states $\varphi, \overline{\varphi}$ or an input state φ and a channel Λ^*, so that it is denoted by $T^{\mathcal{S}}(\varphi; \overline{\varphi})$ or $T^{\mathcal{S}}(\varphi; \Lambda^*)$, which is called a transmitted complexity from φ to $\overline{\varphi}$ or $\Lambda^* \varphi$. Typical examples of these complexities are entropy and mutual entropy, which play essential roles in several fields.

These complexities should satisfy the following properties: Let $(\mathcal{A}_1, \mathfrak{S}_1, \alpha^1(G))$, $(\mathcal{A}_2, \mathfrak{S}_2, \alpha^2(G))$ be two systems and $(\mathcal{A}, \mathfrak{S}, \alpha(G))$ be the compound system such that $\mathcal{A} = \mathcal{A}_1 \otimes \mathcal{A}_2$, and Λ^* be a channel from $\mathfrak{S}_1$ to $\mathfrak{S}_2$.

(i) For any state $\varphi \in \mathcal{S} \subset \mathfrak{S}_1$, $C^{\mathcal{S}}(\varphi) \geq 0$ and $T^{\mathcal{S}}(\varphi; \Lambda^*) \geq 0$.

(ii) For any bijection j from $\mathfrak{S}_1$ to $\mathfrak{S}_1$,

$$C^{j(\mathcal{S})}(j(\varphi)) = C^{\mathcal{S}}(\varphi)$$
$$T^{j(\mathcal{S})}(j(\varphi); \Lambda^*) = T^{\mathcal{S}}(\varphi; \Lambda^*).$$

(iii) For any $\Phi \in \mathcal{S} \subset \mathfrak{S}$, put $\varphi \equiv \Phi \restriction \mathcal{A}_1 \in \mathcal{S}_1 \equiv \mathcal{S} \restriction \mathcal{A}_1$ and $\psi \equiv \Phi \restriction \mathcal{A}_2 \in \mathcal{S}_2 \equiv \mathcal{S} \restriction \mathcal{A}_2$

$$C^{\mathcal{S}}(\Phi) \leq C_1^{\mathcal{S}}(\varphi) + C_2^{\mathcal{S}}(\psi).$$

Moreover

(iv) $0 \leq T^{\mathcal{S}}(\varphi; \Lambda^*) \leq C^{\mathcal{S}}(\varphi)$.

(v) $T^{\mathcal{S}}(\varphi; id) = C^{\mathcal{S}}(\varphi)$, where "id" is an identity channel from $\mathfrak{S}_1$ to $\mathfrak{S}_1$; $id(\varphi) = \varphi$, $\forall \varphi \in \mathfrak{S}_1$.

Definition 1.1: *Information dynamics is a dynamics described by a set*

$$\{\mathcal{A}, \mathfrak{S}, \mathcal{S}, \alpha \, ; \overline{\mathcal{A}}, \overline{\mathfrak{S}}, \overline{\mathcal{S}}, \overline{\alpha} \, ; \Gamma \, ; C^{\mathcal{S}}(\varphi), T^{\mathcal{S}}(\varphi; \Gamma) \, ; R\},$$

where R is a certain relation among above quantities.

Therefore, for a system of interest, we have to

(1) mathematically determine $\mathcal{A}, \mathfrak{S}, \mathcal{S}, \alpha$; $\overline{\mathcal{A}}, \overline{\mathfrak{S}}, \overline{\mathcal{S}}, \overline{\alpha}$

(2) choose Γ and R ;

(3) fix $C^{\mathcal{S}}(\varphi)$ and $T^{\mathcal{S}}(\varphi; \Gamma)$.

By setting the above (1) ~ (3) in general quantum systems (GQS), we can apply this general frames to several topics such as optical communication[1,5,6],fractal theory [8], molecular genetics [7]. We here mainly discuss an application of information dynamics to optical communication processes.

§2. Channel and Lifting

In order to consider an application of information dynamics to optical communication processes, we have to mathematically set all quantities needed. Let both input system $(\mathcal{A}, \mathfrak{S}, \alpha)$ and output system $(\overline{\mathcal{A}}, \overline{\mathfrak{S}}, \overline{\alpha})$ be described by certain C*-algebraic triples. Very often $\mathcal{A} = \overline{\mathcal{A}}$, $\mathfrak{S} = \overline{\mathfrak{S}}$, $\alpha = \overline{\alpha}$.

A map $\Lambda^* : \mathfrak{S} \to \overline{\mathfrak{S}}$ is called a channel. In particular, Λ^* is called completely positive channel if the dual map $\Lambda : \overline{\mathcal{A}} \to \mathcal{A}$ satisfies

$$\sum_{i,j} B_i^* \Lambda(A_i^* A_j) B_j \geq 0 \quad \text{for } \forall B_j \in \mathcal{A} \text{ and } \forall A_i \in \overline{\mathcal{A}}.$$

A certain class of channels may be a lifting $\mathcal{E}^*$ from $\mathcal{A}$ to $\mathcal{A} \otimes \overline{\mathcal{A}}$ [1]; that is, a continuous map such as

$$\mathcal{E}^* : \mathfrak{S}(\mathcal{A}) \to \mathfrak{S}(\mathcal{A} \otimes \overline{\mathcal{A}}).$$

A lifting $\mathcal{E}^*$ is said to be nondemolition for a state φ if

$$(\mathcal{E}^* \varphi)(A \otimes I) = \varphi(A) \text{ for } \forall A \in \mathcal{A}.$$

Given a lifting $\mathcal{E}^*$, we can construct channels Λ^* and $\overline{\Lambda^*}$ such that

$$\Lambda^* : \mathfrak{S} \to \overline{\mathfrak{S}}$$

by $\Lambda^* \varphi(\overline{A}) = (\mathcal{E}^* \varphi)(I \otimes \overline{A}), \forall \overline{A} \in \overline{\mathcal{A}}$, and

$$\overline{\Lambda}^* : \mathfrak{S} \to \mathfrak{S}$$

by $\overline{\Lambda}^* \varphi(A) = (\mathcal{E}^* \varphi)(A \otimes I), \forall \overline{A} \in \mathcal{A}$. Conversely, given a channel $\Lambda^* : \mathfrak{S} \to \overline{\mathfrak{S}}$ and $\varphi \in \mathcal{S} \subset \mathfrak{S}$ and a decomposition of φ ; $\varphi = \int_{\mathcal{S}} \omega d\mu$

$$\mathcal{E}^* \varphi = \int_{\mathcal{S}} \omega \otimes \Lambda^* \omega d\mu \quad \text{(compound state [3])}$$

is a nondemolition lifting. Hence we have

[Channels $\Leftrightarrow$ Liftings]

< Examples of channel and lifting >

Here we give some examples of channel and we construct a quantum communication channel. Let $\rho = \sum \lambda_n \rho_n$ be a certain decomposition of a density operator when $\mathcal{A} = \mathrm{B}(\mathcal{H})$ and $\mathfrak{S} = \mathrm{T}(\mathcal{H})_{+,1}$ on a Hilbert space $\mathcal{H}$.

<u><1> Unitary evolution :</u>

$$\rho \to \Lambda_t^* \rho = \mathrm{Ad}U_t(\rho) \equiv U_t^* \rho U_t, \; t \in R \;\Rightarrow \mathcal{E}^* \rho = \sum \lambda_n \rho_n \otimes \Lambda^* \rho_n$$

where U_t is a unitary operator $U_t = \exp(\mathrm{it}H)$.

<2> Semigroup evolution :

$$\rho \to \Lambda_t^* \rho = \mathrm{Ad}V_t(\rho) \equiv V_t^* \rho V_t, \ t \in R^+ \ \Rightarrow \mathcal{E}^* \rho = \sum \lambda_n \rho_n \otimes \Lambda^* \rho_n$$

where $\{V_t \ ; \ \mathrm{t} \in R^+\}$ is a one parameter semigroup on $\mathcal{H}$.

<3> Measurement : Measure $\mathrm{A} = \sum_n a_n P_n$ (spectral decomposition of A) in a state ρ,

$$\rho \to \Lambda^* \rho = \sum_n P_n \rho P_n \Rightarrow \mathcal{E}^* \rho = \sum \lambda_n \rho_n \otimes \Lambda^* \rho_n.$$

<4> Reduction : When two systems described by Hilbert spaces $\mathcal{H}$ and $\mathcal{K}$ interacts and we look at the state change of the first system.

$$\begin{array}{ccc} \mathfrak{S}(\mathcal{H}) & \xrightarrow{\text{interaction}} & \mathfrak{S}(\mathcal{K}). \\ \Cup & & \Cup \\ \rho & & \sigma \end{array}$$

Then

$$\mathcal{E}_t^* \rho \equiv \theta_t = U_t^* (\rho \otimes \sigma) U_t,$$

$$\rho \to \Lambda_t^* \rho = tr_{\mathcal{K}} \theta_t.$$

More generally for <3> and <4>, $\mathcal{A}$ (resp. $\overline{\mathcal{A}}$) is interpreted as the algebra of observables of a system (resp. a measuring apparatus) and $\mathcal{E}^*$ describes an interaction between the system and the apparatus as well as the preparation of the apparatus. If $\varphi \in \mathcal{S}(\mathcal{A})$ is the preparation of the system, i.e., the state before the interaction with the apparatus, then $\overline{\Lambda}^* \varphi \in \mathcal{S}(\mathcal{A})$ (resp. $\Lambda^* \varphi \in \mathcal{S}(\overline{\mathcal{A}})$) is the state of the system (resp. of the apparatus) after the measurement.

<5> Isometric lifting : For an isometry V ($V^* V = I_{\mathcal{H}}$) from $\mathcal{H}$ to $\mathcal{H} \otimes \mathcal{K}$, a lifting defined by

$$\mathcal{E}^* \rho = V \rho V^*, \quad \forall \rho \in \mathfrak{S}(\mathcal{H}_1)$$

is called an isometric lifting [1].

<6> Quantum Communication Channel :

Let us construct a quantum (communication) channel.

$$\begin{array}{ccc} (Noise)\ \mathfrak{S}(\mathcal{K}_1) & & \\ \downarrow & & \\ \mathfrak{S}(\mathcal{H}_1) \ni \rho & \longrightarrow & \bar{\rho} = \Lambda^* \rho \in \mathfrak{S}(\mathcal{H}_2) \\ & & \downarrow \\ & & (Loss)\ \mathfrak{S}(\mathcal{K}_2) \end{array}$$

Let $\nu \in \mathfrak{S}(\mathcal{K}_1)$ be a state representing the noise and a, π, γ be the following maps: (1) a : $B(\mathcal{H}_2) \rightarrow B(\mathcal{H}_2 \otimes \mathcal{K}_2)$ given by $a(A) = A \otimes I$ for any $A \in B(\mathcal{H}_2)$, (2) $\pi : B(\mathcal{H}_2 \otimes \mathcal{K}_2) \rightarrow B(\mathcal{H}_1 \otimes \mathcal{K}_1)$ completely positive with $\pi(I) = I$, (3) $\gamma : B(\mathcal{H}_1 \otimes \mathcal{K}_1) \rightarrow B(\mathcal{H}_1)$ by $\gamma(Q) = tr_{\mathcal{K}_1} \nu Q$ for any $Q \in B(\mathcal{H}_1 \otimes \mathcal{K}_1)$.

$$\Lambda = \gamma \circ \pi \circ a.$$

Then

$$\Lambda^* = a^* \circ \pi^* \circ \gamma^*$$

or equivalently,

$$\Lambda^* \rho = tr_{\mathcal{K}_2} \pi^*(\rho \otimes \nu), \ \forall \rho \in \mathfrak{S}(\mathcal{H}_1).$$

We call this channel "quantum communication channel".

When $\mathcal{H}_1 = \mathcal{H}_2 = \mathcal{H}$,

$$\mathcal{E}^* : \rho \in \mathfrak{S}(\mathcal{H}) \rightarrow \pi^*(\rho \otimes \nu) \in \mathfrak{S}(\mathcal{H} \otimes \mathcal{K})$$

is a lifting, and

$$\Lambda^* \rho = tr_{\mathcal{K}} \mathcal{E}^* \rho.$$

In <5> and <6>, an input signal is transmitted and received by an apparatus which produces an output signal. Here $\mathcal{A}$ (resp. $\overline{\mathcal{A}}$) is interpreted as the algebra of observables of the input (resp. output) signal and $\mathcal{E}^*$ describes an interaction between the input signal and the receiver as well as the surroundings of the receiver. If $\rho_1 \in \mathcal{S}(\mathcal{A})$ is the input signal, then the state $\Lambda^* \varphi \in \mathcal{S}(\overline{\mathcal{A}})$ is the state of the (observed) output signal.

§3. Entropies describing complexities in GQS

Next we discuss two types of entropy for general quantum states as the complexities of information dynamics which are needed for optical communication.

Let $(\mathcal{A}, \mathfrak{S}, \alpha(R))$ be a C*-dynamical system and $\mathcal{S}$ be a weak* compact and convex subset of $\mathfrak{S}$, ex$\mathcal{S}$ be the set of all extreme points of $\mathcal{S}$.

Every state $\varphi \in \mathcal{S}$ has a maximal measure μ pseudosupported on ex $\mathcal{S}$ such that

$$\varphi = \int_{\mathcal{S}} \omega d\mu.$$

The measure μ giving the above decomposition is not unique unless $\mathcal{S}$ is a Choquet simplex, so that we denote the set of all such measures by $M_{\varphi}(\mathcal{S})$. For the probability measure μ, define

$$H(\mu) = \sup\{-\sum_{A_k \in \tilde{A}} \mu(A_k) \log \mu(A_k); \ \tilde{A} \in P(\mathcal{S})\},$$

where $P(\mathcal{S})$ is the set of all finite partitions of $\mathcal{S}$. Then the entropy of a state $\varphi \in \mathcal{S}$ w.r.t. $\mathcal{S}$ is defined by

$$S^{\mathcal{S}}(\varphi) = \inf\{H(\mu); \ \mu \in M_{\varphi}(\mathcal{S})\}.$$

This entropy does depend on the set $\mathcal{S}$ chosen, and we call it "$\mathcal{S}$ - entropy". Even when $\mathrm{S}(\varphi) = +\infty$, $\mathrm{S}^{\mathcal{S}}(\varphi) < +\infty$ for some $\mathcal{S}$, which is a remarkable property of $\mathcal{S}$ - entropy [4].

A compound state (lifting) $\Phi_\mu^{\mathcal{S}} (\equiv \mathcal{E}_\mu^* \varphi)$ of φ and $(\bar{\varphi} \equiv) \Lambda^* \varphi$ with respect to $\mathcal{S}$ and μ was introduced as

$$\Phi_\mu^{\mathcal{S}} = \int_{\mathcal{S}} \omega \otimes \Lambda^* \omega d\mu.$$

The mutual entropy w.r.t. an initial state $\varphi \in \mathcal{S}$, the decomposition measure μ and a channel Λ^* is defined by [2,6]

$$\mathrm{I}_\mu^{\mathcal{S}}(\varphi; \Lambda^*) = \mathrm{S}(\Phi_\mu^{\mathcal{S}} | \Phi_0),$$

where $\mathrm{S}(\cdot|\cdot)$ is the relative entropy for two states. In some cases, this mutual entropy can be written as

$$\mathrm{I}_\mu^{\mathcal{S}}(\varphi; \Lambda^*) = \int_\Omega \mathrm{S}(\Lambda^* \omega | \Lambda^* \varphi) d\mu.$$

The mutual entropy w.r.t. an initial state $\varphi \in \mathcal{S}$ and a channel Λ^* is then defined by

$$\mathrm{I}^{\mathcal{S}}(\varphi; \Lambda^*) = \lim_{\varepsilon \to 0} \sup\{\mathrm{I}_\mu^{\mathcal{S}}(\varphi; \Lambda^*); \mu \in F_\varphi(\mathcal{S}; \varepsilon)\},$$

where $F_\varphi(\mathcal{S})$ is the subset of the set $M_\varphi(\mathcal{S})$ such that $F_\varphi(\mathcal{S}; \varepsilon) = \{\mu \in M_\varphi(\mathcal{S}); \mathrm{S}^{\mathcal{S}}(\varphi) \leq \mathrm{H}(\mu) < \mathrm{S}^{\mathcal{S}}(\varphi) + \varepsilon < +\infty\}$ or $F_\varphi(\mathcal{S}; \varepsilon) = M_\varphi(\mathcal{S})$ when $\mathrm{S}^{\mathcal{S}}(\varphi) = \infty$. Note that the mutual entropy should be used when the decomposition measure is fixed. In the sequel we use the simple notations $\mathrm{S}(\varphi)$, Φ_μ, $\mathrm{I}_\mu(\varphi; \Lambda^*)$ and $\mathrm{I}(\varphi; \Lambda^*)$ when $\mathcal{S} = \mathfrak{S}$.

Let us write the mutual entropy in usual quantum system, namely, when $\mathcal{A}$ is the full algebra $B(\mathcal{H})$ and any normal state φ is described by a density operator ρ such as $\varphi(\cdot) = tr \rho \cdot$. Then our entropy $\mathrm{S}(\varphi)$ is shown to be equal to that of von Neumann: $\mathrm{S}(\varphi) = \mathrm{S}(\rho) = -tr \rho \log \rho$. Every Schatten decomposition $\rho = \sum_n \lambda_n E_n$, $E_n = |x_n><x_n|$ (i.e., λ_n is the eigenvalue of ρ and x_n is its associated eigenvector) provides every orthogonal measure in $M_\varphi(\mathfrak{S})$ defining the entropy $\mathrm{S}(\varphi)$. Since the Schatten decomposition of ρ is not unique unless every eigenvalue λ_n is nondegenerate, the compound state Φ is expressed as

$$\Phi_E(Q) = tr \sigma_E Q, \ Q \in \mathcal{A} \otimes \overline{\mathcal{A}},$$

with

$$\sigma_E = \sum_n \lambda_n E_n \otimes \Lambda^* E_n,$$

where E represents a Schatten decomposition $\{E_n\}$. Then the mutual entropy for φ and the channel Λ^* is given by

$$\mathrm{I}(\rho; \Lambda^*)(\equiv \mathrm{I}(\varphi; \Lambda^*)) = \sup\{\mathrm{I}_E(\rho; \Lambda^*); \ E = \{E_n\} \ of \ \rho\},$$

with

$$\mathrm{I}_E(\rho; \Lambda^*) = \mathrm{S}(\sigma_E | \sigma_0) = tr \sigma_E (\log \sigma_E - \log \sigma_0),$$

where $\sigma_0 = \rho \otimes \Lambda^* \rho$. This form of the mutual entropy was introduced in [2] to study optical communication processes.

Fundamental properties of $S^{\mathcal{S}}(\varphi)$ and $I^{\mathcal{S}}(\varphi; \Lambda^*)$:

Theorem 3.1: *When $\mathcal{A} = B(\mathcal{H})$ and $\alpha_t = Ad(U_t)$ with a unitary operator U_t, for any state φ given by $\varphi(\cdot) = tr\rho \cdot$ with a density operator ρ, we have the followings:*
(1) $S(\varphi) = -tr\rho \log \rho$.
(2) *If φ is an α-invariant faithful state and every eigenvalue of ρ is nondegenerate, then* $S^I(\varphi) = S(\varphi)$.
(3) *If $\varphi \in K(\alpha)$, then* $S^K(\varphi) = 0$.

Two states φ_1 and φ_2 are said to be orthogonal each other (denoted by $\varphi_1 \perp \varphi_2$) if their supports $s(\varphi_1)$ and $s(\varphi_2)$ are orthogonal, where the support $s(\varphi)$ of φ means the smallest projection E satisfying $\varphi(I\text{-}E) = 0$. The measure $\mu \in M_\varphi(\mathcal{S})$ is said to be orthogonal if $(\int_Q \omega d\mu) \perp (\int_{\Omega/Q} \omega d\mu)$ is satisfied for every Borel set Q in $\mathcal{S}$. A channel is called normal if it sends a normal state to a normal state.

Theorem 3.2: *For a normal C.P. channel Λ^* and a normal state φ, if a measure μ has a discrete support and is orthogonal, then*

$$I_\mu(\varphi; \Lambda^*) = \int_{\mathfrak{S}} S(\Lambda^*\omega | \Lambda^*\varphi) d\mu < S(\varphi) + \varepsilon.$$

Theorem 3.3: *For a density operator ρ*

$$0 \le I(\rho, \Lambda^*) \le \min\{S(\rho), S(\Lambda^*\rho)\}.$$

In the notatins of Section 1, two complexities are

$$C^{\mathcal{S}}(\varphi) = S^{\mathcal{S}}(\varphi)$$
$$T^{\mathcal{S}}(\varphi; \Lambda^*) = I^{\mathcal{S}}(\varphi; \Lambda^*).$$

Without seriously taking the original meaning of the entropy S (i.e., forgetting the correspondence between each element of φ and that of $\Lambda^*\varphi$), and as far as the complexity is concerned, $C^{\mathcal{S}}(\varphi)$ and $T^{\mathcal{S}}(\varphi; \Lambda^*)$ can be given as follows : For any $\varphi \in \mathcal{S}$ and any decomposition of φ

$$\varphi = \int_{\mathcal{S}} \omega d\mu,$$

the complexities are

$$T^{\mathcal{S}}(\varphi; \Lambda^*) \equiv \int_{\mathfrak{S}} S(\Lambda^*\omega | \Lambda^*\varphi) d\mu$$
$$C^{\mathcal{S}}(\varphi) \equiv T^{\mathcal{S}}(\varphi; id).$$

Watanabe [10] discusses the efficiency of a modulation M by

$$r(M) = \frac{\mathrm{I}(\rho_M;\Lambda^*)}{\mathrm{S}(\rho_M)},$$

where ρ_M is a modulated input state.

Remark: The channel capacity is given by

$$\mathrm{C}^{\mathcal{S}}(\Lambda^*) = \sup\{\mathrm{I}^{\mathcal{S}}(\varphi;\Lambda^*)\,;\,\varphi\in\mathcal{S}\},$$

which is also useful to study the efficiency of channel for communication of information.

§4. Applications of Auantum Channel to Optical Communication Processes

As applications of a mathematical expression of quantum channel given in <6> of §2, we here discuss an attenuation optical communication process and derivation of error probability.

4.1 Attenuation process

4.1.1 Conventional expression:

A quantum system composed of photons is described by the Hamiltonian $H = a^*a+1/2$, where a^* and a are creation and annihilation operators of photon, respectively. The Schrödinger equation $Hx(q) = Ex(q)$ is easily solved, whose eigenvalue is $E_n =$ n+1/2 (n $\geq$ 0) and the eigenvector x_n(q) for E_n is $(1/(\pi^{1/2}\mathrm{n}!)^{1/2})H_n(\sqrt{2q})\exp(-q^2/2)$, where $H_n(q)$ is the n-th Hermite function. The Hilbert space of this system is a certain closed linear span of linear combinations of $x_n(q)$ (n=0,1,2 ...).

The model is considered as follows: When n_1 photons are transmitted from the input system, m_1 photons from the noise system add to the signal. Then m_2 photons are lost to the loss system through the channel, and n_2 photons are detected in the output system. In this model, the Hilbert spaces are denoted by $\mathcal{H}_1$, $\mathcal{H}_2$, $\mathcal{K}_1$, $\mathcal{K}_2$ and their coordinates represented wave functions are respectively denoted by

$$x^{(1)}_{n_1}(q_1) = \{2^{n_1}\pi(n_1!)\}^{-\frac{1}{2}}\exp\{\frac{-q_1^2}{2}\}H_{n_1}(q_1),$$

$$y^{(1)}_{m_1}(t_1) = \{2^{m_1}\pi(m_1!)\}^{-\frac{1}{2}}\exp\{\frac{-t_1^2}{2}\}H_{m_1}(t_1),$$

$$x^{(2)}_{n_2}(q_2) = \{2^{n_2}\pi(n_2!)\}^{-\frac{1}{2}}\exp\{\frac{-q_2^2}{2}\}H_{n_2}(q_2),$$

$$y^{(2)}_{m_2}(t_2) = \{2^{m_2}\pi(m_2!)\}^{-\frac{1}{2}}\exp\{\frac{-t_2^2}{2}\}H_{m_2}(t_2).$$

According to the conservation of energy $(n_1+m_1 = n_2+m_2)$, we take a following linear transformation among the coordinates q_1, t_1, q_2, t_2 of the input, noise, output and loss systems, respectively:

$$\begin{cases} q_2 = \alpha q_1 + \beta t_1, \\ t_2 = -\beta q_1 + \alpha t_1. \end{cases} \qquad (\alpha^2+\beta^2 = 1)$$

For simplicity, we put $m_1 = 0$. We define a mappings $U : \mathcal{H}_1 \otimes \mathcal{K}_1 \to \mathcal{H}_2 \otimes \mathcal{K}_2$ and π^* by

$$\begin{aligned} &U(x_n^{(1)} \otimes y_0^{(1)})(q_2, t_2) \\ &= x_n^{(2)} \otimes y_0^{(2)}(\alpha q_2 - \beta t_2, \beta q_2 + \alpha t_2) \\ &= \sum_{j=0}^{n} \sqrt{\frac{n!}{(n-j)!j!}} \alpha^j (-\beta)^{n-j} x_j^{(2)} \otimes y_{n-j}^{(2)}(q_2, t_2). \\ &\pi^* = U(\cdot)U^*. \end{aligned}$$

Thus from the expression Λ^* of quantum channel of <6> in §2, the attenuation channel Λ^* is given by

$$\Lambda^* \rho = tr_{\mathcal{K}_2} U^*(\rho \otimes \nu) U, \tag{4.1}$$

with the noise $\nu = |y_0^{(1)} >< y_0^{(1)}| \in \mathfrak{S}(\mathcal{K}_1)$ due to the "zero point fluctuation" of electromagnetic field ($y_0^{(1)}$ is a vacuum state vector in $\mathcal{K}_1$). Note that we may take $\mathcal{H} = \mathcal{H}_2 = \mathcal{K}_2$.

4.1.2 A simple expression [1]:

The above attenuation process (4.1) can be written by a little simple way.

Let $\mathcal{H} = \mathcal{K} = \Gamma(\mathrm{C})$ (Fock space) and let

$$|\theta >= \exp\left(-\frac{|\theta|^2}{2}\right) \sum_n \frac{\theta^n}{\sqrt{n!}} |n>, \ \theta \in C$$

denote the coherent vector, where $|n>$ is the number state. Define a mapping V from $\Gamma(\mathcal{H})$ to $\Gamma(\mathcal{H}) \otimes \Gamma(\mathcal{K})$ as

$$V|\theta >= |\alpha\theta > \otimes |\beta\theta >$$

with $\alpha, \beta \in \mathrm{C}$, $|\alpha|^2 + |\beta|^2 = 1$.

V represents the interaction of the signal with an apparatus or a receiver and it means that by the effect of the interaction a coherent signal (beam) $|\theta>$ splits into two signals (beams) still coherent but of lower intensity although the total intensity (energy) is preserved by the transformation.

Now, let us show the equivalence of the above operator V and the operator U in the conventional expression.

$$\begin{aligned} &V|\theta >= |\alpha\theta > \otimes |\beta\theta > \\ &= \exp(-\frac{|\alpha\theta|^2}{2}) \sum_n \frac{(\alpha\theta)^n}{\sqrt{n!}} |n > \exp(-\frac{|\beta\theta|^2}{2}) \sum_m \frac{(\beta\theta)^m}{\sqrt{m!}} |m > \\ &= \exp(-\frac{|\theta|^2}{2}) \sum_n \sum_m \frac{(\alpha\theta)^n (\beta\theta)^m}{\sqrt{n!m!}} |n > \otimes |m > \\ &= \exp(-\frac{|\theta|^2}{2}) \{\sum_{N=0}^{\infty} \frac{\theta^N}{\sqrt{N!}}\} \{\sum_{n=0}^{\infty} \alpha^n \beta^{N-n} \sqrt{\frac{N!}{n!(N-n)!}} |n > \otimes |N-n >\}, \end{aligned}$$

which implies, for any nonnegative integer N,

$$V|N>=\sum_{n=0}^{N}\alpha^n\beta^{N-n}\sqrt{\frac{N!}{n!(N-n)!}}|n>\otimes|N-n>.$$

Thus U equals to V by replacing β with $-\beta$.

Therefore the attenuation channel can be written as

$$\Lambda^*\rho=tr_{\mathcal{K}}V\rho V^*.$$

4.2 Error probability

Let ξ_i be the quantum code corresponding to a symbol $c_i\in\mathcal{C}$: For simplicity, take

$$\mathcal{C}=\{0,1\}\Longleftrightarrow\Xi=\{\xi_0,\xi_1\}.$$

One expression of quantum code is due to a state of photon; for instance, ξ_0 is the vacuum state and ξ_1 is another state such as a coherent or a squeesed state.

Two states (quantum mechanical codes) $\xi_0^{(1)}$ and $\xi_1^{(1)}$ in the input system are transmitted to the output system through a channel Λ^*. Consider a Z-type signal transmission, namely, the signal "0" represented by the state $\xi_0^{(1)}$ goes always to "0" represented by $\xi_0^{(2)}$ and the signal "1" represented by the state $\xi_1^{(1)}$ goes to $\xi_0^{(2)}$ or other states. Then the error probability q_e comes from that the signal "1" is recognized as the signal "0", so that it is given by

$$\begin{aligned}q_e&=tr\Lambda^*(\xi_1^{(1)})\xi_0^{(2)}\\&=tr_{\mathcal{H}_2}(tr_{\mathcal{K}_2}\pi^*(\xi_1^{(1)}\otimes\nu))\xi_0^{(2)}\end{aligned}$$

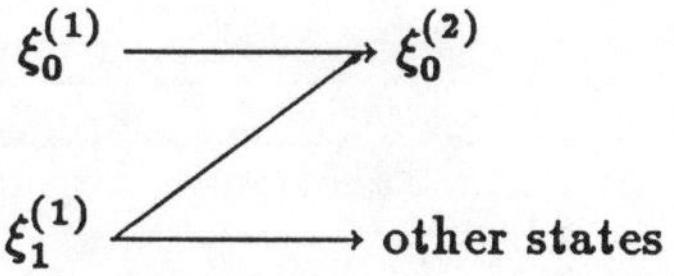

Based on this error probability q_e, the error probability of PCM modulation with the t_0-tuple error correcting code with the weight N and that for PPM modulation are given by

$$\begin{aligned}P_e^{PCM}&=\sum_{j=t_0+1}^{N}{}_NC_jq_e^j(1-q_e)^{N-j},\\P_e^{PPM}&=q_e,\end{aligned}$$

where ${}_NC_j=N!/\{(N-j)!j!\}$.

Concrete computation and physical discussion of error probability for some optical processes are given in the paper [9] of this volume, by which we obtain an interesting observation. A certain input squeezed state gives us a better error probability than the input coherent state; that is, the error probability very much depends on the way of squeezing the coherent state.

References: I here simply give references of mine and see [2] and [6] for a complete reference.

[1] L. Accardi and M. Ohya : "Compoud channels, transition expectations and liftings", preprint.
[2] M. Ohya : "On compound state and mutual information in quantum information theory", IEEE.Trans.Inf.Theory, 29, pp.770-774 (1983).
[3] ________ : "Note on quantum probability", L. Nuovo Cimento, 38,pp.402-404 (1983).
[4] ________ : "Entropy transmission in C*-dynamical systems", J. Math. Anal. Appl., 100, pp.222-235 (1984).
[5] M. Ohya, H. Yoshimi and O. Hirota : "Rigorous derivation of error probability in quantum control communication processes", IEICE of Japan, J71-B, No.4, 533-539 (1988).
[6] M. Ohya : "Some aspects of quantum information theory and their applications to irreversible processes", Rep. on Math. Phys., 27, pp.19-47 (1989).
[7] ________ : "Information theoretical treatment of genes", Trans. IEICE, E70, No.5, pp.556-560 (1989).
[8] ________ : "Fractal dimensions of states", to appear in Quantum Probability and Applications (edited by L. Accardi and W. von Waldenfels), KLUWER Publishing Company.
[9] M. Ohya and H. Suyari : "Rigorous derivation of error probability in coherent optical communication", in this volume.
[10] N. Watanabe : "Efficiency of optical modulations with coherent state", in this volume.

INFORMATION THEORY AND QUANTUM CORRELATIONS

Simon J. D. Phoenix[(1)] and Stephen M. Barnett[(2)]

[(1)]*British Telecom Research Laboratories, Martlesham Heath, Ipswich. IP5 7RE. England.*

[(2)]*Department of Physics, King's College London, The Strand, London. WC2R 2LS. England.*

I. INTRODUCTION

There are many examples in physics of phenomena which require some element of a quantum mechanical description for their understanding. Perhaps the most fascinating of these is the celebrated Einstein-Podolsky-Rosen (EPR) paradox [1] in which one of a pair of correlated quantum systems appears to display an instantaneous response to the observation of its partner. The correlations in such quantum systems are not fully characterized by sets of observation probabilities between pairs of observables. [2] The extra correlation inherent in quantum mechanics can be recovered in classical physics at the expense of the introduction of "hidden variables". Predictions based on local hidden-variable theories must satisfy the Bell inequalities; predictions based on quantum mechanics do not have to satisfy this requirement. [3] Quantum correlations, as opposed to those which can be described using *conventional* classical mechanics, are difficult to quantify in a precise fashion. The fundamental reason for this difficulty is that in quantum mechanics we must deal with probability *amplitudes* rather than probabilities. In quantum mechanics there exists the possibility for a given system to be prepared in a pure state; there is no analogous quantity in conventional classical mechanics.

As we have remarked above the distinction between quantum and classical correlations resides in the appearance of probability amplitudes in a quantum mechanical description. An absolute and precise measure of the strength of a correlation is the information content of that correlation. This measure has been developed independently by Zurek [4] and by ourselves [5] and called the "index of correlation" by us. Conventional

Lecture Notes in Physics, Vol. 378
C. Bendjaballah, O. Hirota, S. Reynaud (Eds.),
Quantum Aspects of Optical Communications

measures of correlation in quantum mechanics, such as the correlation coefficient, strictly only measure correlations between specific observables. The index of correlation can be applied to correlations between quantum *states* and when applied in this fashion is observable-independent. The index of correlation is a general measure and can also be applied to specific observables. In one case we ask for the information content of the correlation between quantum states; in another we could ask for the information content of the correlation between specific observables. It is also legitimate to ask for the information content of a correlation between two classical systems. The index of correlation as applied to all these situations can give us insights into the nature of quantum and classical correlations. It should be strongly emphasized here that by "classical" we imply a particular description of physical phenomena *which does not include hidden variables of any form.* Thus the information-theoretic bounds on the correlation in the classical and quantum cases which we shall derive are less stringent than the bounds placed by the Bell inequalities. Nevertheless, the index of correlation allows us to distinguish between quantum and classical correlations in a way which highlights the fundamental role played by probability amplitudes in quantum mechanics.

II. THE INDEX OF CORRELATION

The index of correlation between two correlated systems or observables is given by

$$I_c = Inf - (Inf_a + Inf_b) \tag{1}$$

where Inf is the total information content associated with the correlation and $Inf_{a(b)}$ is the information content of system, or observable, $a(b)$ considered independently. The information content of an ensemble of systems, or observables of those systems, is given by an expression of the form

$$Inf = S^{max} - S \tag{2}$$

where S^{max} is the maximum possible entropy and S is the actual entropy associated with the system or observable. [6] [7] Let us examine the three specific uses of the index of correlation in quantum and classical mechanics.

A. The Quantum Index of Correlation

The quantum index of correlation, labelled I_c^{qu}, is a direct measure of the strength of correlation between two quantum systems. Using the fact that the maximum entropy is simply the sum of the two maximum individual system entropies we find that the index of correlation takes the simple and appealing form

$$I_c^{qu} = S_a + S_b - S \tag{3}$$

where the individual system entropies have been labelled by the suffix a or b. The entropies in this expression are formed from the density operators or reduced density operators according to the prescription

$$\begin{aligned} S &= - Tr\,\rho \ln \rho \\ S_{a(b)} &= - Tr_{a(b)}\,\rho_{a(b)} \ln \rho_{a(b)} \,. \end{aligned} \tag{4}$$

These entropies obey the Araki-Lieb inequalities [8] which states that for any two quantum systems, labelled a and b, we have

$$| S_a - S_b | \leq S \leq S_a + S_b \,. \tag{5}$$

As we have shown, [5] the optimally correlated quantum state of two systems is a pure state in which each of the component systems is maximally disordered subject to any physical constraints. If the total system is in a pure state then it follows from (5) that each of the component systems have equal entropies and the maximum value that I_c^{qu} can take is just twice this entropy.

The most important, and physically relevant, example of a correlated system, for our purposes, is that of the two-mode squeezed vacuum state (which, for convenience we shall call simply a "squeezed state" throughout this paper). In this state one mode of an electromagnetic field is strongly correlated to another. The squeezed state can be expanded in terms of the number states of the modes and we find

$$|r,\theta\rangle = (\cosh r)^{-1} \sum_{n=0}^{\infty} (\tanh r)^n e^{in\theta} |n\rangle_a |n\rangle_b \quad . \tag{6}$$

It is easy to see from this expression that the photon numbers in each of the modes are perfectly correlated. However, the correlation is more fundamental than this, as we have previously shown [5] *the squeezed state (6) is the most strongly correlated two-mode state of light subject to the constraint of a mean energy.* The index of correlation for this state is given by [5]

$$I_c = 2[\cosh^2 r \ln(\cosh^2 r) - \sinh^2 r \ln(\sinh^2 r)] \quad . \tag{7}$$

B. The Shannon Index of Correlation

The Shannon index of correlation is a direct measure of the strength of correlation between two observables. It is formed from the probability distributions associated with these observables. For instance, in quantum mechanics an observable $\hat{N}$ has a probability distribution on a particular state given by $P(n) = \langle n|\rho|n\rangle$, where $\hat{N}|n\rangle = n|n\rangle$. The entropy associated with this observable is simply given by

$$S(\hat{N}) = -\sum_{n=0}^{\infty} P(n) \ln P(n) \quad . \tag{8}$$

The extension to entropies of joint observables is obvious. Shannon entropies for specific observables have been studied in the context of the Jaynes-Cummings model of optical resonance. [9] However, these entropies enable us to define an index of correlation for two specific observables $\hat{N}_a$, $\hat{M}_b$ of the a and b systems respectively; it is given by

$$I_c(\hat{N}_a,\hat{M}_b) = S_a(\hat{N}_a) + S_b(\hat{M}_b) - S(\hat{N}_a,\hat{M}_b) \quad . \tag{9}$$

These entropies obey a modified Araki-Lieb inequality given by [10]

$$\max\left(S_a(\hat{N}_a),S_b(\hat{M}_b)\right) \leq S(\hat{N}_a,\hat{M}_b) \leq S_a(\hat{N}_a) + S_b(\hat{M}_b) \quad . \tag{10}$$

This inequality tells us that the observables considered jointly must be at least as disordered as either of the observables alone. We find that optimally correlated observables must have equal disorder, or entropy, in the individual observables and the maximum value that the Shannon index can take is just equal to this entropy. From this we deduce the important and fundamental result that for an optimally correlated quantum state *any given pair of observables contains at most only half the information about the correlations.* Formally,

$$I_c(\hat{N}_a,\hat{M}_b) \;\leq\; \frac{1}{2}\, I_c^{qu}(max)\ . \tag{11}$$

Furthermore, it can be shown [10] that the Shannon index for a given pair of observables is always less than or equal to the index of correlation for that state. This expresses the physically reasonable fact that the information content of the correlation between specific observables of a correlated quantum state cannot exceed the total information content of the correlation.

The squeezed state (6) again provides us with an excellent example with which to illustrate these concepts. The Shannon index of correlation can be calculated exactly for a range of observables. Specifically, we list below the index of correlation for number-number correlations, phase-phase correlations, number-phase correlations and quadrature-quadrature correlations. [10] It should be noted that it is the existence, and form, of the Hermitian phase operator which allows us to calculate the phase correlations exactly. [11] [12] Thus we have

$$I_c(\hat{N}_a,\hat{N}_b) = \frac{1}{2}\, I_c \tag{12}$$

$$I_c(\hat{\phi}_{\theta_a},\hat{\phi}_{\theta_b}) = \ln(\cosh^2 r) \;<\; \frac{1}{2}\, I_c \tag{13}$$

$$I_c(\hat{N}_a,\hat{\phi}_{\theta_b}) = 0 \tag{14}$$

$$I_c(\hat{X}_a,\hat{X}_b) = \ln(\cosh 2r) \;<\; \frac{1}{2}\, I_c \tag{15}$$

$$I_c(\hat{X}_a,\hat{Y}_b) = 0 \tag{16}$$

The number of photons in each mode is perfectly correlated and we find that the information content of this correlation is precisely half that of the total information content of the correlation between the modes. Furthermore, we find that

$$\frac{1}{2} I_c > I_c(\hat{X}_a, \hat{X}_b) > I_c(\hat{\phi}_a, \hat{\phi}_b) .$$

C. Quantum vs. Classical ?

Before discussing the classical index of correlation we introduce a figure of merit for the strength of the correlation. This figure of merit, η, is a normalized index of correlation and is defined by

$$\eta = \frac{I_c}{\max(S_a, S_b)} . \tag{17}$$

This quantity will allow us to make a direct comparison between the quantum and classical regimes. We emphasize again that the use of the word "classical" does not include any form of hidden-variable theory whatsoever; any stochastic behaviour is introduced in the normal fashion through distributions on phase space. It should be noted, however, that there are considerable difficulties in defining entropies of continuous variables in classical mechanics and some of these are addressed in the review by Wehrl. [13]

The difference between correlated classical systems and correlated quantum systems arises because a classical system must be at least as disordered as either of its components; a quantum system in a pure state has zero total entropy. Classical systems therefore obey an inequality of the form (8); quantum systems obey inequality (5). This is a consequence of the distinction between pure and mixed states in quantum mechanics. It is the superposition principle for probability *amplitudes* which allows us to make this distinction in quantum mechanics. We find that the figure of merit in the classical and quantum cases is bounded by

$$\eta^{cl} = \frac{I_c^{cl}}{\max(S_a,S_b)} \leq 1$$

$$\eta^{qu} = \frac{I_c^{qu}}{\max(S_a,S_b)} \leq 2\,. \tag{18}$$

Quantum mechanics allows systems to be *twice* as strongly correlated as classical mechanics. Thus if we are to compare correlations in conventional classical and quantum mechanics we must probe the region $1 < \eta \leq 2$ in order to distinguish purely quantum effects. As the modified Araki-Lieb inequalities (10) hold for a given pair of observables, experiments on the correlation between a single pair of observables can never access this region. Indeed, the celebrated Bell inequalities can only be formulated by considering more than one observable for at least one of the correlated systems. The index of correlation has highlighted the fact that it is the pure state entanglement of probability amplitudes, for which there is no classical analogue, which admits this extra correlation in quantum mechanics. This excess correlation can be recovered in classical-like theories at the expense of the introduction of hidden variables.

Leaving aside the difficult question of the precise definition of entropies in classical mechanics the quantum mechanical bound (18) reveals the new and fundamental property of the squeezed state (6). For this state we find that $\eta = 2$ so that the modes in a squeezed state are as strongly correlated as quantum mechanics will allow. There are no two systems in quantum mechanics which can be more strongly correlated than this. This places the squeezed state (6) in a rather privileged position.

ACKNOWLEDGEMENTS

We would like to thank Harvey Brown for helpful comments.

REFERENCES

1. A. Einstein, B. Podolsky and N. Rosen, *Phys. Rev.*, **47** 777 (1935).
2. J. S. Bell, *J. de Physique,* Colloque C2, suppl. 3, **42** 41 (1981).
3. J. S. Bell, *Speakable and Unspeakable in Quantum Mechanics,* (Cambridge University Press, Cambridge, 1987).
4. W. H. Zurek, in *Quantum Optics, Experimental Gravitation and Measurement Theory,* eds. P. Meystre and M. O. Scully, (Plenum Press, New York, 1983).
5. S. M. Barnett and S. J. D. Phoenix, *Phys. Rev. A,* **40** 2404 (1989).
6. C. E. Shannon and W. Weaver, *The Mathematical Theory of Communication,* (University of Illinois Press, Urbana, 1963).
7. L. Brillouin, *Science and Information Theory,* (Academic Press, New York, 1962).
8. H. Araki and E. Lieb, *Commun. Math. Phys.*, **18** 160 (1970).
9. S. J. D. Phoenix and P. L. Knight, *Ann. Phys.* (*NY*), **186** 381 (1988).
10. S. M. Barnett and S. J. D. Phoenix, submitted to *Phys. Rev. A.*
11. S. M. Barnett and D. T. Pegg, *J. Mod. Opt.*, **36** 7 (1989).
12. S. M. Barnett and D. T. Pegg, *Phys. Rev. A,* in press.
13. A. Wehrl, *Rev. Mod. Phys.*, **50** 221 (1978).

PHYSICAL INFORMATION THEORY FOR 30 YEARS: BASIC CONCEPTS AND RESULTS

Lev B. Levitin

Boston University, Boston MA, USA

1 Introduction

Communication channels have been studied from physical point of view for more than 30 years. It is remarkable that quantum detection and estimation theory was first developed in the USA where a comprehensive monograph [1] was published in 1976 by C. W. Helstrom, while physical information theory was first developed in USSR, and the only monograph in this area [2] by V. V. Mityugov appeared there also in 1976. This paper is devoted to the latter area only.

The origin of the physical information theory ascends to L. Boltzmann and L. Szilard [3], who attributed an information meaning to the thermodynamic notion of entropy long before the quantitative measure of information was rigorously introduced by C. E. Shannon [4]. However, the information theory leading off with fundamental Shannon's work developed at first as a pure mathematical branch of science. D. Gabor in his pioneering works of the 50's [5] was, seemingly, the first who pointed out that "the communication theory should be considered as a branch of physics" and attempted to estimate the minimum energy required for transmission of one bit of information. Later, in the remarkable book by L. Brillouin [6], a profound relationship between physical entropy and information was suggested in the form of the "negentropy principle of information". Though the term "negentropy" was not rigorously defined, the principle implied that any information is represented by a certain state of a physical system and associated with its deviation from the thermodynamic equilibrium. The "entropy defect principle" [7–10] implements this basic idea in a mathematically consistent and rigorous form. According to this principle, any information is represented by an ensemble of states of a physical system and associated with the average deviation from the thermodynamic equilibrium caused by the choice of one of those states. Thus the information properties of real systems can be described in purely physical terms, which opens a way to developing a consistent physical information theory.

It should be borne in mind that the physical foundations of information theory must be essentially based on quantum theory. Indeed, in classical statistical physics, the entropy dimension is that of logarithm of action, and entropy includes, as was noted by Planck [11], an infinite additive constant, so that only difference of two values of entropy has a physical meaning, but not the absolute value of entropy. In contrast with this, according to quantum theory, every real physical system has a

Lecture Notes in Physics, Vol. 378
C. Bendjaballah, O. Hirota, S. Reynaud (Eds.),
Quantum Aspects of Optical Communications

well-determined finite ("absolute" by Planck) entropy which leades to a natural absolute measure of information associated with statistical ensembles of quantum macrostates.

On the other hand, the vigorous development of quantum electronics and microelectronics (VLSI), which brought into use the infrared and optical range of electromagnetic waves and allowed us to build systems with extremely low thermal noise, where quantum effects are paramount, calls for consideration of quantum nature of information transmission processes.

The physical information theory developed historically along the following three lines:

1. Investigation of the interrelations between basic concepts of statistical physics and information theory.
2. Investigation of the physical nature of information transmission and processing.
3. Application of information-theoretical concepts and approaches to the problems of statistical physics.

Numerous works have been devoted to the first two areas (e.g. [2,12–22]). In particular, some papers focused on investigation of non-classical (two-photon, squeezed) states of electromagnetic field [23–25], and on approaches to a global quantum description of an information channel [26]. The use of quantum carriers of information brings into consideration the fundamental properties of quantum measurements (e.g. [27-33]). It was found that a consistent quantum-mechanical treatment of information transmission calls for generalization of the basic information-theoretical concepts [9,10,34–37], The effect of the quantum-mechanical irreversibility of the measurement on the information transfer has been revealed in the quantum-mechanical formulation of the entropy defect principle [9,10,34]. The controversial problem of information efficiency of direct and indirect quantum measurements has been solved in [38].

Investigations in the third of above-mentioned areas (e.g. [6,18,20,22,39-44]) promise to shed a new light on the foundations of statistical physics, in particular, regarding the origin and the meaning of the Second Law of Thermodynamics. Results obtained in [10,42] lead to a conjecture that in the case of irreversible processes the change of information rather than that of entropy determines the amount of work produced by a non-equilibrium system, and there exists an equivalence relation between work and information.

Information theoretical concepts seem to be fruitfully applicable to some classical problems of statistical physics, such as the Ising model [45-46].

We describe below some important steps in the development of the basic concepts of physical information theory.

2 Entropy Defect Principle (Quasiclassical Formulation) and Ideal Physical Information Channels

The entropy defect principle was first formulated for quasiclassical case in [7,8], with the view to obtain a general expression for information (in Shannon's sence) in terms of physical entropy. The quasiclassical description of physical system implies, as usual, that all the operators corresponding to physical observables commute, which leads to two important premises:

1. all different microstates ("complexions", by Planck) of the system are inequivocally distinguishable by same (complete) measurement;
2. any macrostate of the system in uniquely represented by a statistical ensemble of microstates.

The entropy defect I_0 of a system is defined as the difference between the entropy of the macrostate of the system, which arises under the action of a signal chosen randomly from an ensemble of signals, on one hand, and the average entropy of the macrostates obtained by the action of determinate (prescribed) signal from the same ensemble. Premises 1 and 2 lead then to the entropy defect principle: the information, I, (in Shannon's sense) carried by an ensemble of states of a physical system is equal to its entropy defect:

$$I = I_0 \tag{1}$$

This principle has been applied to the ideal physical channels, i.e., channels uniquely specifies the microstates of the transmitted signal — the physical agent that carries the information, and the receiver measures the microstate of the received signal. The following results have been obtained in [41].

Theorem 2.1 *The capacity of a nonideal channel does not exceed the capacity of an ideal channel with the same transition probabilities, obeying the same constraints on the ensemble of microstates of the transmitted signal.*

Consider an ideal physical channel with statistically independent additive noise, which is an idealized model of systems where information carriers are particles obeying Bose-Einstein statistics. Let the ensemble average of the energy of the transmitted signal be E_0, that of the noise E_1. Then, by virtue of additivity, the average energy of the received signal is $E_2 = E_0 + E_1$. Assume, moreover, that the noise is thermal, with temperature $T_1 = T(E_1)$, where $T(E)$ is the temperature of the system as a function of its average energy.

Theorem 2.2 *The maximum amount of information that can be transmitted over an ideal physical channel with statistically independent additive noise is*

$$I_{\max} = \int_{E_1}^{E_2} \frac{dE}{kT} = \int_{T_1}^{T_2} \frac{1}{kT} \frac{dE(T)}{dT} dT \tag{2}$$

and this maximum is achieved when the distribution of microstates of the received signal corresponds to thermodynamic equilibrium at the temperature T_2 *determined by*

$$\int_{T_1}^{T_2} \frac{dE(T)}{dT} dT = E_0. \tag{3}$$

Here $E(T)$ *is the average energy of the system as a function of temperature;* $E(T_2) = E_2$.

Corollary 2.1 *The minimum amount of energy necessary to transmit one natural unit of information over a channel with statistically independent additive noise satisfies the inequality*

$$E_{\min} = \frac{E_0}{I} \geq kT_1 \tag{4}$$

where T_1 *is the temperature of the noise. Equality holds asymptotically in case of a weak signal, i.e. when*

$$\frac{T_2 - T_1}{T_2} \ll 1. \tag{5}$$

This corollary is a rigorous specialization of the well-known conjecture by Brillouin [6] that the minimum energy necessary to obtain one bit of information at temperature T_1 is $kT_1 \ln 2$.

In particular, the minimum energy per nat has been calculated in [7,8,13] for a broadband one-dimensional photon channel:

$$E_{\min} = kT + \frac{3}{\pi}\hbar R. \tag{6}$$

Here $\hbar$ is Planck's constant, R is the rate of information transmission (nat/s). This elegant formula clearly shows both the effect of thermal noise and the limitation imposed by the quantum structure of the electromagnetic field: the faster one transmits information, the more energy required per each unit of information. Other types of ideal physical channels have been investigated in [7,8,12-14,48,49].

3 Entropy Defect Principle (Quantum Formulation)

The assumptions 1 and 2 of Section 2 are not valid for a consistent quantum-mechanical description. In this section we generalize the concept of information (by Shannon) for quantum statistical ensembles, obtain the consistently quantum-theoretical formulation of the entropy defect principle and analyze the effect of irreversibility of a quantum measurement on information transmission.

In quantum mechanics a macrostate s of a physical system is described by a density operator (density matrix) ρ which is a self-adjoint nonnegative definite operator with $Tr\,\rho = 1$ in a separable Hilbert space $\mathcal{H}$. A microstate (pure state) of a physical system is described by a state vector (wave function) $\psi \in \mathcal{H}$ of a length equal to 1, or alternatively, by a density operator ρ of a special form, namely such that any matrix element $\rho_{nn'} = \psi_n \bar{\psi}_{n'}$ (the bar denotes complex conjugation). Any complete direct quantum-mechanical measurement is associated with a certain countable orthonormal basis L in the space $\mathcal{H}$, the probability of the k-th outcome being given by the diagonal element $\rho_{kk}(L)$ of the density matrix ρ in the basis L:

$$\rho_{kk}(L) = Tr(\rho P_k) = \sum_n \sum_{n'} \rho_{nn'}(L)\delta_{kn'}\delta_{n'n}. \tag{7}$$

Here P_k is a projection operator, having in the basis L a form:

$$||P_{k,nn'}(L)|| = ||\delta_{kn}\delta_{nn'}||. \tag{8}$$

Obviously, P_k is a nonnegative definite operator, and the set of all P_k ($k = 1, 2, \ldots$) forms an orthogonal resolution of the identity $\mathbf{1}$ in the space $\mathcal{H}$:

$$P_k \geq 0; \quad \sum_k P_k = 1; \quad Tr(P_k P_{k'}) = 0 \quad (k \neq k'). \tag{9}$$

Note that each P_k is a density operator corresponding to a microstate m_k described by a state vector ψ_k which is one of the basis vectors: $\psi_k \in L$. Thus, the physical meaning of the k-th outcome of the measurement is that the system is found in the microstate m_k.

The entropy of a quantum system in a macrostate s described by a density matrix ρ is defined as follows [50,51]

$$H = -Tr\, \rho \ln \rho. \tag{10}$$

A quantum-mechanical measurement changes the state of a system in an irreversible way: the entropy of the ensemble of microstates obtained as a result of a measurement is, generally speaking, larger than the entropy of the initial state. Namely, by Klein's lemma [50,51]

$$-\,Tr\, \rho \ln \rho \leq -\sum \rho_{kk}(L) \ln \rho_{kk}(L), \tag{11}$$

where equality holds if ρ is diagonal in basis L.

Consider now an ensemble $S = \{s_i, p_i\}$ of macrostates of a physical system, each macrostate s_i occuring with probability p_i and being described by a density matrix $\rho^{(i)}$. (The preparation of a certain macrostate s_i can be interpreted as "physical encoding" of a signal x_i taken from an ensemble of signals $X = \{x_i, p_i\}$.) The a priori state of the system s chosen at random from the ensemble S is described by a density matrix ρ:

$$\rho = \sum_i p_i \rho^{(i)} \tag{12}$$

Definition 3.1 *The entropy defect of a system described by the ensemble of macrostates S is the quantity*

$$I_0 = -Tr\, \rho \ln \rho + \sum_i p_i Tr\, \rho^{(i)} \ln \rho^{(i)}. \tag{13}$$

Entropy defect characterizes the average decrease of the entropy of the system, when it becomes known, which macrostate (out of the ensemble S) has been chosen.

Theorem 3.1

$$0 \leq I_0 \leq -\sum_i p_i \ln p_i. \tag{14}$$

The lefthand equality hold iff all $\rho^{(i)}$ are identical, the righthand equality holds iff all $\rho^{(i)}$ are orthogonal (i.e. $Tr\,\rho^{(i)}\rho^{(i')} = 0,\ i \neq i'$).

Thus, entropy defect does not exceed the entropy of the signal $H(X) = -\sum_i p_i \ln p_i$.

We have seen in Section 2 that, in the quasiclassical case, the entropy defect is equal to the information about the macrostate of the system obtained by measuring its microstates. The situation is different and more complex in quantum theory. Information about the macrostate (or about the signal, since the random macrostate S and signal X are in one-to-one correspondence) depends on the choice of the complete set of observables to be measured.

Definition 3.2 *Information about the macrostate S of a physical system obtained in a complete measurement associated with a countable orthonormal basis L in the Hilbert space $\mathcal{H}$ is the quantity*

$$I_L = -\sum_k \rho_{kk}(L) \ln \rho_{kk}(L) + \sum_i \sum_k p_i \rho_{kk}^{(i)}(L) \ln \rho_{kk}^{(i)}(L) \tag{15}$$

Here $\rho_{kk}(L)$ and $\rho_{kk}^{(i)}(L)$ are the diagonal elements of ρ and $\rho^{(i)}$, respectively, in basis L.

Definition 3.3 *Information about the macrostate S of a physical system obtainable by direct measurement is the quantity*

$$I = \sup_L I_L, \tag{16}$$

where the least upper bound is taken over all possible choices of L (or, in other words, over all possible orthonormal resolutions of identity in $\mathcal{H}$).

The quantity I plays the same role in information transmission as information by Shannon in the classical theory. (Note that here we restrict ourselves to direct measurements only.)

Theorem 3.2

$$I \leq I_0, \tag{17}$$

and the equality holds iff all the matrices $\rho^{(i)}$ commute. In this case

$$I = I_{L_0} = I_0 \tag{18}$$

for a basis L_0 in which all $\rho^{(i)}$ are diagonal.

Theorem 3.2 is the quantum-mechanical counterpart of the entropy defect principle. It shows that the irreversibility of a quantum measurement results in inevitable loss of information (except for the case when all the density matrices are diagonal, and the quasiclassical approximation is applicable).

4 Information in Indirect Quantum Measurements

The concept of quantum measurement can be extended by introducing indirect measurements. Consider again a quantum system with an ensemble of macrostates $S = \{s_i, p_i\}$, each macrostate s_i being described by a density matrix $\rho^{(i)}$ in a Hilbert space $\mathcal{H}_1$. Consider an auxiliary quantum system (called in [1] "ancilla") whose state is independent of the state of the system carrying information and described by a density operator $\rho^{(0)}$ in a separable Hilbert space $\mathcal{H}_2$. Then the state of the compound system consisting of the two systems together is described by a tensor-product density operator

$$\tilde{\rho}^{(i)} = \rho^{(i)} \otimes \rho^{(0)} \tag{19}$$

in the tensor-product Hilbert space $\mathcal{H} = \mathcal{H}_1 \otimes \mathcal{H}_2$. Denote also

$$\tilde{\rho} = \rho \otimes \rho^{(0)} = \sum_i p_i \rho^{(i)} \otimes \rho^{(0)} \tag{20}$$

A complete indirect (generalized) measurement is a usual quantum measurement performed over the compound system and associated with a certain countable orthonormal basis K in the space $\mathcal{H}$.

Definition 4.1 *Information about the macrostate S of the quantum system for a given state of the ancilla in the outcome of an indirect measurement associated with a basis K in the space $\mathcal{H}_1 \otimes \mathcal{H}_2$ is the quantity*

$$J_K = -\sum_k \sum_n \rho_{kn,kn}(K) \ln \rho_{kn,kn}(K) + \sum_i \sum_k \sum_n p_i \rho^{(i)}_{kn,kn}(K) \ln \rho^{(i)}_{kn,kn}(K), \tag{21}$$

where $\rho_{kn,kn}$ and $\rho^{(i)}_{kn,kn}$ are the diagonal elements of density matrices $\tilde{\rho}$ and $\tilde{\rho^{(i)}}$, respectively, in the basis K.

Definition 4.2 *Information about the macrostate S of a physical system obtainable by indirect measurement is the quantity*

$$J = \sup_k \sup_{\rho^{(0)}} J_K \tag{22}$$

It has been shown by Naimark in 1940 that any direct measurement, i.e. any orthogonal resolution of identity in $\mathcal{H}$ is equivalent to a resolution of identity (generally speaking, non-orthogonal) in $\mathcal{H}_1$, and, conversely, any non-orthogonal resolution of identity in $\mathcal{H}_1$ can be considered as induced by an orthogonal resolution of identity in a certain tensor-product space $\mathcal{H}$.

Indirect quantum measurements were discussed in a number of works (e.g. [1,2,27–29,31–33,35–38]). It was shown that randomized, successive and adaptive quantum measurements are equivalent to some indirect measurement [32], and, thus, indirect measurements are the most general kind of quantum measurements. It was also shown [36] that

$$J \leq I_0, \tag{23}$$

where the equality holds iff all the density matrices $\rho^{(i)}$ commute.

However it remained a controversial problem for more than ten years, whether indirect measurements can yield more information than the direct ones. In [35] an example was constructed, where, indeed, $J > 1$. This example, however, uses "state vectors" in a two-dimensional Euclidean space and therefore it seem to be not relevant to quantum systems with states described by operators in Hilbert (infinite-dimensional) space. Intuitively, the phenomenon that J can exceed I in a finite-dimensional space is due to the fact that in such cases the cardinality $|S|$ of the set of states exceeds the dimensionality d of the space in which they are embedded. (It has been shown by Davies [37] that the cardinality of the set $\{B_K\}$ of self-adjoint operators used in the optimal indirect measurement obeys an inequality

$$d \leq |\{B_K\}| \leq d^2 \tag{24}$$

but, obviously, this result is not applicable in infinite-dimensional case.) Indeed it was finally proved [38] that the situation is different in an infinite-dimensional Hilbert space.

Theorem 4.1 *For any set of density operators $\rho^{(i)}$ in a separable (infinite-dimensional) Hilberet space, for any set of probabilities p_i such that $H = -\sum_i p_i \ln p_i < \infty$ and for any ancilla with a separable Hilbert or Euclidean space*

$$I\{\rho^{(i)}; p_i\} = J\{\rho^{(i)}; p_i\}. \tag{25}$$

Theorem 4.1 shows that in the case of a quantum system associated with a separable Hilbert space direct measurements are as efficient as more general indirect measurements.

References

[1] Helstrom, C. W., "Quantum Detection and Estimation Theory," Academic Press, New York, 1976.

[2] Mityugov, V. V., "Physical Foundations of Information Thoery," Sovietskoe Radio, Moscow, 1976 (in Russian).

[3] Szilard, L., "Ober die Entropieverminerung in einem thermodynamischen System bei Eingriff intelligenter Wesen," Z. Physik, B. 53, No. 5, 1929, pp. 840–856 (in German).

[4] Shannon, C. E. and W. Weaver, "The Mathematical Theory of Communication," Urbana Univ. Press., Chicago, IL, 1949.

[5] Gabor, D., "Communication Theory and Physics," Phil. Mag., 41, 1950, pp. 1161–1187.

[6] Brillouin, L., "Science and Information Theory," Acad. Press, New York, 1956.

[7] Lebędev, D. S. and L. B. Levitin, "Transfer of Information by Electromagnetic Field," in "Theory of Information Transmission," Nauka, Moscow, 1964, 1-20 (in Russian).

[8] Lebedev, D. S. and L. B. Levitin, "Information Transmission by Electromagnetic Field," Inform. Contr. 9, No. 1, 1966, 1-22.

[9] Levitin, L. B., "On the Quantum Measure of the Amount of Information," Proc. IV National Conf. on Information Theory, Tashkent, 1969, pp. 111–115 (in Russian).

[10] Levitin, L. B., "Information Theory for Quantum Systems," in "Information, Complexity and Control in Quantum Physics," Springer-Verlag, New York, 1987, pp. 15–47.

[11] Planck, M., "Warmestrahlung," Berlin 1913 (Engilsh Translation, "Theory of Heat Radiation," Dover, New York, 1959).

[12] Gordon, J. P., "Quantum Effects in Communication Systems," Proc. IRE 60, No. 9, 1962, pp. 1898–1908.

[13] Lebedev, D. S. and L. B. Levitin, "The Maximum Amount of Information Transmissible by Electromagnetic Field," Soviet Physics-Doklady, 8, 1963, pp. 377–379.

[14] Takahasi, H., "Information Theory of Quantum-Mechanical Channels," in "Advances in Communication Systems," Vol. 1, Acad. Press, New York, 1965, pp. 227–310.

[15] Stratonovich, R. L., "Information Transmission Rate in some Quantum Communication Channels," Probl. Inform. Transm., 2, 1966, pp. 45–57.

[16] Mityugov, V. V., "On Quantum Theory of Information Transmission," Probl. Info. Transm., 2, 1966, pp. 48–58.

[17] Ingarden, R. S., "Quantum Information Theory," Institut of Physics, N. Copernicus Univ., Torun, Poland, 1975.

[18] Vainshtein, V. D. and S. D. Tvorogov, "Some Problems of Measurements of Quantum Observables and Determination of Joint Entropy in Quantum Statistics," Comm. Math. Phys., 43, 1975, pp. 273–278.

[19] Drikker, A. S., "Homodine Reception of a Quantum Electromagnetic Signal," Probl. Inform. Transm., 12, 1976, pp. 57–68.

[20] Poplavskii, R. P., "Thermodynamic Models of Information Processes," Sov. Phys. Uspekhi, 115, 1975, pp. 222–241.

[21] Pierce, J. B., E. C. Posner, and E. C. Rodemich, "The Capacity of the Photon Counting Channel," IEEE Trans. on Inform. Theory, IT-27, 1981, pp. 61–77.

[22] Kosloff, R., "Therodynamics Aspects of the Quantum-Mechanical Measuring Process," in "Advances in Chemical Physics," Wiley and Sons, 1981, pp.153–193.

[23] Vourdas, A., "Superposition of Squeezed Coherent States with Thermal Light," Phys. Rev. A, 34, 1986, pp. 3466–3469.

[24] Walls, D. F., "Squeezed States of Light," Nature, 306, 1983, pp. 141–146.

[25] Yuen, H. P. and J. H. Shapivo, "Optical Communication Theory with Two-Photon Coherent States—Part III," IEEE Trans. on Inform. Theory, IT-26, 1980, pp. 78–92.

[26] Ohya, M., "On Compound State and Mutual Information in Quantum Information Theory," IEEE Trans. on Inform. Theory, IT-29, 1983, pp. 770–774.

[27] Arthurs, E. and J. L. Kelly, "On the Simultaneous Measurement of a Pair of Conjugate Observables," Bell System Tech. J., 44, 1965, pp. 725–729.

[28] Gordon, J. P. and W. J. Louisell, "Simultaneous Measurement of NonCommuting Observables," in "Physics of Quantum Electronics," McGraw-Hill, New York, 1966, pp. 833–840.

[29] Levitin, L. B. and V. V. Mitiugov, "Reception of Coherent Signals by Splitting of the Beam," 1st Conf. on Problems of Information Transmission by Laser Radiation, Kiev, 1968.

[30] Helstrom, C. W., "Detection Theory and Quantum Mechanics," Inform. Contr., 10, 1967, pp. 254–291.

[31] Davis, E. B. and J. T. Lewis, "An Operational Approach to Quantum Probability," Comm. Math. Phys., 17, 1970, pp. 239–260.

[32] Benioff, P. A., "Decision Procedures in Quantum Mechanics," J. Math. Phys., 13, 1972, pp. 908–915.

[33] Helstrom, C. W. and R. S. Kennedy, "Noncommuting Observables in Quantum Detection and Estimation Theory," IEEE Trans. on Inform. Theory, IT-20, 1974, pp. 16–24.

[34] Levitin, L. B., "Amount of Information and the Quantum-Mechanical Irreversibility of Measurement," Proc. of the II Int. Symp. on Inform. Transm. Theory, Yerevan, 1971, pp. 144–147.

[35] Holevo, A. S., "Informational Aspects of Quantum Measurements," Probl. Inform. Transm., 9, 1973, pp. 31–42.

[36] Holevo, A. S., "Certain Estimates of Information Transmissible Over a Quantum Communication Channel," Probl. Inform. Transm., 9, 1973, pp. 3–11.

[37] Davies, E. B., "Information and Quantum Measurement," IEEE Trans. on Inform. Theory, IT-24, 1979, pp. 596–599.

[38] Levitin, L. B., "Direct and Indirect Quantum Measurements Yield Equal Maximum Information," IEEE Int. Symp. on Information Theory, Santa Monica, CA, 1981.

[39] Jaynes, E. T., "Information Theory and Statistical Mechanics," Phys. Rev., Part I, 106, 1957, pp. 620–630; Part II, 108, 1959, pp. 171–190.

[40] Katz, A., "Principles of Statistical Mechanics: The Information Theory Approach," W. H. Freeman, 1967.

[41] Ingarden, R. S, "Information Theory and Thermodynamics," Part I, Torun, Poland, 1974; Part II, Torun, Poland, 1976.

[42] Levitin, L. B., "Quantum Amount of Information and Maximum Work," in "Proc. of the 13th IUPAP Conf. on Statistical Physics," A. Hilger, Bristol, England, 1978.

[43] Poplavskii, R. P., "Maxwell Demon and Correlation Between Information and Entropy," Sov. Phys. Uspekhi, 128, 1979, pp. 165–176.

[44] Mityugov, V. V., "Entropy, Information and Work in Quantum Statistics," Probl. Control and Inform. Theory, 2, 1973, pp. 243–256.

[45] Berger, T., "Communication Theory via Random Fields," IEEE Int. Symp. on Inform. Theory, St. Jovite, Quebec, Canada, 1983.

[46] Levitin, L. B., "Information-Theoretical Approach to Ising Problem," IEEE Int. Symp. on Inform. Theory, Brighton, England, 1985.

[47] Levitin, L. B., "A Thermodynamic Characterization of Ideal Physical Information Channels," Journal of Information and Optimization Sciences, 2, 1981, pp. 259–266.

[48] Levitin, L. B., "Information Transmission in an Ideal Photon Channel," Probl. Inform. Transm., 1, 1965, pp. 55–62.

[49] Levitin, L. B., "Ideal Corpuscular Information Channels," IEEE Int. Symp. on Inform. Theroy, St. Jovite, Quebec, Canada, 1983.

[50] Klein, O. "Zur Quantenmechanischen Begrundung des zweiten Hauptsatzes der Warmelehre," Z. Phys., 72, 1931, pp. 767–775.

[51] Neumann, J., "Mathematische Grundlagen der Quantenmechanik," Springer Verlag, Berlin, 1932 (English Translation, "Mathematical Foundations of Quantum Mechanics," Princeton Univ. Press, Princeton, NJ, 1955).

Quantum Nondemolition Measurement in Optical Communication Theory

Tomohiko UYEMATSU† Osamu HIROTA†† Kohichi SAKANIWA†

†Tokyo Institute of Technology, Tokyo 152, JAPAN

††Tamagawa University, Machida-shi, Tokyo 194, JAPAN

1. Introduction

According to the development of quantum optics, a new measurement scheme called "quantum nondemolition (QND) measurement" has been theoretically discussed[1-5] and some experiments of this scheme have been reported [6,7]. Quantum nondemolition measurement means that the back action noise arising from the measurement may be evaded in its process. This measurement scheme provides a new aspect of quantum optical communication besides the measurement itself. For example, we have already proposed a lossless optical coupler for optical network system[8] by employing a back-action evading (BAE) amplifier[9], which can amplify one of the quadrature amplitudes of light without degradation of signal to quantum noise ratio. However, the proposed optical coupler is essentially a QND measurement of the quadrature amplitude of light. Thus, in order to analyze its characteristics precisely, the theory of QND measurement must be reformulated from the view point of the optical communication theory.

In this paper, we shall describe the theory of QND measurement from the view point of optical communication theory. Namely, the quantum noise caused by the QND measurement is analyzed precisely. First, we show that standard quantum theory enables us to analyze the QND measurement of a quadrature amplitude of light. Next, we obtain the probability densities for quadrature amplitudes in the QND measurement including the state reduction after the measurement. Moreover, we show that the QND measurement is completely described by a specific channel in the communication theory.

2. QND Measurement for Quadrature Amplitude

In this section, we discuss a QND measurement scheme for quadrature amplitudes of lights.

First, we describe a definition of QND measurement.

[Definition] (QND measurement)

Assume that the state of the signal light is in an eigenstate $|a>$ of the observable $\widehat{A}$ which is to be measured. After the measurement of the observable $\widehat{A}$, if the state of signal light remains in $|a>$, namely, the state is conserved, this measurement is called "QND measurement" for $\widehat{A}$.

We consider one of the quadrature amplitudes in the signal light

$$\widehat{X}_1(t) \triangleq a_s e^{i\omega_s t} + a_s^\dagger e^{-i\omega_s t} \tag{1}$$

is measured, where a_s and ω_s denote the photon annihilation operator and the frequency of the signal light. To measure the observable $\widehat{X}_1(t)$, it is necessary to couple the signal

Lecture Notes in Physics, Vol. 378
C. Bendjaballah, O. Hirota, S. Reynaud (Eds.),
Quantum Aspects of Optical Communications

$$\times \frac{1}{\sqrt{2\pi}} \cdot \frac{1}{\sqrt{\sigma_\alpha^2 G^2 + \sigma_\beta^2}} e^{-\frac{1}{2(\sigma_\alpha^2 G^2+\sigma_\beta^2)}(y_2-\beta_2-G\alpha_1)^2}, \tag{16}$$

where G is the time dependent amplitude gain given by eq.(12). Hence, the probability density for $\widehat{Y}_2$ is given by

$$p(y_2,t) = \int_{-\infty}^{\infty} p(x_1,y_2,t)\,dx_1$$
$$= \frac{1}{\sqrt{2\pi}} \cdot \frac{1}{\sqrt{\sigma_\alpha^2 G^2 + \sigma_\beta^2}} e^{-\frac{(y_2-\beta_2-G\alpha_1)^2}{2(\sigma_\alpha^2 G^2+\sigma_\beta^2)}}. \tag{17}$$

This implies that $\widehat{Y}_2$ has a Gaussian distribution with average $\beta_2 + G\alpha_1$ and variance $(\sigma_\alpha^2 G^2 + \sigma_\beta^2)$. Thus, if we read out a value for $\widehat{Y}_2$ and get a result y_2^{out}, the estimated value for $\widehat{X}_1$ is given by

$$\alpha_1' = \frac{1}{G}(y_2^{out} - \beta_2). \tag{18}$$

Moreover, if $\beta_2 = 0$, the signal to quantum noise ratio (SNR) for this QND measurement is given by

$$SNR_m \triangleq \frac{<\widehat{Y}_2(t)>^2}{<\Delta\widehat{Y}_2^2(t)>} = \frac{G^2\alpha_1^2}{\sigma_\alpha^2 G^2 + \sigma_\beta^2}. \tag{19}$$

It should be noted that if the amplitude gain G becomes large, SNR_m approaches to the SNR of the signal, namely

$$\lim_{|G|\to\infty} SNR_m = \frac{\alpha_1^2}{\sigma_\alpha^2} = \frac{<\widehat{X}_1(0)>^2}{<\Delta\widehat{X}_1^2(0)>}. \tag{20}$$

Next, we analyze the effect of probe-state reduction when the measurement on $\widehat{Y}_2$ is made. At time t, we read out a value for $\widehat{Y}_2$ and get a result y_2^{out}. Then, the probability density for $\widehat{X}_1$ is given by

$$p_a(x_1,t|y_2^{out}) = \frac{p(x_1,y_2^{out},t)}{p(y_2^{out},t)}$$
$$= \frac{1}{\sqrt{2\pi}} \cdot \frac{\sqrt{\sigma_\alpha^2 G^2 + \sigma_\beta^2}}{\sigma_\alpha \sigma_\beta} e^{-\frac{\sigma_\alpha^2 G^2+\sigma_\beta^2}{2\sigma_\alpha^2\sigma_\beta^2}\left(x_1 - \frac{\sigma_\alpha^2 G^2 \alpha_1' + \sigma_\beta^2 \alpha_1}{\sigma_\alpha^2 G^2 + \sigma_\beta^2}\right)^2}, \tag{21}$$

where α_1' is defined by eq.(18). Thus, $\widehat{X}_1$ has a Gaussian distribution with average $(\sigma_\alpha^2 G^2 \alpha_1' + \sigma_\beta^2 \alpha_1)/(\sigma_\alpha^2 G^2 + \sigma_\beta^2)$ and variance $\sigma_\alpha^2\sigma_\beta^2/(\sigma_\alpha^2 G^2 + \sigma_\beta^2)$ after the readout of $\widehat{Y}_2$. This implies that the average of the signal light approaches to its estimate α_1' as $\sigma_\alpha^2 G^2$ becomes large. On the other hand, the variance of $\widehat{X}_1$ after the readout of $\widehat{Y}_2$ is smaller than its initial variance σ_α^2. Moreover, the variance is decreased as the amplitude gain $|G|$ becomes large.

light to another light called probe light. Then, we define the quadrature amplitudes $\widehat{Y}_1(t)$ and $\widehat{Y}_2(t)$ of the probe light as follows:

$$\left.\begin{aligned}\widehat{Y}_1(t) &\triangleq a_p e^{i\omega_p t} + a_p^\dagger e^{-i\omega_p t}\\ \widehat{Y}_2(t) &\triangleq -i(a_p e^{i\omega_p t} - a_p^\dagger e^{-i\omega_p t})\end{aligned}\right\}, \tag{2}$$

where a_p and ω_p denote the photon annihilation operator and the frequency of the probe light. We assume that $\widehat{Y}_2$ is the readout observable of the probe light and that the coupling between the signal and probe lights is described by the interaction Hamiltonian given by

$$\widehat{H}_I = \hbar K \widehat{X}_1(t)\widehat{Y}_1(t), \tag{3}$$

where K is the coupling constant. Hamiltonians of the form (3) are known to be consistent with QND measurements and studied in many literatures[5,7,9,10].

Now, we give an analysis of this measurement scheme. If we start the interaction at time $t = 0$, the state vector at time t in the Schrödinger picture satisfies the following equation of motion:

$$i\hbar\frac{\partial}{\partial t}|\psi(t)>= (\widehat{H}_s + \widehat{H}_p + \widehat{H}_I)|\psi(t)>, \tag{4}$$

where $\widehat{H}_s$ and $\widehat{H}_p$ are defined by

$$\left.\begin{aligned}\widehat{H}_s &\triangleq \hbar\omega_s a_s^\dagger a_s\\ \widehat{H}_p &\triangleq \hbar\omega_p a_p^\dagger a_p\end{aligned}\right\}, \tag{5}$$

and denote the free space Hamiltonians for the signal and probe lights. This equation of motion is solved in Appendix A with the result

$$|\psi(t)>= e^{-iKt\widehat{X}_1(t)\widehat{Y}_1(t)}e^{-i(\widehat{H}_s+\widehat{H}_p)t/\hbar}|\psi(0)>. \tag{6}$$

From this state vector, the averages for $\widehat{X}_1(t)$ and $\widehat{Y}_2(t)$ can be evaluated. As for $\widehat{X}_1(t)$, we have

$$\begin{aligned}<\psi(t)|\widehat{X}_1(t)|\psi(t)> &=<\psi(0)|e^{i(\widehat{H}_s+\widehat{H}_p)t/\hbar}\widehat{X}_1(t)e^{-i(\widehat{H}_s+\widehat{H}_p)t/\hbar}|\psi(0)>\\ &=<\psi(0)|\widehat{X}_1(0)|\psi(0)>.\end{aligned} \tag{7}$$

As for $\widehat{Y}_2(t)$, according to the relation

$$[\widehat{Y}_2(t), e^{-iKt\widehat{X}_1(t)\widehat{Y}_1(t)}] = -(2Kt)e^{-iKt\widehat{X}_1(t)\widehat{Y}_1(t)}\widehat{X}_1(t), \tag{8}$$

we have

$$\begin{aligned}<\psi(t)|\widehat{Y}_2(t)|\psi(t)> &=<\psi(0)|e^{i\widehat{H}_p t/\hbar}\widehat{Y}_2(t)e^{-i\widehat{H}_p t/\hbar}|\psi(0)>\\ &\quad-(2Kt)<\psi(0)|e^{i(\widehat{H}_s+\widehat{H}_p)t/\hbar}\widehat{X}_1(t)e^{-i(\widehat{H}_s+\widehat{H}_p)t/\hbar}|\psi(0)>\\ &=<\psi(0)|\widehat{Y}_2(0)|\psi(0)> -(2Kt)<\psi(0)|\widehat{X}_1(0)|\psi(0)>.\end{aligned} \tag{9}$$

These equations are equivalent to the relations of the quadrature amplitudes

$$\widehat{X}_1^{out} = \widehat{X}_1^{in}, \tag{10}$$
$$\widehat{Y}_2^{out} = \widehat{Y}_2^{in} + G\widehat{X}_1^{in}, \tag{11}$$

where the superscripts *in* and *out* refer to the input and output lights, and the amplitude gain G is defined by

$$G \triangleq -2Kt. \tag{12}$$

These equations are exactly the input and output relations for the BAE amplifier[7,9]. Therefore, the system with interaction Hamiltonian shown in eq.(3) can be regarded as a BAE amplifier.

Next, the wave function of this measurement system is obtained in Appendix B, and is given by

$$\Psi(x_1, y_2, t) = < y_2 + (2Kt)x_1|\eta >< x_1|\xi >, \tag{13}$$

where $|\xi>$ and $|\eta>$ denote the initial states of the signal and probe lights, while $|x_1>$ and $|y_2>$ are the eigenstates for $\widehat{X}_1(0)$ and $\widehat{Y}_2(0)$. This wave function implies that $\widehat{X}_1$ can be determined by making measurement on $\widehat{Y}_2$. Moreover, if the initial state of the signal light is in an eigenstate of $\widehat{X}_1(0)$, the state of the signal light remains in its eigenstate. Therefore, the system satisfies the condition of QND measurement.

3. Channel Model for QND Measurement

In this section, we show that a QND measurement can be modeled by a specific channel in communication theory.

Assume that the signal and probe lights are initially in squeezed states of Yuen [11] given by

$$\left.\begin{array}{ll} |\xi>=|\alpha;\mu_\alpha,\nu_\alpha> & (\alpha=\alpha_1+i\alpha_2) \\ |\eta>=|\beta;\mu_\beta,\nu_\beta> & (\beta=\beta_1+i\beta_2) \end{array}\right\}, \tag{14}$$

where the squeezing parameters μ_α, ν_α, μ_β and ν_β are real numbers. The variances for $\widehat{X}_1(0)$ and $\widehat{Y}_2(0)$ are given by

$$\left.\begin{array}{l} \sigma_\alpha^2 \triangleq < \Delta\widehat{X}_1^2(0) >= (\mu_\alpha - \nu_\alpha)^2 \\ \sigma_\beta^2 \triangleq < \Delta\widehat{Y}_2^2(0) >= (\mu_\beta + \nu_\beta)^2 \end{array}\right\}. \tag{15}$$

Then, from eq.(13), the quantum probability density for the system at time t is given by

$$\begin{aligned} p(x_1, y_2, t) &\triangleq |\Psi(x_1, y_2, t)|^2 \\ &= \frac{1}{\sqrt{2\pi}\sigma_\beta} e^{-\frac{(y_2 - Gx_1 - \beta_2)^2}{2\sigma_\beta^2}} \cdot \frac{1}{\sqrt{2\pi}\sigma_\alpha} e^{-\frac{(x_1-\alpha_1)^2}{2\sigma_\alpha^2}} \\ &= \frac{1}{\sqrt{2\pi}} \cdot \frac{\sqrt{\sigma_\alpha^2 G^2 + \sigma_\beta^2}}{\sigma_\alpha\sigma_\beta} e^{-\frac{\sigma_\alpha^2 G^2 + \sigma_\beta^2}{2\sigma_\alpha^2\sigma_\beta^2}\left(x_1 - \frac{\sigma_\alpha^2 G(y_2-\beta_2)+\sigma_\beta^2\alpha_1}{\sigma_\alpha^2 G^2+\sigma_\beta^2}\right)^2} \end{aligned}$$

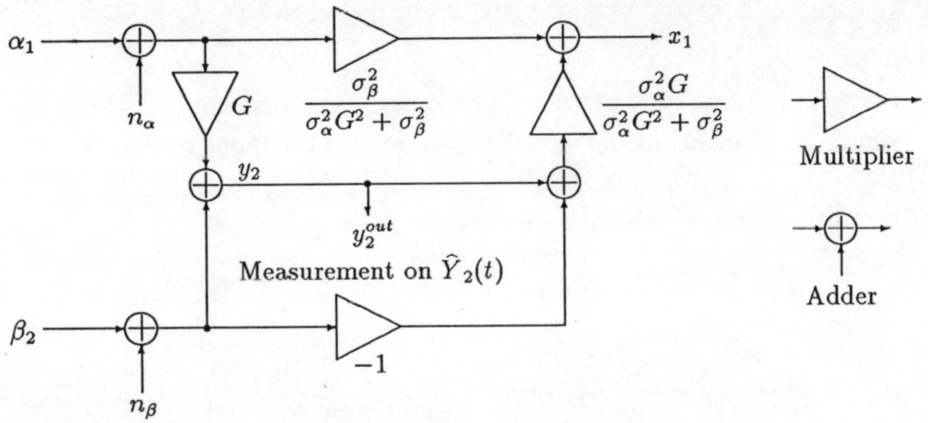

Figure 1: Channel model for QND measurement

In order to show the intuitive meaning of probe-state reduction, consider the probability density for $\widehat{X}_1$ without measurement of $\widehat{Y}_2$. From eq.(16), this density is given by

$$\begin{aligned} p(x_1, t) &= \int_{-\infty}^{\infty} p(x_1, y_2, t)\, dy_2 \\ &= \frac{1}{\sqrt{2\pi}\sigma_\alpha} e^{-\frac{(x_1-\alpha_1)^2}{2\sigma_\alpha^2}} \\ &= p(x_1, 0), \end{aligned} \tag{22}$$

and is equal to that of the initial squeezed state $|\alpha; \mu_\alpha, \nu_\alpha>$. Thus, if we do not measure $\widehat{Y}_2(t)$, the probability density for $\widehat{X}_1$ remains unchanged. Moreover, for the observer who does not know the readout value of $\widehat{Y}_2(t)$, the probability density for $\widehat{X}_1(t)$ is also given by

$$\int_{-\infty}^{\infty} p_a(x_1, t|y_2)\, p(y_2, t)\, dy_2 = p(x_1, t) = p(x_1, 0), \tag{23}$$

whether there is a measurement on $\widehat{Y}_2(t)$ or not. Therefore, we believe that the effect of the state reduction is confined only to the observer who knows the readout value y_2^{out}.

From the above discussions, we can propose that the QND measurement process can be regarded as a communication channel shown in Fig.1, where n_α and n_β denote mutually independent zero-mean white Gaussian noises with variances σ_α^2 and σ_β^2, respectively. In the proposed model, y_2 can be written as

$$y_2 = G\alpha_1 + \beta_2 + Gn_\alpha + n_\beta. \tag{24}$$

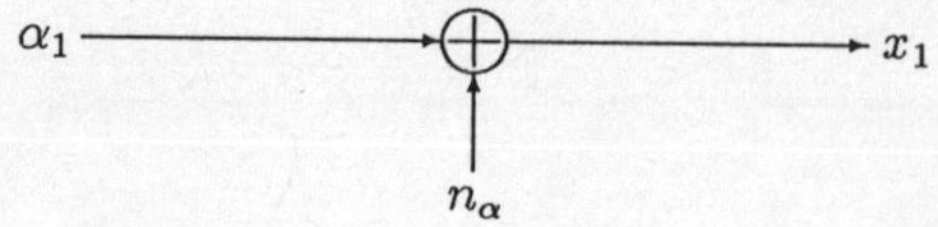

Figure 2: Channel model for QND measurement without readout of y_2

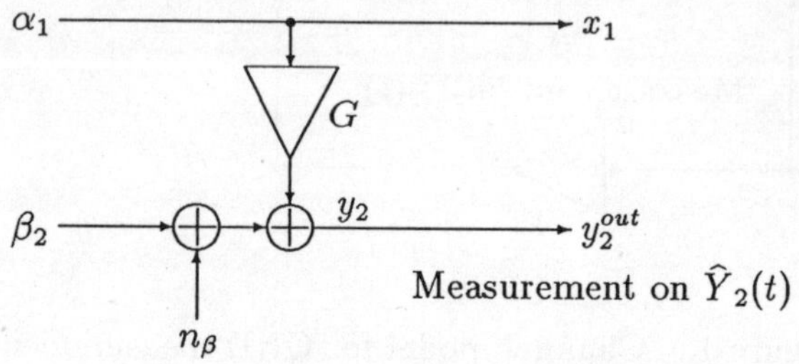

Figure 3: Special case of QND measurement ($\sigma_\alpha = 0$)

Thus, the probability density for y_2 is equal to eq.(17). Next, assume that we know a readout value y_2^{out}. Then, x_1 can be written as

$$x_1 = \frac{\sigma_\alpha^2 G(y_2^{out} - \beta_2) + \sigma_\beta^2 \alpha_1}{\sigma_\alpha^2 G^2 + \sigma_\beta^2} + \frac{\sigma_\beta^2 n_\alpha - \sigma_\alpha^2 G n_\beta}{\sigma_\alpha^2 G^2 + \sigma_\beta^2}. \tag{25}$$

It is easy to see that the probability density for x_1 is equal to eq.(21). On the other hand, if we do not know the value of y_2^{out}, then x_1 can be rewritten as

$$x_1 = \alpha_1 + n_\alpha. \tag{26}$$

This implies a conventional quantum noise model for squeezed states. Actually, if we ignore the output y_2^{out} in the proposed channel model, the model can be simplified to an equivalent one shown in Fig.2, which corresponds to eq.(26). Therefore, the communication channel model shown in Fig.1 can fit the properties of the QND measurement previously mentioned.

Lastly, we note two interesting cases of QND measurement. First, if the signal light is in an eigenstate of $\widehat{X}_1$, namely $\sigma_\alpha = 0$, then the signal light is not affected by the measurement of the probe light as shown in Fig.3. On the other hand, if the probe light is in an eigenstate of $\widehat{Y}_2$, namely $\sigma_\beta = 0$, then the output state of the signal light is completely affected by the measurement of the probe light, and reduced to an eigenstate determined by the readout value y_2^{out} as shown in Fig.4. Therefore, we can conclude that in the QND measurement, the effect of measurement to the signal light is completely specified by the amplitude gain G and the variances of quantum noises σ_α^2 and σ_β^2.

It should be noted that since the BAE amplifier can be modeled by this measurement system, our proposed optical coupler[8] will be precisely analyzed by this channel model.

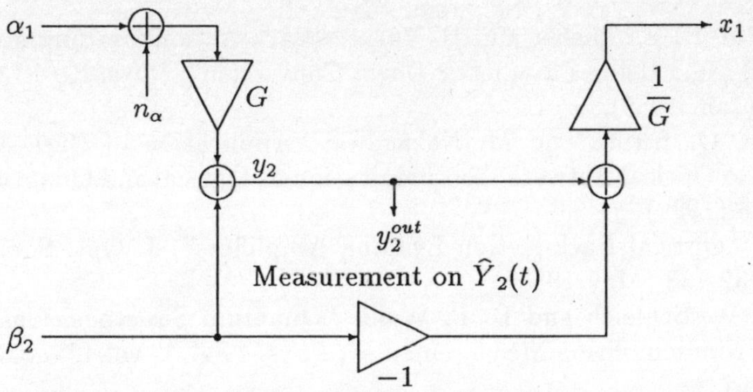

Figure 4: Special case of QND measurement ($\sigma_\beta = 0$)

4. Conclusion

This paper describes the treatment of QND measurement from the view point of the optical communication theory, and obtains the following results:

1. The state vector and the wave function for a QND measurement is obtained by using standard quantum theory.
2. The QND measurement is modeled by a specific channel in communication theory.
3. The effect of QND measurement to the signal light is completely specified by the amplitude gain and the variances of quantum noises of the signal and probe lights.

Acknowledgment

The authors are grateful to Prof. S. Tsujii of Tokyo Institute of Technology for his valuable discussions.

References

[1] C. M. Caves, K. S. Thorne, R. W. P. Drever, V. D. Sandberg and M. Zimmermann: "On the Measurement of a Weak Classical Force Coupled to a Quantum-Mechanical Oscillator. I. Issue of Principle", Rev. Mod. Phys. vol.52, no.2, pp.341-392 (Apr. 1980).

[2] V. B. Braginsky, Y. I. Vorontsov and K. S. Thorne: "Quantum Nondemolition Measurements", Science, vol.209, no.4456, pp.547-557 (Aug. 1980).

[3] G. J. Milburn and D. F. Walls: "Quantum Nondemolition Measurements via Quadratic Coupling", Phys. Rev. A, vol.28, no.4, pp.2065-2070 (Oct. 1983).

[4] N. Imoto, H. A. Haus and Y. Yamamoto: "Quantum Nondemolition Measurement of the Photon Number via the Optical Kerr Effect", Phys. Rev. A, vol.32, no.4, pp.2287-2292 (Oct. 1985).

[5] P. Alsing, G. J. Milburn and D. F. Walls: "Quantum Nondemolition Measurements in Optical Cavities", Phys. Rev. A, vol.37, no.8, pp.2970-2978 (Apr. 1988).

[6] M. D. Levenson, R. M. Shelby, M. Ried and D. F. Walls: "Quantum Nondemolition Detection of Optical Quadrature Amplitudes", Phys. Rev. Let. vol.57, no.20, pp2473-2476 (Nov. 1986).

[7] A. La Porta, R. E. Slusher and B. Yurke: "Back-Action Evading Measurements of an Optical Field Using Parametric Down Conversion", Phys. Rev. Let. vol.62, no.1, pp.28-31 (Jan. 1989).
[8] T. Himori, O. Hirota and M. Nakagawa: "Application of Back-Action Evading Amplifier to Optical Network Communications", Optical and Quantum Elec. vol.21, pp.131-136 (1989).
[9] B. Yurke: "Optical Back-Action-Evading Amplifiers", J. Opt. Soc. Am. B, vol.2, no.5, pp.732-738 (May 1985).
[10] B. Yurke, W. Schleich and D. F. Walls: "Quantum Superpositions Generated by Quantum Nondemolition Measurements", Phys. Rev. A, vol.42, no.3, pp.1703-1711 (Aug. 1990).
[11] H. P. Yuen: "Two-Photon Coherent States of the Radiation Field", Phys. Rev. A, vol.13, no.6, pp.2226-2243 (June 1976).

Appendix A

Define the vector $|\phi(t)>$ by

$$|\phi(t)>= e^{i(\widehat{H}_s+\widehat{H}_p)t/\hbar}|\psi(t)>. \tag{A1}$$

Then eq.(4) can be rewritten as

$$i\hbar\frac{\partial}{\partial t}|\phi(t)>= e^{i(\widehat{H}_s+\widehat{H}_p)t/\hbar}\widehat{H}_I e^{-i(\widehat{H}_s+\widehat{H}_p)t/\hbar}|\phi(t)>. \tag{A2}$$

On the other hand, it is easy to see that the relations

$$\left.\begin{aligned} e^{i\widehat{H}_st/\hbar}\widehat{X}_1(t) &= \widehat{X}_1(0)e^{i\widehat{H}_st/\hbar} \\ e^{i\widehat{H}_pt/\hbar}\widehat{Y}_1(t) &= \widehat{Y}_1(0)e^{i\widehat{H}_pt/\hbar} \end{aligned}\right\} \tag{A3}$$

hold. Combining eq.(3) and (A3), we have

$$\begin{aligned} e^{i(\widehat{H}_s+\widehat{H}_p)t/\hbar}\widehat{H}_I e^{-i(\widehat{H}_s+\widehat{H}_p)t/\hbar} &= e^{i(\widehat{H}_s+\widehat{H}_p)t/\hbar}\hbar K\widehat{X}_1(t)\widehat{Y}_1(t)e^{-i(\widehat{H}_s+\widehat{H}_p)t/\hbar} \\ &= \hbar K e^{i\widehat{H}_st/\hbar}\widehat{X}_1(t)e^{-i\widehat{H}_st/\hbar}e^{i\widehat{H}_pt/\hbar}\widehat{Y}_1(t)e^{-i\widehat{H}_pt/\hbar} \\ &= \hbar K\widehat{X}_1(0)\widehat{Y}_1(0). \end{aligned} \tag{A4}$$

Hence, we can rewrite eq.(A2) as

$$i\hbar\frac{\partial}{\partial t}|\phi(t)>= \hbar K\widehat{X}_1(0)\widehat{Y}_1(0)|\phi(t)>. \tag{A5}$$

Since $\widehat{X}_1(0)\widehat{Y}_1(0)$ does not depend on time, the solution of this equation is easily obtained and given by

$$|\phi(t)>= e^{-iKt\widehat{X}_1(0)\widehat{Y}_1(0)}|\phi(0)>. \tag{A6}$$

From eq.(A1),(A3),(A6) and the relation $|\phi(0)>=|\psi(0)>$, we have

$$\begin{aligned}|\psi(t)> &= e^{-i(\widehat{H}_s+\widehat{H}_p)t/\hbar}|\phi(t)> \\ &= e^{-i(\widehat{H}_s+\widehat{H}_p)t/\hbar}e^{-iKt\widehat{X}_1(0)\widehat{Y}_1(0)}|\psi(0)> \\ &= e^{-iKt\widehat{X}_1(t)\widehat{Y}_1(t)}e^{-i(\widehat{H}_s+\widehat{H}_p)t/\hbar}|\psi(0)> \end{aligned} \tag{A7}$$

which is the solution.

Appendix B

For $t = 0$, the state vector of the system is given by

$$|\psi(0)>=|\xi>|\eta>. \tag{B1}$$

As basis states, we adopt the eigenstate of $\widehat{X}_1(t)$ for the signal light, and that of $\widehat{Y}_2(t)$ for the probe light, that is

$$\left.\begin{aligned}\widehat{X}_1(t)|x_1,t> &= x_1|x_1,t> \\ \widehat{Y}_2(t)|y_2,t> &= y_2|y_2,t>\end{aligned}\right\}. \tag{B2}$$

Then, the wave function $\Psi(x_1,y_2,t)$ is given by

$$\begin{aligned}\Psi(x_1,y_2,t) &=< x_1,t|<y_2,t|\psi(t)> \\ &=< x_1,t|<y_2,t|e^{-iKt\widehat{X}_1(t)\widehat{Y}_1(t)}e^{-i(\widehat{H}_s+\widehat{H}_p)t/\hbar}|\xi>|\eta> \\ &=< y_2,t|e^{-iKtx_1\widehat{Y}_1(t)}e^{-i\widehat{H}_pt/\hbar}|\eta>< x_1,t|e^{-i\widehat{H}_st/\hbar}|\xi>. \end{aligned} \tag{B3}$$

It is easy to see that the relations

$$\left.\begin{aligned}\widehat{X}_1(0)(e^{i\widehat{H}_st/\hbar}|x_1,t>) &= x_1\, e^{i\widehat{H}_st/\hbar}|x_1,t> \\ \widehat{Y}_2(0)(e^{i\widehat{H}_pt/\hbar}|y_2,t>) &= y_2\, e^{i\widehat{H}_pt/\hbar}|y_2,t>\end{aligned}\right\} \tag{B4}$$

hold. Since $|x_1>$ and $|y_2>$ are the eigenstates for $\widehat{X}_1(0)$ and $\widehat{Y}_2(0)$, we have

$$\left.\begin{aligned}|x_1> &= e^{i\widehat{H}_st/\hbar}|x_1,t> \\ |y_2> &= e^{i\widehat{H}_pt/\hbar}|y_2,t>\end{aligned}\right\}. \tag{B5}$$

Substituting eq.(A3) and (B5) to eq.(B3), we have

$$\begin{aligned}\Psi(x_1,y_2,t) &=< y_2,t|e^{-i\widehat{H}_pt/\hbar}e^{-iKtx_1\widehat{Y}_1(0)}|\eta>< x_1,t|e^{-i\widehat{H}_st/\hbar}|\xi> \\ &=< y_2|e^{-iKtx_1\widehat{Y}_1(0)}|\eta>< x_1|\xi> \\ &=< y_2+(2Kt)x_1|\eta>< x_1|\xi>, \end{aligned} \tag{B6}$$

which is the wave function.

INFORMATION IN QUANTUM OPTICAL COMMUNICATIONS

A. VOURDAS
Department of Electrical Engineering and Electronics, The University of Liverpool, Brownlow Hill, P.O. Box 147, Liverpool, L69 3BX.

Abstract

The maximum amount of information that a beam can carry is calculated with the use of quantum mechanical methods. Binary communication systems that use these beams are considered and the transmitted information is calculated.

1. Introduction

Communication at optical frequencies is an exciting area of research with many potential applications. At these frequencies, quantum effects are important and they should be taken into account in the calculations. Refs.[1,2] review the basic formalism for quantum optical communications. Refs.[3,4,5] have used quantum mechanical methods for information calculations. In refs.[6-9] we have performed information calculations in various contexts and here we review them briefly and discuss their relative merits. We define various entropy based quantities and discuss their physical interpretation and their role in characterizing the information in optical communication systems.

In section 2 we consider various signals and assess their potential as information carriers. We subtract the entropy of the signal from the maximum entropy that this signal could have had, with the average number of photons been kept fixed. This provides the maximum information that this signal can carry. The information loss during the measurement process is also considered and a measurement-dependent information is calculated.

In section 3 we consider binary communication systems that use these signals. Two different models are used for the description of the propagation of the signal between the transmitter and the receiver. In the first one, it is assumed that the arriving signal in the receiver is a mixture of the transmitted signal with thermal noise. In the second one, a Hamiltonian that describes the interaction between the signal and the non-linear medium, is used to describe explicitly the propagation. Using these two models, we calculate the density matrix of the arriving signal in the receiver and use it to calculate the transmitted information.
We conclude in section 4 with a discussion of the results.

Lecture Notes in Physics, Vol. 378
C. Bendjaballah, O. Hirota, S. Reynaud (Eds.),
Quantum Aspects of Optical Communications

2. Information of a quantum signal

In this section we introduce an entropy based quantity that characterizes the potential of a quantum signal as information carrier. This calculation is independent of the coding technique of the communication system and provides the maximum information that this beam can carry.

Let ρ be the density matrix describing a signal with total average number of photons $\langle N_{TOT}\rangle$. We define the information carried by this beam as (k_B=h=1):

$$I(\rho) = S_{max} - S(\rho) \tag{1}$$

$$S(\rho) = - Tr\rho\ln\rho \tag{2}$$

$$S_{max} = \ln(1+\langle N_{TOT}\rangle) + \langle N_{TOT}\rangle\ln(1 + \langle N_{TOT}\rangle^{-1}) \tag{3}$$

where $S(\rho)$ is the entropy of the beam and S_{max} is the maximum entropy that this beam could have had; i.e. the entropy of a thermal beam with average number of thermal photons $\langle N_{TOT}\rangle$. The following comments help to clarify the physical meaning of this quantity:

(a) if the density matrix ρ describes a pure state, then $S(\rho) = 0$ and the $I(\rho)$ is equal to the result given in equ.(3), which depends on $\langle N_{TOT}\rangle$ and is independent of the type of the signal (i.e. coherent state, squeezed coherent state, number eigenstate etc.)

(b) the emmitted signals usually arrive at the receiver mixed with thermal noise. In [6,7] we have considered mixtures of various signals (coherent, squeezed coherent and number eigenstates) with thermal noise and we have presented numerical calculations of the quantity of equ.(1). In this case we found that for given numbers of signal photons $\langle N_S\rangle$ and thermal photons $\langle N_T\rangle$, the $I(\rho)$ does depend on the type of the signal.

(c) For thermal beams equ.(1) gives I=0, as it should.

The quantity of equ.(1) expresses the information content of the beam, before the measurement process at the receiver. It is well known, that during the measurement some information might be lost. Let us consider a measurement (Hermitian operator) Λ with eigenvalues λ_N

$$\Lambda = \sum \lambda_N \pi_N \tag{4}$$

$$\pi_N \pi_M = \delta_{NM} \pi_N \tag{5}$$

A system with density matrix ρ will be described after the measurement Λ by the

$$\rho_\Lambda = \sum_N \pi_N\rho\pi_N \tag{6}$$

It is known that [10]

$$S(\rho_\Lambda) \geq S(\rho) \tag{7}$$

With the equality occuring in the case that the ρ,Λ commute. So we can define the following (measurement dependent) information

$$I(\rho,\Lambda) = S_{max} - S(\rho,\Lambda) \tag{8}$$

$$S(\rho,\Lambda) = - Tr\rho_\Lambda \ln\rho_\Lambda \tag{9}$$

$$I(\rho,\Lambda) \leq I(\rho) \tag{10}$$

where the S_{max} has been given in equ.(3). The equality in equ.(10) occurs for commuting ρ,Λ.

3. <u>Information in binary communication systems</u>

In the previous section we have considered the information carried by a quantum signal. A communication system that uses this signal will usually exploit a part or at the very best all this information. We consider now a binary system where a signal is transmitted for each "1" and nothing is transmitted for each "0". The probability of transmitting the symbol i(i=0,1) is known to be $p_i(p_o+p_1=1)$. Let ρ_1,ρ_o be the density matrices describing the arriving signals 1,0 at the receiver. The information per symbol is

$$I = S_{max} - S_{ave} \tag{11}$$

$$S_{ave} = p_o S(\rho_o) + p_1 S(\rho_1) \tag{12}$$

$$S_{max} = S(p_o\rho_o+p_1\rho_1) \tag{13}$$

where S_{ave} is the average of the entropies of the two signals with weights p_o,p_1; and S_{max} is the maximum entropy that this signal could have had and is equal to the entropy of the density matrix $p_o\rho_o+p_1\rho_1$ (concavity property). In the cases $\rho_1=\rho_2$ or $p_o=0$ or $p_1=0$ this definition gives I=0 as it should.

The density matrices entering equs.(11),(12),(13) are those at the receiver. A model is required for the description of the propagation and the evaluation of the density matrices at the receiver, from the density matrices at the transmitter. We have used two such models: In the first one, we simply assumed that noise is added to the emitted signal. In this case the density matrix ρ_1 describes a mixture of the emitted signal with thermal noise. Such density matrices have been studied in detail in [7]. The density matrix ρ_o describes only thermal noise. Numerical results for this model have been presented in [8].

In the second model, the propagation of the signal in a non-linear medium (e.g. optical fiber) has been described with the Hamiltonian

$$H = \omega_1 a_1^+ a_1 + \omega_2 a_2^+ a_2 + \lambda a_1^+ a_2 + \lambda^* a_1 a_2^+ + g a_2^{+2} a_2^2 \tag{14}$$

where a_1^+,a_1 and a_2^+,a_2 are creation and annihilation operators for the electromagnetic signal and the non-linear medium correspondingly. The non-linear medium is described here, as an anharmonic oscillator. This model is good for short propagation distances and low light intensities because it ignores losses and saturation. Using this Hamiltonian we can calculate the density matrices of the arriving signals at the receiver; and then from (11), (12), (13) we can calculate the information. Numerical results have been presented in [9]. It is worth mentioning that the transmitted information has been found to be an oscillatory function of the transmission time T; naively it is expected to be a decreasing function of T. This result might be related to the so-called "collapse and revival of oscillations" which the model of equ.(14) (and also other models) are known to exhibit. We have also calculated the signal-to-noise ratio and found that it is an oscillatory function of T. These results indicate that in certain non-linear media "collapse and revival of the transmitted signal" might occur; and this could be exploited in optical communications.

4. Discussion

A quantum mechanical method has been used to calculate the transmitted information in optical communications. The maximum information that a beam can transmit has been calculated in equs.(1),(2),(3). The effect of a quantum measurement has been taken into account in equs.(8),(9),(10). Binary communication systems have also been considered and the information has been calculated with the use of equs.(11),(12),(13). The systems considered have been intentionally chosen to be simple so that a fully quantum mechanical method can be applied. The ideas of quantum mechanics open a new dimension in the areas of information and coding and this paper is a small contribution in this direction.

References

1. R.O. Harger (Ed.) "Optical Communication Theory" (Dowden, Hutchinson, Ross; Stroudsbourg PA, 1977).
2. C.W. Helstrom "Quantum Detection and Estimation Theory" (Academic, London, 1976).
3. A. Holevo, Probl. Peredachi. Inform. 9 (part 3), 3 (1973).
4. R.S. Ingarden, Rep. Math. Phys. 10, 43 (1976).
5. Y. Yamamoto, H.A. Haus. Rev. Mod. Phys. 58, 1001 (1984).
6. A. Vourdas, Phys. Rev. A37, 3890 (1988).
7. A. Vourdas, Phys. Rev. A39, 206 (1989).
8. A. Vourdas, IEEE Trans. Info. Theo. IT-36, 208, (1990).
9. A. Vourdas, J. Mod. Opt. 37, 1661, (1990).
10. A. Wehrl, Rev. Mod. Phys. 50, 221 (1978).

Session III

– Quantum Stochastic Processes –

Chairman : C.W. Helstrom

STATISTICAL INFERENCE FOR QUANTUM PROCESSES

A.S.Holevo
Steklov Mathematical Institute
Academy of Science of the USSR
Vavilova 42, Moscow, USSR

This is an expository paper containing brief survey of basic notions of generalized statistical model of quantum mechanics, needed, in particular, to formulate a consistent theory of continuous quantum measurement processes. Stochastic representation for such processes are given, which allow explicit description of statistics of observed trajectories and of the corresponding posterior states. The relation between posterior dynamics and nonlinear wave equations is discussed.

§1. Generalized statistical model of quantum mechanics

This model was introduced in the early 70's and considered by many authors (see Davies [1], Holevo [2], Kraus [3] for detailed presentation and further references). In this model there are three levels of description of quantum measurement, to each level corresponding the appropriate mathematical object in the underlying Hilbert space $\mathcal{H}$ of the system.

Level I. The statistics of measurement is given. This means that to each state, described by density operator S in $\mathcal{H}$, there corresponds probability distribution $\mu_S(dx)$ on the space $\mathcal{X}$ of outcomes of the measurement. Under the only natural requirement that the correspondence $S \to \mu_S(dx)$ is affine, i.e. preserves statistical mixtures, one proves [2] that there is a unique generalized observable, i.e. operator-valued measure $M(dx)$ in $\mathcal{H}$, such that $M(dx) \geq 0$, $\int_{\mathcal{X}} M(dx) = I$, satisfying

$$\mu_S(dx) = \operatorname{Tr} S M(dx). \tag{1}$$

The statistics of the measurement is thus adequately described by the generalized observable $M(dx)$. The standard notion of observable is recovered if $M(dx)$ is projection-valued measure, i.e. $M(dx)^2 = M(dx)$. There is a number of problems in which this notion is insufficient, and which require introduction of generalized obser-

Lecture Notes in Physics, Vol. 378
C. Bendjaballah, O. Hirota, S. Reynaud (Eds.),
Quantum Aspects of Optical Communications

vables. Let us mention here quantum detection and estimation theory [2], [4] and the problem of correspondence ("nonexistence" of certain observables) in quantum mechanics [5].

<u>Level II.</u> The statistics of measurement and the state changes after measurement are given. On this level the measurement is adequately described by the notion of instrument, i.e. operation-valued measure $\mathcal{M}(dx)[\cdot]$ in $\mathcal{H}$ [1]. Given a state S, the statistics of outcomes is described by the formula

$$\mu_S(dx) = \mathrm{Tr}\,\mathcal{M}(dx)[S], \tag{2}$$

and posterior state, given an outcome $x \in \mathcal{X}$, is defined by relation

$$S_x = \frac{\mathcal{M}(dx)[S]}{\mathrm{Tr}\,\mathcal{M}(dx)[S]}. \tag{3}$$

Mathematical study of posterior states was made by Ozawa [6], who had shown the existence of the Radon-Nikodym derivatives (3) under mild regularity conditions. The instrument $\mathcal{M}(dx)$ is recovered from (2) and (3) by the relation

$$\mathcal{M}(B)[S] = \int_B S_x\,\mu_S(dx)\ ; \quad B \in \mathcal{B}(\mathcal{X}).$$

The relation (2) can be rewritten in the form (1), where $M(dx) = \mathcal{M}(dx)^*[I]$. Thus to any instrument there corresponds a generalized observable, but the correspondence is not one-to-one, since the level II gives more detailed description of measurement than level I.

<u>Level III.</u> In the scheme of indirect measurement one has the probe system (part of measuring apparatus) described by the Hilbert space $\mathcal{H}_0$, which is initially in the state S_0. The probe system interacts with the observed system via unitary operator U in $\mathcal{H} \otimes \mathcal{H}_0$; then a standard observable $P_0(dx)$ in $\mathcal{H}_0$ is measured. The resulting relation

$$\mathcal{M}(dx)[S] = \mathrm{Tr}_{\mathcal{H}_0}\, U(S \otimes S_0)U^*(I \otimes P_0(dx)) \tag{4}$$

determines an instrument in $\mathcal{H}$. This instrument is completely positive (c.p.) in the sense that operations $\mathcal{M}(dx)^*$ are completely positive [3]. Conversely, Ozawa [7] had shown that every c.p. instrument $\mathcal{M}(dx)$ can be represented by the relation (4) via some indirect measurement. Since $P_0(dx)$ can always be choosen projection-valued measure, this theorem also displays agreement between generalized and standard statistical models of quantum mechanics. Again the correspondence between instruments and indirect measurements is not one-to-one since level III is more detailed than level II.

The formula (4) is related to the following Stinespring type representation for c.p. instrument:

$$\mathcal{M}(dx)[S] = \sum_{k=1}^{\infty} A_k(x) S A_k(x)^* \mu(dx),$$

where μ is a positive measure, $A_k(x)$ - operator-valued functions such that $\sum_{k=1}^{\infty} A_k(x)^* A_k(x) \mu(dx) = I$. Especially simple is the case

$$\mathcal{M}(dx)[S] = A(x) S A(x)^* \mu(dx) \tag{5}$$

Then formulas (2), (3) become

$$\mu_S(dx) = \operatorname{Tr} A(x) S A(x)^* \mu(dx), \tag{6}$$

$$S_x = A(x) S A(x)^* / \operatorname{Tr} A(x) S A(x)^* . \tag{7}$$

§ 2. Theory of continuous quantum measurement processes.

This theory, together with solution of the repeatability problem [1], [7] is the most important application of the notion of instrument. It was gradually developed in the works of Davies [1], Barchielli, Lanz and Prosperi [8], and Holevo [9]. For simplicity we restrict here to measurements of real-valued quantities. Let $\mathcal{Y}$ be the space of all real-valued function (trajectories) $y(t)$ on $\mathbb{R}$, $\mathcal{B}_{a,b}$ - σ-algebra, generated by increments $y(s) - y(t)$, where $a \le t \le s \le b$. By instrumental process with independent increments (i.-process) we call a family of c.p. instruments $\{\mathcal{M}_{a,b}(E);\ a \le b,\ a,b \in \mathbb{R}\}$, where $E \in \mathcal{B}_{a,b}$, such that

$$\mathcal{M}_{b,c}(F)[\mathcal{M}_{a,b}(E)[S]] = \mathcal{M}_{a,c}(E \cap F)[S]; \quad a \le b \le c,$$

for all states S, all $E \in \mathcal{B}_{a,b}$, $F \in \mathcal{B}_{b,c}$. Any continuous homogeneous i.-process is uniquely determined by the generator $\mathcal{L}(\lambda)$ which is a function of $\lambda \in \mathbb{R}$ with values in the complex hull of all operations, described by a noncommutative generalization of the Levy-Khinchine formula [9]. The relation between the i.-process and its generator is

$$\int_{\mathbb{R}} e^{i\lambda x} \mathcal{M}_t(dx) = \exp t\,\mathcal{L}(\lambda),$$

where

$$\mathcal{M}_t(B) = \mathcal{M}_{a,a+t}(y(\cdot):\ y(a+t) - y(a) \in B);\quad B \in \mathcal{B}(\mathbb{R}),$$

is a c.p. instrument on $\mathbb{R}$.

<u>Example 1.</u> Let R , $H = H^*$ be bounded operators in $\mathcal{H}$, $\gamma > 0$, then

$$\mathcal{L}^{(1)}(\lambda)[S] = \mathcal{L}[S] + \frac{i\lambda}{2}(RS + SR^*) - \frac{\gamma\lambda^2}{2}S, \tag{8}$$

where

$$\mathcal{L}[S] = i[S, H] + \frac{1}{4\gamma}(RSR^* - R^*R \circ S) \tag{9}$$

is a generator of i.-process which can be interpreted as a continuous approximate measurement of observable $(R+R^*)/2$ over the system evolving with the hamiltonian H [8]. Parameter γ reflects accuracy of the measurement.

<u>Example 2.</u> Let R be a bounded operator, $z \in \mathbb{C}$. Then

$$\mathcal{L}^{(2)}(\lambda)[S] = i[S, H] + \alpha^2(RSR^* e^{i\lambda} - R^*R \circ S) \tag{10}$$

is a generator of a quantum counting process [10], [11]. Parameter $\alpha^2 = 1/4\gamma$ reflects intensity of counts.

Using quantum stochastic calculus developed in [12], [13], Barchielli and Lupieri [14] found a dilation for continuous homogeneous i.-processes, which can be regarded as a dynamical generalization of the indirect measurement scheme (4). We give it here for the two most important examples described above.

Let $\mathcal{H}_0 = \Gamma(L^2(\mathbb{R}))$ be the Boson Fock space over $L^2(\mathbb{R})$; $A^{\dagger}(t)$, $A(t)$, $\Lambda(t)$ be the creation, annihilation and number processes in $\mathcal{H}_0$, and let $S_0 = |0\rangle\langle 0|$ be the vacuum state in $\mathcal{H}_0$ [12].

Then for continuous measurement process of Example 1 one has the representation

$$\mathcal{M}^{(1)}_{a,b}(E)[S] = \mathrm{Tr}_{\mathcal{H}_0} U_{a,b}(S \otimes S_0) U^*_{a,b}(I \otimes P^{(1)}_{a,b}(E));\ E \in \mathcal{B}_{a,b}, \tag{11}$$

where $U_{a,b}$ in the family of unitary operators in $\mathcal{H} \otimes \mathcal{H}_0$ satisfying the quantum stochastic differential equation

$$d\,U_{a,t} = \left[\frac{1}{2\sqrt{\gamma}}\left(R\,dA^{\dagger}(t) - R^{*}dA(t)\right) - \left(iH + \frac{1}{8\gamma}R^{*}R\right)dt\right]U_{a,t}\,; \qquad (12)$$

$$t \geq a\,;\; U_{a,a} = I\,,$$

and $P^{(1)}_{a,b}(E)$ is the joint spectral measure of the family of compatible observables $\sqrt{\gamma}\,Q(t) = \sqrt{\gamma}\,(A^{\dagger}(t) + A(t))\,;\; a \leq t \leq b$ in $\mathcal{H}_0$. By using time-ordered exponentials defined in [15] one can write the solution of (12) in the form

$$U_{a,b} = \overleftarrow{\exp}\int_a^b \left[\frac{1}{2\sqrt{\gamma}}\left(R\,dA^{\dagger}(t) - R^{*}dA(t)\right) - iH\,dt\right],$$

which gives the representation (11) a tranparent interpretation: the observed system, evolving with the Hamiltonian H, interacts with Boson quantum noise, the interaction Hamiltonian being singular and equal to $\frac{i}{2\sqrt{\gamma}}\left(R\dot{A}^{\dagger}(t) - R^{*}\dot{A}(t)\right)$. Simultaneously, on quantum noise, which plays the role of the probe system, a continuous measurement of compatible family of field operators $\sqrt{\gamma}\,Q(t)\,;\; a \leq t \leq b$, is performed.

For example 2 the dilation has the form

$$\mathcal{M}^{(2)}_{a,b}(E)[S] = \mathrm{Tr}_{\mathcal{H}_0}\, U_{a,b}\,(S \otimes S_0)\, U^{*}_{a,b}\,\left(I \otimes P^{(2)}_{a,b}(E)\right);\quad E \in \mathcal{B}_{a,b}, \qquad (13)$$

where $U_{a,b}$ and S_0 are the same as above and $P^{(2)}_{a,b}(E)$, $E \in \mathcal{B}_{a,b}$, is the joint spectral measure of the family of compatible number observables $\Lambda(t)\,;\; a \leq t \leq b$.

To prove relations (11), (13) one has only to check that the function $\Phi^{(j)}_t(\lambda) = \int e^{ix\lambda}\,\mathcal{M}^{(j)}_t(dx)\,;\; j = 1, 2$, where

$$\mathcal{M}^{(j)}_t(B)[S] = \mathrm{Tr}_{\mathcal{H}_0}\, U_{0,t}\,(S \otimes S_0)\, U^{*}_{0,t}\,\left(I \otimes P^{(j)}_{0,t}\left(y(\cdot) : y(t) - y(0) \in B\right)\right)$$

satisfies the equation

$$d\,\Phi^{(j)}_t(\lambda)[S] = \mathcal{L}^{(j)}(\lambda)\left[\Phi^{(j)}_t(\lambda)[S]\right]dt,$$

i.e. has the form $\Phi^{(j)}_t(\lambda) = \exp t\,\mathcal{L}^{(j)}(\lambda)$. This is accomplished by using quantum Ito formula [12], [13] and the equation (12).

§ 3. Stochastic representations of continuous measurement processes.

Here we give stochastic representations for continuous measurement processes from § 2 via usual Wiener and Poisson process, which will enable easy calculation of the probability distributions of the observed trajectories and the corresponding posterior states. Let us first consider Example 1.

<u>Proposition 1.</u> The i.-process $\{\mathcal{M}^{(1)}_{a,b}(E)\}$ is absolutely continuous with respect to the Wiener measure μ_1 with the variance γ in the space of continuous functions $\mathcal{C}$ in the sense that

$$\mathcal{M}^{(1)}_{a,b}(E)[S] = \int_E \Psi^{(1)}_{a,b}(y)\, S\, \Psi^{(1)}_{a,b}(y)^* d\mu_1(y), \quad E \in \mathcal{B}(\mathcal{C}), \tag{14}$$

where $\Psi^{(1)}_{a,b}(y)$ is the family of bounded operators in $\mathcal{H}$ determined by the stochastic differential equation

$$d\Psi^{(1)}_{a,t}(W) = \left[\frac{R}{2\gamma}\, dW(t) - \left(iH + \frac{1}{8\gamma} R^*R\right)dt\right]\Psi^{(1)}_{a,t}(W); \quad \Psi_{a,a} = I \tag{15}$$

Once written, the relation (14) can be proved just as (11) by using Ito stochastic calculus. However it is instructive to derive it from (11) by using a trick suggested in [15]. To prove (14) let us introduce the family of isometric operators $\Phi_{a,b} = U_{a,b}|0\rangle$ from $\mathcal{H}$ to $\mathcal{H}\otimes\mathcal{H}_0$. From (12) it satisfies the equation

$$d\Phi_{a,t} = \left[\frac{1}{2\sqrt{\gamma}}\left(R\,dA^\dagger - R^* dA\right) - \left(iH + \frac{1}{8\gamma}R^*R\right)dt\right]\Phi_{a,t}. \tag{16}$$

Now $dA(t)\Phi_{a,t} = dA(t)U_{a,t}|0\rangle = U_{a,t}\,dA(t)|0\rangle = 0$, therefore (16) can be rewritten as

$$d\Phi_{a,t} = \left[\frac{1}{2\sqrt{\gamma}} R\,dQ - \left(iH + \frac{1}{8\gamma}R^*R\right)dt\right]\Phi_{a,t}. \tag{17}$$

Let us apply the Segal's duality transform [12], [15], according to which $\Gamma(L^2(\mathbb{R}))$ is isometrically isomorphic to $L^2(\mu_1)$ in the way that $|0\rangle$ corresponds to 1 and $Q(t)$ corresponds to $W(t)/\sqrt{\gamma}$, so that $P^{(1)}_{a,b}(E)$ corresponds to the characteristic function of the set $E \in \mathcal{B}_{a,b} \cap \mathcal{C}$. Denoting by $\Psi^{(1)}_{a,b}$ the image of $\Phi_{a,b}$ under duality transform, we have that $\Psi^{(1)}_{a,t}$ satisfies the stochastic differential equation (15). Moreover (17) converts into (15) and the proposition is proved.

The solution of the equation (15) can be written in the form (see [15]):

$$\Psi^{(1)}_{a,b}(W) = \overleftarrow{\exp}\int_a^b \left[\frac{R}{2\gamma}\, dW(t) - \left(iH + \frac{1}{8\gamma}(R^* + R)R\right)dt\right].$$

<u>Proposition 2.</u> The i.-process $\{\mathcal{M}^{(2)}_{a,b}(E)\}$ is absolutely continuous with respect to the measure μ_2, corresponding to the

Poisson process $N(t)$ with intensity α^2 in the sense that

$$\mathcal{M}^{(2)}_{a,b}(E)[S] = \int_E \Psi^{(2)}_{a,b}(y) S \Psi^{(2)}_{a,b}(y)^* d\mu_2(y), \tag{18}$$

where $\Psi^{(2)}_{a,b}$ is the family of bounded operators in the Hilbert space $\mathcal{H}$, determined by the stochastic differential equation

$$d\Psi^{(2)}_{a,t}(N) = \{(R-I)dN(t) - [iH + \frac{\alpha^2}{2}(R^*R - I)]dt\}\Psi^{(2)}_{a,t}(N) \tag{19}$$
$$a < t; \ \Psi^{(2)}_{a,a} = I.$$

Derivation of (18) from (11) requires some transformations. Without loss of generality we take $a = 0$. The unitary Weyl operator $W_t(\alpha) = exp(\alpha A(t) - \alpha A^+(t))$ has the property

$$W_t(\alpha)\Lambda(s)W_t(\alpha)^* = \Lambda(s) + \alpha A(s) + \alpha A^+(s) + \alpha^2 s \equiv \Pi(s); \quad s \leq t,$$

where $\Pi(t)$ is the Poisson process of intensity in the Fock space [12]. Denoting $\tilde{U}^{(2)}_{0,t} = W_t(\alpha) U_{0,t}$, we have from (11)

$$\mathcal{M}^{(2)}_{0,t}(E)[S] = \mathrm{Tr}_{\mathcal{H}_0} \tilde{U}_t (S \otimes S_0) \tilde{U}_t^* (I \otimes \mathrm{P}^{(2)}_{0,t}(E)), \quad E \in \mathcal{B}_{0,t}, \tag{20}$$

where $\tilde{P}^{(2)}_{0,t}(E)$ is the spectral measure of the family of compatible observables $\Pi(s)$, $0 \leq s \leq t$. By using quantum Ito formula, we find the equation for $\tilde{U}^{(2)}_{0,t}$:

$$d\tilde{U}^{(2)}_{0,t} = \{\alpha(R-I)dA^+ - \alpha(R-I)^* dA - [iH + \frac{\alpha^2}{2}(R^*R + I - 2R)]dt\}\tilde{U}^{(2)}_{0,t}.$$

Introducing isometric operators $\tilde{\Phi}_t = \tilde{U}^{(2)}_{0,t}|0\rangle$ and taking into account that also $d\Lambda(t)\tilde{U}^{(2)}_{0,t}|0\rangle = 0$, we can rewrite it as

$$d\tilde{\Phi}_t = [(R-I)d\Pi - (iH + \frac{\alpha^2}{2}(R^*R - I))dt]\tilde{\Phi}_t. \tag{21}$$

As shown in [12], [15], there is a unitary isomorphism between the Fock space and the space $L^2(\mu_2)$, such that $|0\rangle$ corresponds to 1 and $\Pi(t)$ corresponds to $N(t)$. Under this correspondence the relation (20) becomes (18) and the equation (21) becomes (19).

Note that by [15] solution of the equation (19) can be written in the form

$$\Psi^{(2)}_{a,b} = \overleftarrow{exp}\int_a^b [\ln R\, dN(t) - (iH + \frac{\alpha^2}{2}(R^*R - I))dt].$$

§4. Statistics and posterior states.

Thus in both cases considered, the instruments describing continuous measurement are represented in the form (5). From formula (6) we obtain the probability distribution of the observed trajectories

$$d\mu_S^{(j)}(y) = p_{a,b}^{(j)}(y)\, d\mu_j(y) \; ; \quad j = 1,2,$$

where the density

$$p_{a,b}^{(j)}(y) = \operatorname{Tr} \Psi_{a,b}^{(j)}(y)\, S\, \Psi_{a,b}^{(j)}(y)^* \tag{22}$$

is positive almost everywhere with respect to μ_j. According to (7) the posterior states are

$$S_{a,b}^{(j)}(y) = p_{a,b}^{(j)}(y)^{-1}\, \Psi_{a,b}^{(j)}(y)\, S\, \Psi_{a,b}^{(j)}(y)^* . \tag{23}$$

In particular, if the initial state is pure, $S = |\psi\rangle\langle\psi|$, the posterior state is also pure, $S_{a,b}^{(j)}(y) = |\psi_{a,b}^{(j)}(y)\rangle\langle\psi_{a,b}^{(j)}(y)|$, where

$$\psi_{a,b}^{(j)}(y) = \Psi_{a,b}^{(j)}(y)\psi \Big/ \| \Psi_{a,b}^{(j)}(y)\psi \| . \tag{24}$$

Let us derive stochastic differential equations, describing observed trajectories and posterior states in the continuous measurement process of Example 1. From (22) and the equation (15) it follows that the density $p_{a,t}^{(1)}$ satisfies the stochastic differential equation

$$d p_{a,t}^{(1)}(W) = \gamma^{-1} m_{a,t}(W)\, p_{a,t}^{(1)}(W)\, dW(t),$$

where

$$m_{a,t}(W) = \operatorname{Tr} S_{a,t}^{(1)}(W)\, \frac{R + R^*}{2} \tag{25}$$

is the posterior mean of the observable $(R+R^*)/2$. It follows (see, e.g. [16], §7.2) that

$$p_{a,t}^{(1)}(W) = \exp \gamma^{-1} \Big[\int_a^t m_{a,t}(W)\, dW(t) - \frac{1}{2} \int_a^t m_{a,t}(W)^2 dt \Big]$$

and that the observed process $y(t)$, which has the distribution $\mu_S^{(1)}$, is a diffusion type process satisfying the stochastic differential equation

$$dy(t) = m_{a,t}(y)\, dt + dW(t), \quad t \geqslant a . \tag{26}$$

Moreover,

$$dp^{(1)}_{a,t}(y)^{-1} = -\gamma^{-1} m_{a,t}(y) p^{(1)}_{a,t}(y)^{-1} [dy(t) - m_{a,t}(y)\,dt].$$

Therefore, from classical Ito formula one obtains the following stochastic differential equation for the posterior state (23)

$$\begin{aligned} &d S^{(1)}_{a,t}(y) - \mathcal{L}[S^{(1)}_{a,t}(y)]\,dt = \\ &=[(R - m_{a,t}(y)) S^{(1)}_{a,t}(y) + S^{(1)}_{a,t}(y)(R - m_{a,t}(y))^*] \cdot \\ &\cdot [dy(t) - m_{a,t}(y)\,dt], \end{aligned} \tag{27}$$

where $\mathcal{L}$ is given by (9). This equation was discovered by Belavkin from consideration of quantum analog of the filtering problem for stochastic processes [17], [18]. In the case of pure states the equation reduces to

$$\begin{aligned} d\psi^{(1)}_{a,t}(y) = (R - m_{a,t}(y))\psi^{(1)}_{a,t}(y)[dy(t) - m_{a,t}(y)\,dt] - \\ -[iH + \tfrac{1}{2}(R^*R - 2R\, m_{a,t}(y) + m_{a,t}(y)^2)]\psi^{(1)}_{a,t}(y)\,dt. \end{aligned} \tag{28}$$

Nonlinearity of the equations (27), (28) is due to normalization of the posterior states (23), (24); the background for them is the linear equation (15).

The equation (27) describes posterior dynamics of a quantum object under continuous approximate measurement of certain observable. The fundamental importance of this equation is stressed by the fact that similar equations, but with totally different motivation and interpretation, appeared in the works of several authors, searching for "realistic" foundations for the theory of quantum measurement. In the paper [19] Ghirardi, Rimini, Weber posed the problem of finding equations, giving a unified description of micro- and macro-objects. Various solutions to this problem were proposed. In the works of Gisin [20], Ghirardi, Pearle, Rimini [21] the nonlinear wave equation was introduced, which is of the type (28) but with the differential of an apriori given Wiener process $dW(t)$ in place of $[dy(t) - m_{a,t}(y)\,dt]$. If $H=0$ and $R^* = R$ is an observable with purely point spectrum, then the resulting equation

$$d\psi_t = (R - \langle R\rangle_t)\psi_t\, dW(t) - \tfrac{1}{2}(R - \langle R\rangle_t)^2 \psi_t\, dt, \tag{29}$$

where $\langle R\rangle_t = \langle \psi_t | R\psi_t\rangle$, provides a dynamical description of the projection postulate in the sense that for $t \to +\infty$ ψ_t

tends to one of the eigenvectors of R . In the paper of Gatarek and Gisin [22] a rigorous study of the equation (29) with an unbounded self-adjoint R is given. To show the existence of weak solution of (29) they used Girsanov transformation for the Wiener process $W(t)$, which is formally equivalent to the transformation (26) having transparent statistical meaning in the framework of continuous measurement processes.

In example 2 the stochastic equation for the density is

$$dp_{a,t}^{(2)}(N) = \left[\frac{\lambda_{a,t}(N)}{\alpha^2} - 1 \right] p_{a,t}^{(2)}(N)\,[dN(t) - \alpha^2 dt],$$

where

$$\lambda_{a,t}(N) = \mathrm{Tr}\, S_{a,t}^{(2)}(N) R^* R$$

is the posterior intensity of counts. It follows that

$$p_{a,t}^{(2)}(N) = \exp\left[\int_a^t \ln \frac{\lambda_{a,t}(N)}{\alpha^2}\, dN(t) - \int_a^t [\lambda_{a,t}(N) - \alpha^2]\, dt\right].$$

The resulting equation for the posterior state is

$$dS_{a,t}^{(2)}(y) - \mathcal{L}[S_{a,t}^{(2)}(y)]\,dt =$$

$$= \left[\frac{R S_{a,t}^{(2)}(y) R^*}{\lambda_{a,t}(y)} - S_{a,t}^{(2)}(y) \right] [dy(t) - \lambda_{a,t}(y)\,dt], \qquad (30)$$

where $\mathcal{L}$ is defined as in (9) with $\frac{1}{4\gamma}$ replaced by α^2 (see Belavkin [17] , Barchielli and Belavkin [23]). In the last paper the equation (27) is also derived as a continuous limit of equations (30).

Acknowledgements. The author is greateful to Dr. A.Barchielli for hospitality in the Institute of Physics of the Milan University, for the interest which stimulated appearance of this paper and for drawing the author's attention to the recent activity in the field of nonlinear wave equations. He is also grateful to Prof. G.Ghirardi for the hospitality and stimulating discussion at the ICTP, Trieste.

References

1 Davies E.B. Quantum theory of open system. - London: AP, 1976.
2 Holevo A.S. Probabilistic and statistical aspects of quantum theory. - Amsterdam: North Holland, 9 - 1982.
3 Kraus K. States, effects and operations//LNP - 1983 - v. 190.
4 Helstrom C.W. Quantum detection and estimation theory. - New York: AP, - 1976.

5 Holevo A.S. Covariant measurements and imprimitivity systems. In: Quantum probability and applications to the quantum theory of irreducible processes (ed. L.Accardi, A.Frigerio, V.Gorini)// LNP - Berlin: Springer-Verlag - 1984 - v. 1055. - p. 163-172.
6 Ozawa M. Conditional probability and a posteriori states in quantum mechanics //Publ. RIMS - 1985 - v. 21, N 2. - p. 279-295.
7 Ozawa M. Quantum measuring processes of continuous observables // J. Math. Phys. - 1984 - v. 25. - p. 79-87.
8 Barchielli A., Lanz L., Prosperi G.M. A model for macroscopic description and continuous observations in quantum mechanics //Nuovo Cimento. - 1982. - v. 72B - p. 79-121.
9 Holevo A.S. Conditionally positive definite functions in quantum probability //Proc. Intern. Congr. of Math., Berkeley, Calif. USA, 1986. - 1987. - p. 1011-1020.
10 Srinivas M.D., Davies E.B. Opt. Acta - 1981 - v. 28 - p. 981-996.
11 Holevo A.S. Quantum probabilistic study of counting statistics with application to the "Dolinar receiver" //Izv. vuzov. Matematika. - 1982 - v. 8 - p. 3-19 (Russian).
12 Hudson R.L., Parthasarathy K.R. Quantum Ito's formula and stochastic evolutions //Commun. Math. Phys. - 1984. - v. 93, N 3. - p. 301-323.
13 Gardiner C.W., Collet M.J. Input and output in damped quantum systems: quantum statistical differential equations and the master equation //Phys. Rev. - 1985. - A31. - p. 3761-3774.
14 Barchielli A., Lupieri G. Dilations of operation-valued stochastic processes. In: Quantum probability and applications II (ed. L.Accardi, W. von Waldenfels) // LNM - 1985 - v. 1136 - p. 57-66.
15 Holevo A.S. Quantum stochastic calculus. In: VINITI. Itogi nauki i techniki. ser. Sovr. probl. mat. v. 36 - 1990 - p. 3-28 (Russian).
16 Liptzer R.S., Shiryaev A.N. Statistics of random processes. - Moscow, Nauka. - 1974 (Russian).
17 Belavkin V.P. Nonlinear filtering of quantum continuous signal. In: 9th All-Union conference on coding theory and information transmission. Abstracts. v. 2 - Odessa - 1988 - p. 342-345. (Russian).
18 Belavkin V.P. In: Stochastic methods in mathematics and physics. Proc. 24 Karpacz winter school on theoretical physics. (ed. R.Gielerak and W.Karwowski) - Singapore: World Scientific. - 1989 - p. 310.
19 Chirardi G., Rimini A., Weber T. A model for unified quantum description of macroscopic and microscopic system. In: Quantum probability and applications II (ed. L.Accardi, W. von Waldenfels) //LNM - Berlin: Springer - Verlag - 1985 - v. 1136 - p. 223-233.
20 Gisin N. Stochastic quantum dynamics and relativity //Helv. Phys. Acta - 1989 - v. 62 - p. 363-371.
21 Chirardi G., Pearle P., Rimini A. Markov processes in Hilbert space and continuous spontaneous localization of systems of identical particles //Preprint ICTP. IC/89/44 - 1989.
22 Gatarek D., Gisin N. Continuous quantum jumps and infinitedimensional stochastic equation //Preprint Université de Genève - 1990.
23 Barchielli A., Belavkin V.P. Measurement continuous in time and a posteriori states in quantum mechanics //Preprint IFUM 383/FT - 1990.

A new class of quantum states: examples and applications

Luigi Accardi

Centro Matematico V.Volterra

Università di Roma II

Roma, Italy

1 CHANNELS, LIFTINGS AND TRANSITION EXPECTATIONS

A **system** , either classical or quantum, is supposed to be characterized by its **observables** $\mathcal{A}$, and its **states** $\mathcal{S}$. In the present note, the mathematical model for the observables of a system will be the set of self-adjoint elements of a C*-algebra $\mathcal{A}$, and the mathematical model for its states, the convex set $\mathcal{S}(\mathcal{A})$ of its states (of its normal states, if $\mathcal{A}$ is a von Neumann algebra). The set of extremal states is denoted $\mathcal{S}(\mathcal{A})_{extr}$. Both $\mathcal{S}(\mathcal{A})$ and $\mathcal{S}(\mathcal{A})_{extr}$ are measurable spaces with their Borel structure and the set of probability measures on $\mathcal{S}(\mathcal{A})$ ($\mathcal{S}(\mathcal{A})_{extr}$) is denoted $Prob\mathcal{S}(\mathcal{A})$ ($Prob\mathcal{S}(\mathcal{A})_{extr}$). All C*- and W*-algebras will be realized on some Hilbert space and the tensor products are those induced by the tensor products of the corresponding Hilbert spaces.

Consider two systems, denoted respectively 1, 2 and their algebras of observables, $\mathcal{A}_1, \mathcal{A}_2$. Suppose that the interaction between the two systems is switched on at a sharp time t_o before which the two systems are independent. During the interaction the two systems merge into a larger system denoted $(1,2)$ whose algebra of observables $\mathcal{A}$ contains both $\mathcal{A}_1$ and $\mathcal{A}_2$, in the sense that there are embeddings

$$j_1 : \mathcal{A}_1 \to \mathcal{A} \; ; \; j_2 : \mathcal{A}_2 \to \mathcal{A}$$

and that any physical information on the state of system 1 or of system 2 after the interaction can be obtained by choosing a state φ on $\mathcal{A}$, i.e. a state of the composite system $(1,2)$, and considering its restriction on the algebra $j_1(\mathcal{A}_1)$ (resp. $j_2(\mathcal{A}_2)$). In most applications one chooses

$$\mathcal{A} = \mathcal{A}_1 \otimes \mathcal{A}_2 \; ; \; \varphi = \varphi_1 \otimes \varphi_2$$
$$j_1(a_1) = a_1 \otimes 1_2 \; ; \; j_2(a_2) = 1_1 \otimes a_2$$

We shall be concerned with three basic problems:

i) **The measurement problem**: One supposes that the state φ_2, of system 2 **after** the interaction (e.g., an output signal, a pointer in a measurement apparatus, ...) and the form of the interaction, are known. One wants to know the state φ_1 of system 1 **before** the interaction (e.g., an input signal, the state of a microsystem which has interacted with the apparatus, ...).

ii) **The first prediction problem**: One supposes that the state φ_2, of system 2 **before** the interaction (e.g., the preparation of a microsystem or the preparation

Lecture Notes in Physics, Vol. 378
C. Bendjaballah, O. Hirota, S. Reynaud (Eds.),
Quantum Aspects of Optical Communications

of a measurement apparatus, ...) and the form of the interaction, are known. One wants to know the state of system 2 **after** the interaction.

iii) **The second prediction problem**: Given the state φ_2, of system 2 **before** the interaction, and the form of the interaction, one wants to know the state of the composite system $(1,2)$ **after** the interaction.

iv) **The third prediction problem**: Given the states φ_1, φ_2, of both systems 1 and 2 **before** the interaction, and the form of the interaction, one wants to know the state of the composite system $(1,2)$ **after** the interaction.

The notion of **channel** is introduced to deal with the first two questions; the notion of **lifting** to deal with the third one; the notion of **compound channel** to deal with the fourth one. In the contemporary probabilistic literature, the notion of compound channel is called **coupling** and plays an important role in the theory of interacting particle systems.

A lifting is a channel from a sub-system to a compound system. Channels (resp. liftings) map states into states. Their *dual objects* are the **Markovian operators** (resp. **transition expectations**), which map observables into observables. Transition expectations are a quantum generalization of the notion of **instrument**, introduced in the operational approach to quantum measurement. A compound channel is a map which to a pair of states φ_1, φ_2, of the systems 1 and 2, associates a state φ, of the composite system $(1,2)$, whose marginal states, on the systems 1 and 2, are precisely φ_1 and φ_2. In order to define these notions precisely, we introduce some notations.

If $\mathcal{A}$ and $\mathcal{B}$ are C*-algebras, a **channel** from $\mathcal{A}$ to $\mathcal{B}$ is a map $P^* : \mathcal{S}(\mathcal{A}) \to \mathcal{S}(\mathcal{B})$.

There are various interpretations of the notion of channel, e.g.:

i) A channel as a quantum dynamics: if $\mathcal{A} = \mathcal{B}$ is the algebra of observables of a system at a given time, then $\rho \in \mathcal{S}(\mathcal{A})$ represents the state of the system at time t and $P^*\rho \in \mathcal{S}(\mathcal{B})$ the state of the system a unit time later (discrete time).

ii) A channel as a input-output system: in this case $\rho \in \mathcal{S}(\mathcal{A})$ represents the input and $P^*\rho \in \mathcal{S}(\mathcal{B})$ the output.

iii) A channel as a communication system (this is a particaular case of (*ii*) above: let I (resp. O) be a set, interpreted as a **input** (resp. **output**) alphabet. To each symbol $i \in I$ (resp. $o \in O$) we associate a state $\rho_i \in (\mathcal{S}(\mathcal{A})$ (resp. a positive operator $A_o \in \mathcal{B}$). This correspondence defines the **coding** for the input (resp. output) alphabet. An **error free channel** is an injective map $F : I \to O$ (i.e. to every input symbol there corresponds one output symbol and different inputs correspond to different outputs) with the property that

$$\sum_{i\in I} A_{F(i)} = 1$$

A real channel is never error free and we represent it through a map $P^* : \mathcal{S}(\mathcal{A}) \to \mathcal{S}(\mathcal{B})$. If F is a subset of the output alphabet O, the quantity

$$\sum_{o\in F} P^*(\rho_i)(A_o)$$

represents *the probability that, having sent the symbol i of the input alphabet, one receives one of the symbols $o \in F$ in the output alphabet.* In particular the quantity

$$1 - P^*(\rho_i)(A_{F(i)}) = \sum_{o \neq F(i)} P^*(\rho_i)(A_o)$$

is called **the probability of error relative to the symbol i.** If a channel P^* is affine we speak of a **linear channel** .

A linear channel P^* can be extended by linearity to a linear map (still denoted P^*) from $\mathcal{A}^*$ to $\mathcal{B}^*$ and its dual $P : \mathcal{B} \to \mathcal{A}$ is a positive map. If it is completely positive, we call it a **Markovian operator.**

Notice that the terminology adopted is *neutral*, in the sense that it can be used both for the classical and quantum case. If some of the algebras involved are non commutative, then one often speaks of a quantum channel, a quantum Markovian operator, ...

Remark There is an interesting connection between quantum and classical Markovian operators which has not yet been completely clarified. Roughly speaking, to every quantum Markovian operator, one can associate a classical Markovian kernel $p(d\omega' \,|\omega)$, giving the probability that the state ω' of system 2 is in a certain set of states $d\omega'$ given that the state of system 1 is in state ω. More precisely: to any classical Markovian kernel on the measurable space $\mathcal{S}(\mathcal{B}) \times \mathcal{S}(\mathcal{A})$, i.e. a measurable map

$$p : \omega \in \mathcal{S}(\mathcal{A}) \longrightarrow p(\,\cdot\,|\omega) \in Prob(\mathcal{S}(\mathcal{B}))$$

one can naturally associate a linear quantum channel in the following way: for any state $\varphi \in \mathcal{S}(\mathcal{A})$ one fixes a convex decomposition

$$\varphi = \int_{\mathcal{S}(\mathcal{A})} \omega d\mu_\varphi(\omega)$$

and defines the channel $P^* : \mathcal{S}(\mathcal{A}) \to \mathcal{S}(\mathcal{B})$ through the identity

$$P^*\varphi := \int_{\mathcal{S}(\mathcal{A})} d\mu_\varphi(\omega) \cdot \int_{\mathcal{S}(\mathcal{B})} \omega' p(d\omega'|\omega)$$

The channel P^* is usually nonlinear since the map $\varphi \to \mu_\varphi$ is affine if and only if $\mathcal{S}(\mathcal{A})$ is a symplex ([BR] Theorem (4.1.15) and Corollary (4.1.17)) and this is the case if and only if $\mathcal{A}$ is Abelian ([BR] Example (4.)). On the other hand, given a linear channel P^* from $\mathcal{A}$ to $\mathcal{B}$ one can fix, for each $\omega \in \mathcal{S}(\mathcal{A})$, a convex decomposition

$$P^*\omega := \int_{\mathcal{S}(\mathcal{B})} \omega' p(d\omega'|\omega)$$

In order to prove that $p(\,\cdot\,|\omega)$ is a Markovian kernel on the measurable space $\mathcal{S}(\mathcal{B}) \times \mathcal{S}(\mathcal{A})$, one has to prove the measurability of the map

$$p : \omega \in \mathcal{S}(\mathcal{A}) \longrightarrow p(\,\cdot\,|\omega) \in Prob(\mathcal{S}(\mathcal{B}))$$

which, in many cases can be done .

Definition 1.1: Let $\mathcal{A}_1, \mathcal{A}_2$ be C*-algebras and let $\mathcal{A}_1 \otimes \mathcal{A}_2$ be a fixed C*-tensor product of $\mathcal{A}_1$ and $\mathcal{A}_2$. A **lifting** from $\mathcal{A}_1$ to $\mathcal{A}_1 \otimes \mathcal{A}_2$ is a continuous map

$$\mathcal{E}^* : \mathcal{S}(\mathcal{A}_1) \to \mathcal{S}(\mathcal{A}_1 \otimes \mathcal{A}_2) \tag{1.1}$$

If $\mathcal{E}^*$ is affine, we call it a linear lifting; if it maps pure states into pure states, we call it **pure** .

In the applications it is often important to consider nonlinear liftings and several examples of these objects arise naturally.

To every lifting from $\mathcal{A}_1$ to $\mathcal{A}_1 \otimes \mathcal{A}_2$ we can associate two channels: one from $\mathcal{A}_1$ to $\mathcal{A}_1$, defined by

$$P_1^*\rho_1(a_1) := (\mathcal{E}^*\rho_1)(a_1 \otimes 1) \quad ; \quad \forall a_1 \in \mathcal{A}_1 \tag{1.2}$$

another from $\mathcal{A}_1$ to $\mathcal{A}_2$, defined by

$$P_2^*\rho_1(a_2) := (\mathcal{E}^*\rho_1)(1 \otimes a_2) \quad ; \quad \forall a_2 \in \mathcal{A}_2 \tag{1.3}$$

In general, any state $\varphi \in \mathcal{S}(\mathcal{A}_1 \otimes \mathcal{A}_2)$ such that

$$\varphi \mid_{\mathcal{A}_1 \otimes 1} = \rho_1 \quad ; \quad \varphi \mid 1 \otimes \mathcal{A}_2 = \rho_2 \tag{1.4}$$

will be called a **compound state** of the states $\rho_1 \in \mathcal{S}(\mathcal{A}_1)$ and $\rho_2 \in \mathcal{S}(\mathcal{A}_2)$.

The following problem is important in several applications: Given a state $\rho_1 \in \mathcal{S}(\mathcal{A}_1)$ and a channel $P^* : \mathcal{S}(\mathcal{A}_1) \to \mathcal{S}(\mathcal{A}_2)$, find a standard lifting $\mathcal{E}^* : \mathcal{S}(\mathcal{A}_1) \to \mathcal{S}(\mathcal{A}_1 \otimes \mathcal{A}_2)$ such that $\mathcal{E}^*\rho_1$ is a compound state of ρ_1 and $P^*\rho_1$. Several particular solutions of this problem have been proposed in [O2 ~ O4], [C1 ~ C3], however an explicit description of all the possible solutions to this problem is still missing.

Definition 1.2: A lifting from $\mathcal{A}_1$ to $\mathcal{A}_1 \otimes \mathcal{A}_2$ is called **nondemolition** for a state $\rho_1 \in \mathcal{S}(\mathcal{A}_1)$ if ρ_1 is invariant for P_1^* i.e., if for all $a_1 \in \mathcal{A}_1$

$$(\mathcal{E}^*\rho_1)(a_1 \otimes 1) = \rho_1(a_1) \tag{1.5}$$

Remark: The physical idea of this definition being that the interaction with system 2 does not alter the state of system 1. The notion of "nondemolition lifting", discussed here is essentially (i.e., up to minor technicalities) included in the more abstract notion of "state extension" introduced by Cecchini and Petz [C2,C3] (cf. also Cecchini and Kümmerer [C3]).

Definition 1.3: Let $\mathcal{A}_1, \mathcal{A}_2$ be C*-algebras and let $\mathcal{A}_1 \otimes \mathcal{A}_2$ be a fixed C*-tensor product of $\mathcal{A}_1$ and $\mathcal{A}_2$. A **transition expectation** from $\mathcal{A}_1 \otimes \mathcal{A}_2$ to $\mathcal{A}_1$ is a completely positive linear map $\mathcal{E} : \mathcal{A}_1 \otimes \mathcal{A}_2 \to \mathcal{A}_1$ satisfying

$$\mathcal{E}(1_{\mathcal{A}_1} \otimes 1_{\mathcal{A}_2}) = 1_{\mathcal{A}_1}$$

It is clear that a transition expectation $\mathcal{E} : \mathcal{A}_1 \otimes \mathcal{A}_2 \to \mathcal{A}_1$ defines by duality a linear lifting $\mathcal{E}^*$ from $\mathcal{A}_1$ to $\mathcal{A}_1 \otimes \mathcal{A}_2$. In some cases the converse is also true. For example, if $\mathcal{A}_1, \mathcal{A}_2$ are W*-algebras and $\mathcal{A}_1 \otimes \mathcal{A}_2$ denotes their W*-tensor product, then any linear lifting $\mathcal{E}^*$ from $\mathcal{A}_1$ to $\mathcal{A}_1 \otimes \mathcal{A}_2$ defines by duality a positive linear map $\mathcal{E} : \mathcal{A}_1 \otimes \mathcal{A}_2 \to \mathcal{A}_1$ characterized by

$$\rho_1(\mathcal{E}(a_1 \otimes a_2) := (\mathcal{E}^* \rho_1)(a_1 \otimes a_2) \quad ; \quad \rho_1 \in \mathcal{S}(\mathcal{A}_1) \ , \ a_1 \in \mathcal{A}_1 \ , \ a_2 \in \mathcal{A}_2$$

Since $\mathcal{E}^*$ maps states into states, it follows that

$$\mathcal{E}(1_{\mathcal{A}_1} \otimes 1_{\mathcal{A}_2}) = 1_{\mathcal{A}_1}$$

Transition expectations play a crucial role in the construction of quantum Markov chains and they arise naturally within the framework of measurement theory in the following way: the composite system $(1,2)$ undergoes an evolution $u_t : \mathcal{A} \to \mathcal{A}$ $(t \in R)$, which is a one-parameter group of $*-$automorphisms of $\mathcal{A}$. This means that the state φ of $(1,2)$ evolves according to the law

$$\varphi_t := \varphi \circ u_t$$

and the state φ_1 of the system 1 evolves according to the **reduced evolution**:

$$\varphi_{1,t}(a_1) := \varphi_1(E_2 \circ u_t \circ j_1(a_1)) \ ; \ a_1 \in \mathcal{A}_\infty$$

where $E_2 : \mathcal{A} = \mathcal{A}_1 \otimes \mathcal{A}_2 \to \mathcal{A}_1$ is the linear map characterized by

$$E_2(a_1 \otimes a_2) = a_1 \varphi_2(a_2) \ ; \ a_1 \in \mathcal{A}_1 \quad ; \ a_2 \in \mathcal{A}_2$$

Let us fix a time T; representing the moment when the experiment ends (ideally $T = +\infty$) and consider the linear map $\mathcal{E} : \mathcal{A}_1 \otimes \mathcal{A}_2 \to \mathcal{A}_1$, characterized by

$$\mathcal{E}(a_1 \otimes a_2) = E_2(u_T(a_1 \otimes a_2)) \ ; \ a_1 \in \mathcal{A}_1 \ ; \ a_2 \in \mathcal{A}_2 \tag{1.6}$$

Then $\mathcal{E}$ is a transition expectation.

Transition expectations are also called **quantum instruments** for the following reasons: an instrument in the sense of the operational approach to quantum measurment is obtained by taking the restriction of a transition expectation $\mathcal{E}$ to a subalgebra $\mathcal{C}_1 \otimes \mathcal{A}_2$ of $\mathcal{A}_1 \otimes \mathcal{A}_2$ where $\mathcal{C}_1$ is an abelian von Neumann algebra of $\mathcal{A}_1$. In this case it is known that there exists a probability space $(\Omega, \mathcal{F}, P)$ such that $\mathcal{C}_1$ is isomorphic to $L^\infty(\Omega, \mathcal{F}, P)$ and the points $\omega \in \Omega$ are interpreted as macroscopic parameters of the apparatus. Moreover, the isomorphism (cf. [S1])

$$\mathcal{A} = \mathcal{A}_1 \otimes \mathcal{A}_2 = \mathcal{A}_1 \otimes L^\infty(\Omega, \mathcal{F}, P) \cong L^\infty(\Omega, \mathcal{F}, P; \mathcal{A}_1)$$

implies that the elements of $\mathcal{A}$, i.e., the observables of the composite system $(1,2)$ can be interpreted as functions $(\Omega, \mathcal{F}, P) \to \mathcal{A}_1$, i.e., as **operator valued** random variables. Thus interpreting $(\Omega, \mathcal{F}, P)$ as the sample space of a classical stochastic process, the operational scheme becomes equivalent to the theory of **operator valued classical processes** (cf. [Hol], [Bar]).

In this sense an instrument in the operational sense is an object which is only "half-quantum". The physical motivations for this choice go back to some ideas of

Ludwig according to which the measurement apparatus is usually a macroscopic body so that classical probability should be sufficient for its description. However, since all the examples of physical interest of instruments in the operational sense, produced up to now, are restrictions of quantum instruments, we feel that the latter notion plays a more natural and fundamental role.

The following examples of liftings appear frequently in the applications:

Example 1: Isometric liftings. Let $V : \mathcal{H}_1 \to \mathcal{H}_1 \otimes \mathcal{H}_2$ be an isometry

$$V^*V = 1_{\mathcal{H}_1}. \tag{1.7}$$

Then the map

$$\mathcal{E} : x \in \mathcal{B}(\mathcal{H}_1) \otimes \mathcal{B}(\mathcal{H}_2) \to V^*xV \in \mathcal{B}(\mathcal{H}_1) \tag{1.8}$$

is a transition expectation, and the associated lifting maps a density matrix $w_1 \in T(\mathcal{H}_1)$ into $\mathcal{E}^*w_1 = Vw_1V^*$. Liftings of this type are called isometric. Every isometric lifting is a pure lifting.

Example 1a: The attenuation (or beam splitting) lifting. It is the particular isometric lifting characterized by the properties.

$$\mathcal{H}_1 = \mathcal{H}_2 =: \Gamma(\mathbf{C}) = \text{the Fock space over } \mathbf{C}$$

$$V : \Gamma(\mathbf{C}) \to \Gamma(\mathbf{C}) \otimes \Gamma(\mathbf{C}) \tag{1.9}$$

is characterized by the expression

$$V|\theta >= |\alpha\theta > \otimes |\beta\theta > \tag{1.10}$$

where $|\theta >$ is the normalized coherent vector parametrized by $\theta \in \mathbf{C}$ and $\alpha, \beta \in \mathbf{C}$ are such that

$$|\alpha|^2 + |\beta|^2 = 1 \tag{1.11}$$

Notice that this lifting maps coherent states into products of coherent states. So it maps the simplex of the so called "classical states" (i.e., the convex combinations of coherent vectors) into itself. Finally, one easily verifies that this lifting is not nondemolition.

This lifting has a natural interpretation: $\Gamma(\mathbf{C})$ is the space of a 1-mode coherent EM field (signal); V represents the action, on the signal, of an apparatus (e.g., a receiver or a semi–transparent mirror). Equation (1.10) means that, by the effect of the interaction, a coherent signal (beam) $|\theta >$ splits into 2 signals (beams) still coherent, but of lower intensity. Because of (1.11), the total intensity (energy) is preserved by the transformation.

Example 1b: Superposition beam splitting.
The only difference with Example (1a.) is the form of V, which in this case is

$$V|\theta >= \frac{1}{\sqrt{2}}(|\alpha\theta > \otimes |\beta\theta > -i|\beta\theta > \otimes |\alpha\theta >) \tag{1.12}$$

One easily checks that V extends linearly to an isometry of the form (1.9).

This isometric lifting does not map states into convex combinations of product states, neither it is a nondemolition lifting.

Example 2: The compound lifting.

Let $P^* : \mathcal{S}(\mathcal{A}_1) \to \mathcal{S}(\mathcal{A}_2)$ be a channel. For any $\rho_1 \in \mathcal{S}(\mathcal{A}_2)$, fix a decomposition of ρ_1 as a convex combination of extremal states in $\mathcal{S}(\mathcal{A}_1)$

$$\rho_1 = \int_{\mathcal{S}(\mathcal{A}_1)} \omega_1 p(d\omega_1 \mid \rho_1)$$

where $p(\,\cdot\, \mid \rho_1)$ is a Borel measure on $\mathcal{S}(\mathcal{A}_1)$ with support in the extremal states, and define

$$\mathcal{E}^* \rho_1 := \int_{\mathcal{S}(\mathcal{A}_1)} \omega_1 \otimes P^* \omega_1 p(d\omega_1 \mid \rho_1)$$

Then $\mathcal{E}^* : \mathcal{S}(\mathcal{A}_1) \to \mathcal{S}(\mathcal{A}_2 \otimes \mathcal{A}_2)$ is a lifting, nonlinear even if P^* is linear, and nondemolition for ρ_1.

2.) QUANTUM MARKOV CHAINS

In this Section we briefly recall how to associate, in a natural way, a quantum process to a given transition expectation and a given initial distribution.

If $(\mathcal{E}_n)_{n\geq 0}$ is a sequence of transition expectations from $\mathcal{B} \otimes \mathcal{B}_o$ to $\mathcal{B}_o$. then there exists a unique completely positive identity preserving map $E_{o]} : \otimes_{\mathbf{N}} \mathcal{B} \to \mathcal{B}_o$ such that for each integer n and for each $a_o, a_1, \ldots, a_n \in \mathcal{B}$, one has

$$E_{o]}\left(a_o \otimes a_1 \otimes \ldots \otimes a_n \otimes 1 \otimes \ldots\right) = \mathcal{E}_o\left(a_o \otimes \mathcal{E}_1(a_1 \otimes \ldots \otimes \mathcal{E}_n(a_n \otimes 1))\right) \tag{2.1}$$

In fact, since $\mathcal{E}_n$ is completely positive, then for each $n \in \mathbf{N}$ we can define a completely positive map $\bar{\mathcal{E}}_n$ characterized by

$$\bar{\mathcal{E}}_n\left(a_o \otimes a_1 \otimes \ldots \otimes a_n\right) := a_o \otimes a_2 \otimes \ldots \otimes \mathcal{E}_n(a_{n-1} \otimes a_n)$$

The map $\bar{\mathcal{E}}_{n]} : \mathcal{A}_{[0,n]} \to \mathcal{A}_{[0,n-1]}$ defined by

$$\bar{\mathcal{E}}_{n]} := \bar{\mathcal{E}}_o \circ \ldots \circ \bar{\mathcal{E}}_{n-1} \circ \bar{\mathcal{E}}_n$$

is a completely positive identity preserving map with the property that for each integer n and for each $a_o, a_1, \ldots, a_n \in \mathcal{B}$ the limit

$$\lim_{N\to\infty} \bar{\mathcal{E}}_{N]}(a_o \otimes a_1 \otimes \ldots \otimes a_n \otimes 1 \otimes \ldots)$$

exists and is equal to the right hand side of (2.1). By linearity and continuity this limit defines a unique map $\mathcal{E}_{o]} : \mathcal{A} := \otimes_{\mathbf{N}} \mathcal{B} \to \mathcal{B}_o$.

The map $E_{o]}$ can be used to lift any state φ_o on $\mathcal{B}_o$ to a state φ on $\otimes_{\mathbf{N}}\mathcal{B}$ defined by

$$\varphi = \varphi_o \circ E_{o]} \tag{2.2}$$

By construction φ is characterized by the property :

$$\varphi\left(a_o \otimes a_1 \otimes \ldots \otimes a_n \otimes \ldots\right) = \varphi_o\left\{\mathcal{E}_o\left(a_o \otimes \mathcal{E}_1(a_1 \otimes \ldots \otimes \mathcal{E}_n(a_n \otimes 1))\right)\right\} \tag{2.3}$$

If $\mathcal{E}(a \otimes b) = a \cdot P(b)$, where P is a Markovian operator and the algebra $\mathcal{B}$ is abelian, then the joint expectations (2.3) above reduce to the well known formula for the joint expectations of a classical Markov chain.

The state φ , characterized by (2.2) will be called the **generalized Markov chain** associated to (or determined by) the pair $\{\varphi_o, (\mathcal{E}_n)\}$. If for each n

$$\mathcal{E}_n = \mathcal{E}_o =: \mathcal{E} \tag{2.4}$$

then we speak of an **homogeneous** generalized Markov chain . The completely positive, identity preserving , normal maps $\mathcal{E}_n$ are called the transition expectations of the generalized Markov chain φ . The state φ_o (or sometimes its density matrix w_o) is called the **initial distribution** of the generalized Markov chain .

PROPOSITION The generalized Markov chain φ , determined by the pair $\{\varphi_o, (\mathcal{E}_n)\}$ is stationary if

i) it is homogeneous (i.e. $\mathcal{E}_n = \mathcal{E}$ independently of n)

ii) Denoting P : $\mathcal{B} \to \mathcal{B}$ the backward Markovian operator associated to $\mathcal{E}$ in the sense of (17.6), one has :

$$\varphi_o \circ P = \varphi_o \tag{2.5}$$

PROOF A straightforward verification.

Recall that a bounded linear operator T on a real Banach space V, is called a **Perron operator** with maximal eigenvalue λ , eigenvector b and invariant state ψ if there exists a nonzero vector $b \in V$ and a nonzero continuous linear functional ψ on V such that the operator

$$\frac{T}{\lambda} - b \otimes \psi$$

has norm strictly less than 1 (if $x \in V$, then by definition $(b \otimes \psi)(x) = b\psi(x)$).

Notice that if T is a Perron operator, then for any $x \in V$,

$$\| (\frac{T}{\lambda})^n x - \psi(x)b \| \leq e^{-cn} \| x \| \tag{2.6}$$

for some constant $c > 0$. Moreover it is known (Dunford, Schwartz (1963)) that, if T is a Perron operator and λ, ψ and b are as above, then

$$Tb = \lambda b \quad ; \quad \psi \circ T = \lambda\psi \tag{2.7}$$

$$\psi(b) = \lambda \tag{2.8}$$

Moreover λ is a simple eigenvalue and any eigenvalue of T, different from λ, has modulus strictly less than λ and, if c is the corresponding eigenvector, then

$$\psi(c) = 0 \tag{2.9}$$

PROPOSITION (2.6) Let $\varphi = \{\varphi_o, \mathcal{E}\}$ be a homogeneous Markov chain. If the forward transition operator P , associated to $\mathcal{E}$ is a Perron operator , then φ is exponentially mixing and φ is a factor state .

PROOF (cf.[A3])

3.) CONVEX COMBINATIONS OF PRODUCT STATES

One of the main differences between classical and quantum probability is that, while all the measures on a product space are in the closed convex hull (for the weak topology) of product measures, it is not true that all the states on the tensor product $\mathcal{A}_1 \otimes \mathcal{A}_2$ of two general C^*–algebras are limits (in some topology) of convex combinations of product states.

In particular, the image under a general lifting $\mathcal{E}^*$ of a state φ will usually not be a convex combination of product states.

However the class of liftings with this porperty is particulary interesting because we expect that in this class some features of quantum probability will mix with some features of classical probability. This class is defined as follows:

Definition 3.1: Let $\mathcal{A}_1$ and $\mathcal{A}_2$ be W^*–algebras. A lifting $\mathcal{E}^* : \mathcal{S}(\mathcal{A}_1) \to \mathcal{S}(\mathcal{A}_1 \otimes \mathcal{A}_2)$ will be called **of convex product type**, or shortly a **convex product lifting** , if any state $\omega \in \mathcal{S}(\mathcal{A}_1)$ is mapped by $\mathcal{E}^*$ into a convex combination of states on $\mathcal{A}_1 \otimes \mathcal{A}_2$. If this property holds only for any state ω in a subset $\mathcal{F} \subseteq \mathcal{S}(\mathcal{A}_1)$ then $\mathcal{E}^*$ is called a convex product lifting with respect to the family $\mathcal{F}$.

For any von Neumann algebra $\mathcal{A}$, the set $\mathcal{S}(\mathcal{A})$ of all its states has a natural structure of measurable space with its Borel σ–algebra. In the following any probability measure on $\mathcal{S}(\mathcal{A})$ will be meant with respect to this σ–algebra.

Definition 3.2: A **convex decomposition** of $\varphi \in \mathcal{S}(\mathcal{A})$ is a probability measure μ on $\mathcal{S}(\mathcal{A})$ satisfying

$$\varphi = \int_{\mathcal{S}(\mathcal{A})} \omega d\mu(\omega) \tag{3.1}$$

If μ is pseudosupported in the set of extremal states of $\mathcal{S}(\mathcal{A})$, we speak of an **extremal convex decomposition** of φ.

Proposition 3.1: To every lifting of convex product type $\mathcal{E}^* : \mathcal{S}(\mathcal{A}_1) \to \mathcal{S}(\mathcal{A}_1 \otimes \mathcal{A}_2)$, one can associate a pair

$$\{p_\rho(d\omega_1),\ p_\rho(d\omega_2|\omega_1)\}$$

with the following properties:

(i) $p_\rho(d\omega_1)$ is a probability measure on $\mathcal{S}(\mathcal{A}_1)$

(ii) $p_\rho(d\omega_2|\omega_1)$ is a Markovian kernel from $\mathcal{S}(\mathcal{A}_1)$ to $\mathcal{S}(\mathcal{A}_2)$.

Conversely every pair satisfying (i) and (ii) above determines (via (3.3) and (3.4) below) a unique convex product lifting.

Proof: For $\mathcal{E}^*$ as in Definition (3.1), let us fix a state $\rho_1 \in \mathcal{S}(\mathcal{A}_1)$ and also a decomposition of $\mathcal{E}^*\rho_1$ as a convex combination of product states

$$\mathcal{E}^*\rho_1 = \int_{\mathcal{S}(\mathcal{A}_1)\times\mathcal{S}(\mathcal{A}_2)} \omega_1 \otimes \omega_2 dp(\omega_1, \omega_2|\rho_1) \tag{3.2}$$

Denoting $p_{\rho_1}(d\omega_2|\omega_1)$ the conditional probability of $p(\,\cdot\,|\rho_1)$ on the σ–algebra of the first factor and $dp_{\rho_1}(\omega_1)$ the marginal of $p(\,\cdot\,|\rho_1)$ on the first factor, we obtain

$$\mathcal{E}^*\rho_1 = \int_{\mathcal{S}(\mathcal{A}_1)}\int_{\mathcal{S}(\mathcal{A}_2)} \omega_1 \otimes \omega_2 dp_{\rho_1}(\omega_1)p_{\rho_1}(d\omega_2|\omega_1) = \int_{\mathcal{S}(\mathcal{A}_1)} \omega_1 \otimes P^*_\rho\omega_1 dp_{\rho_1}(\omega_1) \tag{3.3}$$

$$P^*_{\rho_1}\omega_1 := \int_{\mathcal{S}(\mathcal{A}_2)} \omega_2 p_{\rho_1}(d\omega_2|\omega_1) \tag{3.4}$$

Thus any lifting $\mathcal{E}^* : \mathcal{S}(\mathcal{A}_1) \to \mathcal{S}(\mathcal{A}_1 \otimes \mathcal{A}_2)$, of convex product type, has the form (3.3), where p_{ρ_1} is a probability measure on $\mathcal{S}(\mathcal{A}_1)$ and the map $P^*_{\rho_1} : \mathcal{S}(\mathcal{A}_1) \to \mathcal{S}(\mathcal{A}_2)$, is given by (3.4). Notice that $P^*_{\rho_1}$ is a channel in the sense of Section 1 and it is usually nonlinear both in ω_1 and ρ_1.

Conversely, given p_{ρ_1} and $P^*_{\rho_1}$ as above, if we define $\mathcal{E}^*$ by (3.2), then clearly $\mathcal{E}^*$ is a convex product lifting from $\mathcal{S}(\mathcal{A}_1)$ to $\mathcal{S}(\mathcal{A}_1 \otimes \mathcal{A}_2)$. Finally it is clear that the map

$$(\omega_1, S_2) \in \mathcal{S}(\mathcal{A}_1) \times \text{Borel}(\mathcal{S}(\mathcal{A}_2)) \to p_{\rho_1}(S_2|\omega_1) \in [0,1]$$

is a classical Markovian kernel on the Borel space $\mathcal{S}(\mathcal{A}_1) \times \mathcal{S}(\mathcal{A}_2)$.

Remark: If in (3.2) one conditions on the σ-algebra of the second factor rather than on the first one, the resulting lifting is

$$\mathcal{E}^*\rho_1 = \int_{\mathcal{S}(\mathcal{A}_1)}\int_{\mathcal{S}(\mathcal{A}_2)} \omega_1 \otimes \omega_2 dq_{\rho_1}(\omega_2) dq_{\rho_1}(d\omega_1|\omega_2) \tag{3.5}$$

where $dq_{\rho_1}(\omega_2)$ is a probability measure on $\mathcal{S}(\mathcal{A}_2)$ and $dq_{\rho_1}(d\omega_1|\omega_2)$ a Markovian kernel from $\mathcal{S}(\mathcal{A}_2)$ to $\mathcal{S}(\mathcal{A}_1)$.

The construction of quantum Markov chains can be extended to the case of a lifting of convex product type (not necessarily linear) as follows: if $\mathcal{E}^* : \mathcal{S}(\mathcal{A}_2) \to \mathcal{S}(\mathcal{A}_1 \otimes \mathcal{A}_2)$ is a lifting of convex product type, then it has the form:

$$\mathcal{E}^*\rho_2 = \int_{\mathcal{S}(\mathcal{A}_1)}\int_{\mathcal{S}(\mathcal{A}_2)} \omega^1 \otimes \omega^2 p(d\omega^1, d\omega^2|\rho_2) \tag{3.6}$$

Notice that $p(d\omega^1, d\omega^2|\rho_2)$ can be considered as a Markovian Kernel on the space

$$\mathcal{S}_{12} := \mathcal{S}(\mathcal{A}_1) \times \mathcal{S}(\mathcal{A}_2)$$

which is constant on the first conditioning, i.e.,

$$p(d\omega^1, d\omega^2|\rho_1, \rho_2) = p(d\omega^1, d\omega^2|\rho_2) \quad ; \quad \omega_1, \rho_1, \in \mathcal{S}(\mathcal{A}_1) \ , \ \omega_2, \rho_2 \in \mathcal{S}(\mathcal{A}_2) \tag{3.7}$$

Clearly (3.6) is a state on $\mathcal{A}_1 \otimes \mathcal{A}_2$. If we apply $\mathcal{E}^*$ to ω^2 in (3.6), we obtain the following state on $(\mathcal{A}_1 \otimes \mathcal{A}_2) \otimes \mathcal{A}_2$:

$$\int_{S_{12}} p(d\omega_1^1, d\omega_1^2, |\rho_2)\omega_1^1 \otimes \mathcal{E}^*\omega_1^2 =$$

$$= \int_{S_{12}} \int_{S_{12}} p(d\omega_1^1, d\omega_1^2|\rho_2)p(d\omega_2^1, d\omega_2^2|\omega_1^2)\omega_1^1 \otimes \omega_2^1 \otimes \omega_2^2$$

where

$$\omega_i^1 \in \mathcal{S}(\mathcal{A}_1) \; and \; \omega_i^2 \in \mathcal{S}(\mathcal{A}_2).$$

Applying again $\mathcal{E}^*$ to ω_2^2 we find

$$\int_{S_{12}} \int_{S_{12}} \int_{S_{12}} p(d\omega_1^1, d\omega_1^2|\rho_2)p(d\omega_2^1, d\omega_2^2|\omega_1^2)p(d\omega_3^1, d\omega_3^2|\omega_2^2)\omega_1^1 \otimes \omega_2^1 \otimes \omega_3^1 \otimes \omega_3^2$$

At the n–th iteration we obtain the state $\mathcal{E}_{n]}^*\rho_2$ on $(\otimes\mathcal{A}_1)^n \otimes \mathcal{A}_2$, defined by:

$$\mathcal{E}_{n]}\rho_2 := \int_{S_{12}^n} (\otimes_{i=1}^n \omega_i^1) \otimes \omega_n^2 \cdot \Pi_{i=2}^n p(d\omega_i^1, d\omega_i^2|\omega_{i-1}^2)p(d\omega_1^1, d\omega_1^2, |\rho_2) \tag{3.8}$$

This suggests to introduce the classical Markov process

$$\xi_n := (\xi_n^1, \xi_n^2) : (\Omega, \mathcal{F}, P) \to \mathcal{S}(\mathcal{A}_1) \times \mathcal{S}(\mathcal{A}_2) = S_{12} \tag{3.9}$$

with the transition function given by (3.7) and initial distribution $p(\cdot|\rho_2)$. Now let $\mathcal{A} := \otimes_{\mathbf{N}}\mathcal{A}_1$. The identification

$$a_1 \otimes \ldots \otimes a_n \cong a_1 \otimes a_2 \otimes \ldots \otimes a_n \otimes 1 \otimes 1 \otimes \cdots$$

induces a natural identification of $(\otimes\mathcal{A}_1)^n$ with a sub-algebra $\mathcal{A}_{[2,n]}$ of $\mathcal{A} = \otimes_{\mathbf{N}}\mathcal{A}_1$ (the product of the first n-factors).

In particular, if $\rho_2 \in \mathcal{S}(\mathcal{A}_2)$ is a state on $\mathcal{A}_2$, the restriction of $\mathcal{E}_{n]}^*\rho_2$ on $(\otimes\mathcal{A}_1)^n$ is a state on $(\otimes^n\mathcal{A}_1)$ and, with the above identification, we can consider it a state $\rho_{[1,n]}$ on $\mathcal{A}$. Following from all above, in particular (3.8), we obtain

Proposition 3.2: For any $\rho_2 \in \mathcal{S}(\mathcal{A}_2)$ the limit

$$\lim_{n\to\infty} \quad \rho_{[1,n]} =: \varphi \tag{3.10}$$

exists pointwise on $\mathcal{A}$. Moreover, if E_ξ denotes the mean with respect to the process $\{\xi_n\}$, defined by (3.9), then one has

$$\varphi = E_\xi \left(\otimes_{n\in\mathbf{N}}\xi_n^1\right). \tag{3.11}$$

BIBLIOGRAPHY

[A1] Accardi L.; Quantum Kalman filtering. Contribution to the memorial volume for the 60-th birthday of R.E. Kalman

[A2] Accardi L.; Noncommutative Markov Chains with preassigned evolution: an application to the quantum theory of measurement. 29, (1978), 226–243.

[A3] Accardi L.; Six Lectures on Quantum Probability World Scientific, to appear

[A4] Accardi L., Watson G.S.; Quantum random walks and coherent quantum chains, in : Quantum Probability and Applications IV Springer LNM 1396(1987) 73-88.

[A5] L.Accardi; An outline of quantum probability to appear in Uspekhi Matematicheskikh Nauk. (in russian) 1991.

[Al1] Alicki R., Frigerio A.; Quantum Poisson noise and linear quantum Boltzmann equation. preprint, March 1989.

[Ba1] Bach A.; The Simplex Structure of the Classical States of the Quantum Harmonic Oscillator. Commun. Math. Phys. 107 (1986) 553-561.

[Bar1] Barchielli A.; Stochastic Processes and Continual Measurements in Quantum Mechanics. Preprint. [BR] O. Bratteli and D. V. Robinson; Operator algebras and quantum statistical mechanics I and II, Springer Verlag, Berlin, 1981.

[C1] Cecchini C.; Stochastic Couplings for von Neumann Algebras. in: Quantum Probability and Applications III Springer LNM n. 1303 (1988),1–5.

[C2] Cecchini C., Petz D.; Classes of conditional expectations over von Neymann algebras. Journal of functional analysis 91, 1990.

[C3] Cecchini C., Kúmmerer B.; Stochastic Transitions on Preduals of von Neumann Algebras. in: Quantum Probability and Applications V, Springer LNM 1442.

[C4] Cecchini C., Petz D.; State Extention and a Radon-Nikodym Theorem for Conditional Expectations on von Neumann Algebras. Pacific Journal of Mathematics, 138, (1989) 9–24.

[Hi1] Hirota O., Yamazaki K., Nakagawa M., Ohya M.; Properties of error correcting code using photon pulse, Transaction of IECE Japan, E69, (1986), 917–919.

[Ho1] Holevo A.S.; Probabilistic and statistical aspects of quantum theory. North Holland (1982).

[Hol1] Hollenhorst J.N.; Quantum limits on resonant-mass gravitational-radiation detectors, Physical Review, D19,(1979) 1669–1679.

[Mc1] McEliece R.J.; Practical codes for photon communication, IEEE Transaction of Information Theory, 27 (1981),393–397.

[O1] Ohya M.; Quantum Ergodic Channels in Operator Algebras. Reprinted from Journal of Mathematical Analysis and Applications 84, (1981), 318–327.

[O2] Ohya M.; On Compound State and Mutual Information in Quantum Information Th, IEEE Transactions on Information Theory, 29 (1983), 770–774.

[O3] Ohya M.; Note on Quantum Probability. Lettere al Nuovo Cimento 38 (1983) 402–404.

[O4] Ohya M.; State Change and Entropies in Quantum Dynamical Systems. Springer LNM 1136 (1988), 397–408 (ed) Accardi L., von Waldenfels W.

[O5] Ohya M.; Some Aspects of Quantum Information Theory and Their Applications, Physics, 27, (1989), 19–47.

[P1] Petz D.; Characterization of Sufficient Observation Channels. Preprint.

[P2] Petz D.; Sufficient Subalgebras and the Relative Entropy of States of a von Neumann Algebra. Commun. Math. Phys. 105 (1986) 123–131.

[S1] Sakai S.; C^*-algebras and W^*-algebras, Springer, (1971).

[St1] Stoler D.; Equivalence class of minimum uncertainty packets, Physical Review, D4,(1971) 1925.

[Y1] Yuen H.P.; Two-photon coherent states of the radiation field, Physical Review, A13, (1976),2226–2243.

Continuous non–demolition observation quantum filtering and optimal estimation

V. P. Belavkin

Moscow Institute of Electronic Mashinedesign
B. Vusovski 3/12, Moscow 109028, USSR

Abstract. A quantum stochastic model for an open dynamical system (receiver) with output channel of observation is given. An equation for the moment generating operator of the corresponding instrument is derived and reduced equations for the wave function and density matrix of the system under the observation are found. The dynamical problem of quantum filtering for a noncomutative output process is solved and a quantum stochastical equation for the optimal dynamical estimate of an input Markovian process is found. The results are illustrated on an example of optimal estimation of the Langevin force in a quantum oscilator (optical or Weber's antenna).

Introduction. The time evolution of quantum system under a continuous observation can be obtained in the frame work of quantum stochastic (QS) calculus of output nondemolition processes, firstly introdused in [1] and recently developed in a quite general form in [2,3,4,5,6]. A stochastic posterior Schrödinger wave equation for an observed spinless particle derived in [4] by using the quantum filtering method [6], provided an explanation of the quantum Zeno paradox [7,8]. In the paper we give a derivation of the reduced wave equation for Markovian system in a Bose field under a continuous observation which is supposed to be nondemolition in the sense [4,5,6] of the commutativity $[Y(s),\ X(t)] = 0$ of observables $Y(s)$, $s \leq t$, where $X(t)$ is the arbitrary variable of the system in the Heisenberg picture at time t. We shall obtain it by a non–unitary dilation of the characteristic operator of an instrument for the observable output process, but in contrast to [9] we restrict ourselves to the diffusion observation, i.e. to a continu-

Lecture Notes in Physics, Vol. 378
C. Bendjaballah, O. Hirota, S. Reynaud (Eds.),
Quantum Aspects of Optical Communications

ous nondemolition measurement of a quantum Brownian motion. This gives the possibility to solve the dynamical problems of the quantum detection and estimation theory [10].

1 The dynamical model

Let us suppouse that the nondemolition observation describes the continuous in time indirect measurement of some non–commuting selfadjoint operators X_i, $i = 1, \dots, d$, acting in the Hilbert space $\mathcal{H}_0$ of an open quantum system (antenna), with additive errors $\hat{e}_j(t)$, $j = 1, \dots, d$ being white noises. In general the noise $\hat{e}(t) = \{\hat{e}_1, \dots, \hat{e}_d\}(t)$ is given as a quantum Gaussian nondegenerated δ–correlated real vector process $\hat{e}(t) = 2\Re\dot{\mathtt{A}}(t)$, defined in a Fock space $\mathcal{F}$ by the first and second moments

$$\langle \hat{e}_j(t) \rangle = 0, \quad \langle \hat{e}_j(t')\hat{e}_{j'}(t) \rangle = \gamma_{jj'}\delta(t' - t) \tag{1}$$

with respect to the vacuum state $|0\rangle \in \mathcal{F}$. Here $2\Re\dot{\mathtt{A}}(t) = \dot{\mathtt{A}}(t) + \dot{\mathtt{A}}(t)^*$ is the generalized derivative of a real–vector Bose field $\hat{v}(t) = \{\hat{v}_1, \dots, \hat{v}_d\}(t)$ with $\hat{v}_j(t) = \mathtt{A}_j(t) + \mathtt{A}_j^*(t)$, $j = 1, \dots, d$, where $\mathtt{A}_j(t)$ are the annihilation operators, satisfying the commutation relations

$$[\hat{v}_j(t'), \ \hat{v}_{j'}(t)] = 2i \min\{t, \ t'\}\Im\gamma_{jj'} \ , \ 2i\Im\gamma_{jj'} = \gamma_{jj'} - \gamma_{jj'}^*.$$

One can realize the operators $\mathtt{A}_j(t)$ as the linear combinations $\mathtt{A}_j(t) = \sum \gamma_{jk}\mathtt{A}^k(t)$ of the canonical Bose annihilation operators in the space $\mathcal{F}$ of simmetrical tensors over the copmlex vector–functions $\alpha(t) = \{\alpha_j\}(t)$ with

$$(\alpha, \ \alpha) = \int \alpha_i^*(t)\gamma^{ik}\alpha_k(t)dt \equiv \|\alpha\|^2 < \infty \, .$$

Here the positive–definite metric tensor γ^{ik} defines the inverse $d \times d$–matrix $\gamma^{-1} = [\gamma^{jj'}]$, $\sum_{j=1}^d \gamma^{ij}\gamma_{jk} = \delta_k^i = \sum_{j=1}^d \gamma_{ij}\gamma^{jk}$, $\forall i, k = 1, \dots, d$ for the nondegenerated correlation matrix $\gamma = [\gamma_{jj'}]$, $jj' = 1, \dots, d$,

$$[\mathsf{a}(\alpha^*), \ \mathsf{a}^*(\alpha')] = (\alpha, \alpha') \, , \ [\mathsf{a}(\alpha), \ \mathsf{a}(\alpha')] = 0 \tag{2}$$

for $\mathsf{a}(\alpha^*) = \int \alpha_i^*(t) d\mathtt{A}^i(t)$, $\mathsf{a}^*(\alpha) = \mathsf{a}(\alpha^*)^*$ (we use Einstein notations for the convolutions $\alpha_i^*\beta^i = \sum \alpha_i^*\beta^i$ over the indeces $i = 1, 2, \dots$ in contrast to the scalar product notations $\alpha^*\beta$ for the finite sums $\sum_{j=1}^{d}\alpha_j^*\beta^j$, and omit the identity operator I is the Fock space $\mathcal{F}$).

The output vector–process $\dot{Y}(t) = X(t)+I\otimes\hat{e}(t)$, defined by the integrals

$$Y_j(t) = \int_0^t X_j(r)dr + I \otimes 2\Re A_j(t), \;\; j = 1, \ldots, d \tag{3}$$

of the Heisenberg operators $X_j(t) = U(t)^*X_jU(t)$, can be realized for a singular coupling of the system with the Bose field (channel) $\{A^i(t)\}$ by the observables $\hat{v}_j = 2\Re(\gamma_{jk}A^k)$ in the interaction picture

$$Y_j(t) = U(t)^*\hat{V}_j(t)U(t) = 2\Re B_j(t), \;\; \hat{V}_j = I \otimes \hat{v}_j, \tag{4}$$

where $B_j(t) = U(t)^*(I \otimes A_j(t))U(t)$ are the annihilation output processes, introduced in [2,3,4,5], and I is the identity operator in $\mathcal{H}_0$. The unitary evolution $U(t)$ is described in the tensor product $\mathcal{H} = \mathcal{H}_0 \otimes \mathcal{F}$ by a Schrödinger–Ito quantum stochastic equation [11]

$$dU(t) + KU(t)dt = i\left(\frac{\kappa}{\hbar}\, Q \otimes d\vartheta_t - 2\Im(L_k^* \otimes dA^k(t))\right)U(t), \tag{5}$$

where $K = \frac{1}{2}\left(\sigma^2Q^2/\hbar^2 + L_i^*\gamma^{ik}L_k\right)$, $H = H^*$ is a Hamiltonian of the system, L_i, $i = 1, 2, \ldots$ are the operators in $\mathcal{H}_0$ with $L_j + L_j^* = X_j$, $j = 1, \ldots, d$, and $Q = Q^*$ is an operator in $\mathcal{H}_0$ of a generalized coordinate for an input stochastic signal ϑ_t with $(d\vartheta_t)^2 = \sigma^2 dt$ as the conjugated momentum of an external force. The solution $t \mapsto U(t)$ of the equation (4) for $U(0) = I$ is adaptive $U(s) = U^s \otimes \mathrm{I}_{[s}$ with recpect to the tensor decomposition $\mathcal{H} = \mathcal{H}_s \otimes \mathcal{F}_{[s}$, where $\mathcal{H}_s = \mathcal{H}_0 \otimes \mathcal{F}_s$ and $\mathcal{F}_s$, $\mathcal{F}_{[s}$ are the Fock spaces, generated by the vector–function $\alpha_s(t) = 0$, $\forall t \geq s$ and $\alpha_{[s}(t) = 0$, $t < s$. Hence $(I \otimes \hat{v}(s))U(t) = U(t)Y(s)$ for any $s \leq t$, and

$$\begin{aligned} [Y_j(s),\, Y_k(t)] &= U(t)^*(I \otimes [2\Re A_j(s),\, 2\Re A_k(t)])U(t) = 2is\Im\gamma_{jk}I \\ [Y_j(s),\, X(t)] &= U(t)^*[I \otimes \hat{v}_j(s),\, X \otimes \mathrm{I}]U(t) = 0, \;\; \forall s \leq t \end{aligned} \tag{6}$$

for any operator $X(t) = U(t)^*(X \otimes \mathrm{I})U(t)$ of the system in the Heisenberg picture. This proves the nondemolition property of any observable process $Y_j(t)$ with respect to the system, introduced as the princeple for the output quantum processes in [1,2]. Using the quantum Ito formula (4) for [4] with $d\hat{v}_j = 2\Re dA_j$ and multiplication table

$$dA^i dA_k^* = \delta_k^i dt, \; dA_i^* dA^k = 0, \; dA^i dA^k = 0 = dA_i^* dA_k^*, \; X(t) = \{X_j\}(t), \tag{7}$$

one can derive the equation (3): $dY(t) = X(t)dt + I \otimes d\hat{v}(t)$, where $X(t) = \{X_j\}(t)$, $X_j(t) = U(t)^*(L_j + L_j^*)U(t)$.

Theorem 1 *Let us suppose that the input real–valued signal ϑ_t satisfies the stochastic differential equation*

$$d\vartheta_t + v(\vartheta_t)dt = \sigma d\mathtt{W}_t, \tag{8}$$

where $\mathtt{W}_t$ is an independent standart Wiener process, defined by the moments: $\langle \mathtt{W}_t \rangle = 0$, $\langle \mathtt{W}_s \mathtt{W}_t \rangle = s$ for $s \leq t$ with respect to the vacuum state as $2\Re\mathtt{A}^0(t)$ for a canonical annihilation process $\mathtt{A}^0 = \mathtt{A}_0$ in Fock space $\mathcal{F}$, corresponding to $\gamma^{00} = 1 = \gamma_{00}$. Then any twicely differentiable in the strong operator topology function $X : \vartheta \in \mathbf{R} \mapsto X(\vartheta) \in \mathcal{B}(\mathcal{H}_0)$ satisfies in the Heisenberg picture $X(t) = U_t X(\vartheta_t) U_t^* = X^t(\vartheta_t)$ the quantum Langevian equation

$$dX(t) - U_t \Lambda\{X\}(\vartheta_t)U_t^* dt = \sigma U_t(X'(\vartheta_t) - \frac{\kappa}{\hbar}\,[iQ,\ X(\vartheta_t)])U_t^* \otimes d\mathtt{W}_t$$
$$+ U_t[i\Re L_k^*,\ X(\vartheta_t)]U_t^* \otimes 2d\Im\mathtt{A}^k(t) + U_t[i\Im L_k^*,\ X(\vartheta_t)]U_t^* \otimes 2d\Re\mathtt{A}^k(t). \tag{9}$$

Here $U_t^* = U^t = U_t^{-1}$ is the unitary operator $U(t)$ on $\mathcal{H}_t = \mathcal{H}_0 \otimes \mathcal{F}_t$, the stochastic differntials $d\mathtt{W}_t = \mathtt{W}_{t+dt} - \mathtt{W}_t$, $d\mathtt{A}_i^*$, $d\mathtt{A}_t$ are represented in $\mathcal{F}_{[t}$,

$$\Lambda\{X\}(\vartheta) = \frac{1}{2}\,\sigma^2\delta^2 X(\vartheta) - v(\vartheta)X'(\vartheta) +$$
$$\frac{1}{2}\,(L_k^*\gamma^{kk'}[X(\vartheta),\ L_{k'}] + [L_k^*,\ X(\vartheta)]\gamma^{kk'}L_{k'}) + \frac{i}{\hbar}\,[H,\ X(\vartheta)], \tag{10}$$

and $\delta X(\vartheta) = X'(\vartheta) - \frac{\kappa}{\hbar}\,[iQ, X(\vartheta)]$, $X'(\vartheta) = \frac{d}{d\vartheta}\,X(\vartheta)$.

2 The reduced evolution

Let $\mathcal{A}$ be the algebra of all essentially bounded operator–valued functions $X : \vartheta \in \mathbf{R} \mapsto X(\vartheta) \in \mathcal{L}(\mathcal{H}_0)$ and $\mathcal{B}_t$ be the von Neumann algebra, generated by the observables $\boldsymbol{Y}_t = \{\boldsymbol{Y}(s) | s \leq t\}$. The quantum stochastic dynamics (9) of the system, described by the algebra $\mathcal{A}$, should be reduced for each $t \in \mathbf{R}_+$ to a standart representation of the output algebra $\mathcal{B}_t$. Such the representation gives the equivalence $U_t^* \boldsymbol{Y}(s) U_t = I \otimes \hat{v}(s)$, $s \leq t$ of the algebra $\mathcal{B}_t$ and the algebra of observables $\{\hat{v}_1, \ldots, \hat{v}_d\}''_{s\leq t}$, generating a Fock subspace $\mathcal{G}_t \subseteq \mathcal{F}_t$ on the faithful vector $|0\rangle \in \mathcal{F}$. It corresponds to the description of the reduced dynamics in the Schrödinger picture by the (unnormalized) conditional states [12] as the completely positive maps $\check{\phi}_t : \mathcal{A} \to \check{\ell}_t$ into the operator coalgebra $\check{\ell}_t$ of the bounded commutant $\check{\ell}'_t = \{\hat{v}_1, \ldots, \hat{v}_d\}''_{s\leq t}$. The coalgebra $\check{\ell}$ of a von Neumann algebra $\check{\ell}'$ on $\mathcal{G}$ with a faithful cyclic vector $|0\rangle \in \mathcal{G}$ is defined as the space of continuous operators $\check{p} : \ell' \to \ell$, $\|\check{p}\|_* < \infty$ on $\ell' = \check{\ell}'|0\rangle$ into the predual space ℓ of $\check{\ell}'$, satisfying the condition $\check{p}z = \hat{z}p$, $\forall \hat{z} = \check{\ell}'$, where $z = \hat{z}|0\rangle$, $p = \check{p}|0\rangle$, $\|\check{p}\|_* = \sup\{\|\check{p}z\|_1/\|z\|_\infty\}$. Here the pre-Hilbert subspace $\ell' \subset \mathcal{G}$ is an achieved Tomita's algebra [13] with the identity $1 = |0\rangle$, involution $z \mapsto z^* = \hat{z}^*|0\rangle$ and the $*$-norm $\|z\|_\infty = \|\hat{z}\| = \|z^*\|_\infty$, the predual coalgebra $\ell \supset \mathcal{G}$ is the completion of the space of all functionals

$$\rho(z) = \langle 0|\check{p}\hat{z}|0\rangle = \langle z^*, p\rangle, \tag{11}$$

associated with the scalar product $\langle q, p\rangle$ in $\mathcal{G} \ni q, p$ with respect to the norm

$$\|p\|_1 = \sup\{\rho(z)|z \in \ell', \ \|z\|_\infty \leq 1\} = \|\check{p}\|_*,$$

and the operator $\hat{z}$ is continued on ℓ as the preconjugated to $\hat{z}^*$ on ℓ' with respect to the pairing (11): $\langle q, \hat{z}p\rangle = \langle \hat{z}^*q, p\rangle = \langle z^*q, p\rangle$, $\forall q \in \ell'$.

The coalgebra $\check{\ell}$ with the involution $\check{p} \mapsto \check{p}^*$, defined by $\check{p}^*z = \hat{z}p^*, \forall p \in \ell$, $\langle z, p^*\rangle = \langle z^*, p\rangle^*$, $\forall z \in \ell'$, and with the coidentity $\check{1} = \mathrm{I} = \hat{1}$, corresponding to the vacuum state $v(z) = \langle 0|\hat{z}|0\rangle \equiv \langle z, 1\rangle$, $1 = |0\rangle$, commute with $\widehat{\ell'} = \check{\ell}'$ in the sense $\check{p}\hat{q} = \hat{q}\check{p}$ for all $\hat{q} \in \widehat{\ell'}$, $\check{p} \in \check{\ell}'$:

$$\check{p}\hat{q}z = \check{p}qz = \widehat{qz}p = \hat{q}\hat{z}p = \hat{q}\check{p}z, \quad \forall z \in \ell'.$$

Usually the density operator $\check{p} \in \check{\ell}$ of a form (11) has the range $\check{p}\ell$ in $\mathcal{G}$, as it is in the case of a majorized positive form $\rho(z^*z) \leq cv(z^*z)$ for a $c > 0$, corresponding to the bounded $\check{p}$ on $\mathcal{G}$: $\|\check{p}\| \leq c$; moreover, any operator $\check{p} \in \check{\ell}$ can be approximated by the density operators from the bounded commutant of $\check{\ell}'$ in the norm $\|\check{p}\|_* = \|p\|_1$, where $p = \check{p}|0\rangle$.

The unnormalized conditional state $\check{\phi}_t$ on the initial algebra $\mathcal{A}$ with respect to the observable algebra $\check{\ell}'_t$ and the vacuum state $\langle\hat{z}\rangle = \langle 0|\hat{z}|0\rangle$ on $\check{\ell}'_t$ can be defined as the composition of an initial normal state ϕ on $\mathcal{A}$ and a linear normal map $\check{\Phi}_t : X \in \mathcal{A} \to \check{\Phi}_t\{X\}$ into the unbounded commutant $\check{\mathcal{L}}'_t$ of $I \otimes \check{\ell}'_t$-valued essentially bounded functions $\hat{Z} : \vartheta \in \mathbf{R} \mapsto I \otimes \hat{z}(\vartheta)$. Having in mind the composition $\hat{\phi}_t\{X\} = \phi\{\check{\Phi}_t\{X\}\}$ with $\phi :\ \mathcal{A} \to \mathrm{C}$ defined as $(\phi \otimes \mathrm{Id}) \circ \check{\Phi}_t$, one can write the condition

$$\langle \Psi,\ \check{\Phi}_t\{X\}(\vartheta)\hat{Z}_t\Psi\rangle = \langle U(t)\Psi, (X(\vartheta_t) \otimes \hat{z}_t)U(t)\Psi\rangle,$$

$\forall \Psi \in \psi \otimes |0\rangle$, $\hat{Z}_t = I \otimes \hat{z}_t$ for the definition of $\check{\Phi}_t\{X\}$ with $\psi \in \mathcal{H}_0$, $\hat{z}_t \in \check{\mathcal{E}}_t'$ in the form

$$\langle 0|\check{\Phi}_t\{X\}\hat{Z}_t|0\rangle = \langle 0|X(t)Z(t)|0\rangle, \ \ \forall X \in \mathcal{A},\ Z(t) \in \mathcal{B}_t, \tag{12}$$

where $X(t) = U_tX(\vartheta_t)U_t^* = X^t(\vartheta_t)$ is defined by the solution $\vartheta_t = \varphi_t^w(\vartheta)$ of the equation (8) for a fixed ϑ and $\hat{Z}_t = U_t^*Z(t)U_t$.

In order to derive a quantum stochastic equation for the reduced dynamics $\check{\Phi}_t$, let us find the differential evolution for the factorial generating map $\Phi_t^{(\beta)} :\ \mathcal{A} \to \mathcal{A}$,

$$\Phi_t^{(\beta)}\{X\} = \langle 0|\check{\Phi}_t\{X\}(I \otimes \hat{e}_t^{(\beta)})|0\rangle = \langle 0|X^{(\beta)}(t)|0\rangle. \tag{13}$$

Here $Z_t = I \otimes \hat{e}_t^{(\beta)}$ is defined by the Wick (normal) exponent

$$\hat{e}_t^{(\beta)} =: e^{\hat{v}_t(\beta)} := e^{\mathrm{a}^*(\beta_t)}e^{\mathrm{a}(\beta_t)}, \tag{14}$$

of the observable $\hat{v}_t(\beta) = \int_0^t \beta(r)d\hat{v}(r) = (\mathrm{a} + \mathrm{a}^*)(\beta_t)$, where $\beta(t) = \{\beta^1, \ldots, \beta^d\}(t)$ are locally square–integrable functions, $\boldsymbol{\beta}_s(t) = 0$ for $t \geq s$, $\boldsymbol{\beta}_s(r) = \boldsymbol{\beta}(r)$ for $r < s$, and $X^{(\beta)}(t) = U_t(X(\vartheta_t) \otimes \hat{e}_t^{(\beta)})U_t^* = X(t)Z(t)$. Taking into account the equation (9) and

$$dZ(t) = \sum_{j=1}^{d} \beta^j(t)(U_tX_jU_t^*dt + I \otimes d2\Re\mathtt{A}_j(t)) \tag{15}$$

for $Z(t) = U_t\hat{e}_t^{(\beta)}U_t^*$ one can obtain by quantum Ito formula [11]

$$d(X(t)Z(t)) = dX(t)Z(t) + X(t)dZ(t) + dX(t)dZ(t) =$$
$$U_t(\Lambda\{X\}(\vartheta_t) + (\boldsymbol{L}^*X_t^{(\beta)} + X_t^{(\beta)}\boldsymbol{L})\boldsymbol{\beta}(t))U_t^*dt + \sigma U_t\delta X_t^{(\beta)}U_t^* \otimes d\mathtt{W}_t +$$
$$U_t([L_k^*, X_t^{(\beta)}] + \beta_k(t))U_t^* \otimes d\mathtt{A}^k(t) + d\mathtt{A}^k(t)^* \otimes U_t([X_t^{(\beta)},\ L_k] + \beta_k(t))U_t^*,$$

where $X_t^{(\beta)} = X(\vartheta_t) \otimes e_t^{(\beta)}$, $\delta = \partial/\partial\vartheta_t - \frac{\kappa}{\hbar}\ [iQ, \cdot]$. Hence the map $\Phi_t^{(\beta)}$: $X \in \mathcal{A} \mapsto \langle 0|U_tX_t^{(\beta)}U_t^*|0\rangle$ satisfies the equation

$$\frac{d}{dt}\,\Phi_t^{(\beta)}\{X\} = \Phi_t^{(\beta)}\{\Lambda\{X\} + (\boldsymbol{L}^*X + X\boldsymbol{L})\boldsymbol{\beta}(t)\},\ \ \Phi_0^{(\beta)}\{X\} = X. \tag{16}$$

In the same way, using the Ito formula for the product $\check{G}_t \hat{Z}_t$ of $\check{G}_t = \check{\Phi}_t\{X\}$ and the Wick exponent (14), one can obtain the equation for (13), if

$$d\check{\Phi}_t\{X\} - \check{\Phi}_t\{\Lambda\{X\}\}dt = \sum_{j=1}^{d} \check{\Phi}_t\{L_j^* X + X L_j\} \otimes d\check{v}^j(t), \tag{17}$$

where the operators $\check{v}(s) = 2\Re \mathbb{A}(s) = \{\check{v}^1, \ldots, \check{v}^d\}(s)$, $s \leq t$ satisfying the canonical commutation relations

$$[\check{v}^j(t'), \check{v}^{j'}(t)] = 2i \min\{t, t'\} \Im \gamma^{jj'},$$

generate the predual coalgebra as unbounded commutant $\check{\ell}_t = \{\hat{v}(s) | s \leq t\}'$. It proves the following theorem.

Theorem 2 *Let the initial state $\phi : \mathcal{A} \to \mathbf{C}$ be a normal one, described by a density $f : \vartheta \mapsto f(\vartheta)$ with values in trace–class operators on $\mathcal{H}_0$ and $\int \mathrm{Tr} f(\vartheta) d\vartheta = 1$. Then the conditional state*

$$\check{\Phi}_t\{X\} = \int \mathrm{Tr}_{\mathcal{H}_0}\{f(\vartheta)\check{\Phi}_t\{X\}(\vartheta)\} d\vartheta, \tag{18}$$

is described in the standard representation by the operator–function $\check{f}_t(\vartheta)$ as $\check{\phi}_t\{X\} = \int \mathrm{Tr}_{\mathcal{H}_0}\{\check{f}_t(\vartheta) X(\vartheta)\} d\vartheta$, satisfying the quantum stochastic equation

$$d\check{f}_t(\vartheta) - \Lambda[\check{f}_t](\vartheta)dt = \sum_{i=1}^{d} (L_j \check{f}_t(\vartheta) + \check{f}_t(\vartheta) L_j^*) \otimes d\check{v}^j(t). \tag{19}$$

Here $\check{f}_0(\vartheta) = f(\vartheta)$, $[\check{v}^j(t'), \hat{v}_{j'}(t)] = 0$, $\forall t, t' \in \mathbf{R}_+, j, j' = 1, \ldots, d$, and

$$\Lambda[\check{f}_t](\vartheta) = \partial v(\vartheta)\check{f}_t(\vartheta)/\partial\vartheta + \frac{\sigma^2}{2}\, \delta^2 \check{f}_t(\vartheta) +$$

$$\frac{i}{\hbar}\, [\check{f}_t(\vartheta), H] + \frac{1}{2}\, \gamma^{kk'}([L_{k'}, \check{f}_t(\vartheta) L_k^*] + [L_{k'} \check{f}_t(\vartheta), L_k^*]), \tag{20}$$

where $\delta^2 \check{f}_t(\vartheta) = \partial^2 \check{f}_t(\vartheta)/\partial\vartheta^2 - 2\frac{\kappa}{\hbar}\, [iQ, \partial\check{f}_t(\vartheta)/\partial\vartheta) + \frac{\kappa^2}{\hbar^2}\, [iQ, [iQ, \check{f}_t(\vartheta)]]$. It is normalized on the density operator $\check{p}_t = \int \mathrm{Tr}_{\mathcal{H}_0} \check{f}_t(\vartheta) d\vartheta \in \check{\ell}_t$ defining by (11) the output state $\rho_t(z) = \langle Z(t) \rangle = \langle p_t, z \rangle$ on $\ell' \ni z$ for $Z(t) = U(t)^*(I \otimes \hat{z})U(t)$, $\langle Z \rangle = \phi\{\langle 0|Z|0\rangle\}$ with respect to the vacuum state $v(z) = \langle 0|\hat{z}|0\rangle$.

3 The optimal measurement.

The quantum filter defines a quantum measurement on the output of the system at a time instant $t \in \mathbf{R}_+$. In general it is described by a $\mathcal{B}_t$–valued positive measure $M(t, dx)$ on a Borel space $\mathcal{X}$, normalized on the identity operator in $\mathcal{H} = \mathcal{H}_0 \otimes \mathcal{F}$: $\int M(t, dx) = I$. The problem of optimal quantum filtering is the problem of finding of optimal measurement $M^0(t)$, giving the minimal value of the mean

$$\int \langle S(t, x) M(t, dx) \rangle = \int \langle m_t(dx),\ c_t(x) \rangle \tag{21}$$

for a $\mathcal{A}$–valued measurable function $S : x \in \mathcal{X} \mapsto S(x) = S(x)^*$ in the Heisenberg picture $S(t, x) = U_t S(x, \vartheta_t) U_t^*$ with respect to the initial state $\langle Z \rangle = \phi\{\langle 0|Z|0 \rangle\}$. Due to (12) the mean (21) is given in the standard representation $\mathcal{B}_t \to \check{\ell}'_t$ as the integral of the pairing (11) for

$$m_t(dx) = \hat{m}_t(dx)|0\rangle, \quad c_t(x)|0\rangle = \check{c}_t(x)|0\rangle, \tag{22}$$

where $\hat{m}_t$ defines the measure $M(t) = U_t(I \otimes \hat{m}_t)U_t^*$ in the Schrödinger picture, and

$$\check{c}_t(x) = \int \mathrm{Tr}_{\mathcal{H}_0}\{\check{f}_t(\vartheta) S(x, \vartheta)\} d\vartheta = \check{\phi}_t\{S_t(x)\}. \tag{23}$$

The duality princeple gives the necessary and sufficient conditions [14] of optimality

$$\int \langle m_t^0(dx), c_t(x) \rangle = \inf_{\hat{m} \geq 0} \{ \int \langle m_t(dx), c_t(x) \rangle | \int \hat{m}_t(dx) = \mathrm{I} \} \tag{24}$$

of a positive $\check{\ell}'_t$–valued measure $m_t^0 \geq 0$ with $\int \hat{m}_t^0(dx) = \mathrm{I}$ as the conditions for the problem

$$\sup\{\langle 1, l_t \rangle | \check{l}_t \leq \check{c}_t(x), x \in \mathcal{X}\} = \langle 1, l_t^0 \rangle, \tag{25}$$

where l_t^0 defines by $\check{l}_t^0 z = \hat{z} l_t^0$ the operator $\check{l}_t^0 \leq \check{c}_t(x)$, $\forall x \in \mathcal{X}$, for which $\int \langle m_t^0(dx), c_t(x) \rangle = \langle 1, l_t^0 \rangle$. The last condition can be written in the form of the equation

$$(\check{c}_t(x) - \check{l}_t^0) m_t^0(dx), \text{ or } \hat{m}_t^0(dx)(c_t(x) - l_t^0) = 0.$$

Let us consider now the problem of optimal estimation [10] of a selfadjoint operator $X(t) = X^t(\vartheta_t)$, given in the Schrödinger picture by a $\mathcal{B}(\mathcal{H}_0)$-valued function $X(\vartheta) = X(\vartheta)^*$, with $x = \lambda \in \mathbf{R}$, $S(\lambda, \vartheta) = (X(\vartheta) - \lambda I)^2$. One can treat in such a way the problem of filtering of a real measurable function $x(\vartheta_t)$ of the input diffusion signal ϑ_t, if $X(\vartheta) = x(\vartheta)I$.

Theorem 3 *The solution of the optimisation problem (24) for the quadratic cost function*

$$c_t(\lambda) = \lambda^2 p_t - 2\lambda\phi_t\{X\} + \phi_t\{X^2\},\ \phi_t\{\cdot\} = \check{\phi}\{\cdot\}|0\rangle$$

is given by the spectral measure $\hat{m}_t^0$ of a selfadjoint operator

$$\hat{x}_t = \int \mathrm{Tr}_{\mathcal{H}_0}\{\hat{g}_t(\vartheta)\}d\vartheta = \int \lambda \hat{m}_t^0(d\lambda). \tag{26}$$

The posterior density operator–function $\hat{g}_t = \hat{g}_t^* \in \mathcal{A}_* \otimes \ell_t'$ defined by the solution $g_t(\vartheta) = \hat{g}_t(\vartheta)|0\rangle$ of the equation

$$\Re[P_t g_t(\vartheta)] = f_t(\vartheta),\quad P_t = I \otimes \check{p}_t,\quad f_t(\vartheta) = \check{f}_t(\vartheta)|0\rangle, \tag{27}$$

satisfies the nonlinear quantum stochastic equation

$$d\hat{g}_t(\vartheta) - \Gamma_t[\hat{g}_t](\vartheta)dt = \sum_{j=1}^{d} \Xi_t^j[\hat{g}_t] \otimes d\hat{v}_j(t),\quad \hat{v}_0(\vartheta) = f(\vartheta), \tag{28}$$

where $\Gamma_t(\vartheta)$ and $\Xi^j(\vartheta)$ are given by the equations

$$\Re\sum_{j=1}^{d} \gamma_{ij} P_t \Xi_t^j[\hat{g}](\vartheta) = \Re\{P_t(L_i g(\vartheta) + g(\vartheta)L_i^*) - R_{it} g(\vartheta)\},\ i = 1, \ldots, d$$

$$\Re P_t \Gamma_t[\hat{g}](\vartheta) = \Re\{P_t \Lambda[\hat{g}](\vartheta) - \sum_{j=1}^{d} R_{jt} \Xi_t^j[\hat{g}]\}, \tag{29}$$

with $R_{jt} = I \otimes \check{r}_{jt}$, and $\check{r}_{jt} z = \hat{z} r_{jt}$, $\forall z \in \ell_t'$ is defined by $X_j = L_j + L_j^*$ and g_t as

$$r_{jt} = \int \mathrm{Tr}_{\mathcal{H}_0}\{X_j(\vartheta)f_t(\vartheta)\}d\vartheta = \Re[\check{p}_t x_{jt}],\ x_{jt} = \int \mathrm{Tr}_{\mathcal{H}_0}\{g_t(\vartheta)X_j(\vartheta)\}d\vartheta. \tag{30}$$

Proof. Denoting $\hat{u}_t = \int \lambda \hat{m}_t(d\lambda)$ for an orthogonal projective–valued measure $\hat{m}_t(d\lambda) \in \check{\ell}'_t$, one obtains

$$\int \langle m_t(d\lambda), c(\lambda)\rangle = \langle u_t^2, p_t\rangle - 2\langle u_t, \phi_t\{X\}\rangle + \langle 1, \phi_t\{X^2\}\rangle =$$
$$\langle u_t^2, p_t\rangle - 2\langle u_t, \Re\check{p}_t x_t\rangle + \langle 1, \phi_t\{X^2\}\rangle \geq \langle 1, l_t^0\rangle,$$

where $\Re\check{p}_t x_t = \phi_t\{X\}$ is equivalent to (26), (27) in terms of the density function $g_t \in \mathcal{A}_* \otimes \ell'_t$, defining the posterior mean value $x_t = \hat{x}_t|0\rangle$, and $l_t^0 \in \ell_t$ is defined by $\langle x_t z x_t, p_t\rangle = \langle z, l_t^0\rangle$ for all $z \in \ell'_t$. The inequality is due to $2\langle u_t, \Re\check{p}_t x_t\rangle = 2\Re\langle u_t, \check{p}_t x_t\rangle = \langle u_t x_t + x_t u_t, p_t\rangle$ and positivity of $p_t \in \ell_t$ and $(u - x)^2 \in \ell'_t$:

$$\int \langle m_t(d\lambda), c_t(\lambda)\rangle - \langle 1, l_t^0\rangle = \langle (u_t - x_t)^2,\ p_t\rangle \geq 0.$$

Hence the optimal $\hat{m}_t^0$ should be the spectral measure of the operator $\hat{x}_t \in \check{\ell}'_t$, uniqually defined on the faithful vector $|0\rangle$ by $x_t = \int \mathrm{Tr}_{\mathcal{H}_0}\{g_t(\vartheta)X(\vartheta)\}d\vartheta$. Looking for the operator $\hat{x}_t = \hat{x}_t^*$ as for the solution of some quantum stochastic equation $d\hat{x}_t - \hat{\gamma}_t dt = \sum_{j=1}^d \hat{\xi}_t^j \otimes d\hat{v}_j(t)$ with some $\hat{\gamma}_t = \hat{\gamma}_t^*$, $\hat{\xi}_t^j = \hat{\xi}_t^{j*}$, one can obtain, applying the quantum Ito formula

$$d\Re\check{p}_t x_t = \Re(d\check{p}_t x_t + \check{p}_t dx_t + d\check{p}_t dx_t) = d\phi_t\{X\} =$$
$$\Re\sum_{j=1}^d \check{\phi}_t\{X_j\}(x_t \otimes dv^j + \xi_t^j dt) + \Re\check{p}(\sum_{j=1}^d \xi_t^j \otimes dv_j + \gamma_t dt),$$

where $dp_t = \sum_{j=1}^d \phi_t\{X_j\} \otimes dv^j(t)$ for $p_t = \phi_t\{I\}$ due to $\Lambda\{I\} = 0$, $X_j = L_j + L_i^*$, and

$$v_j = (\mathtt{A}_j^* + \mathtt{A}_j)|0\rangle = \mathtt{A}_j^*|0\rangle = \mathtt{A}^{i*}|0\rangle\gamma_{ij} = (\mathtt{A}^{i*} + \mathtt{A}^i)|0\rangle\gamma_{ij} = v^i\gamma_{ij}.$$

Comparing it with the equation (19) for $\phi_t\{X\} = \check{\phi}_t\{X\}|0\rangle$, one can find

$$\Re[\check{\phi}_t\{X_i\}x + \check{p}_t\gamma_{ij}\xi_t^j] = \phi_t\{L_j^* X + XL_j\} = \Re\check{p}_t\pi_t\{L_j^* X + XL_j\},$$
$$\Re[\check{\phi}_t\{X_j\}\xi_t^j + \check{p}_t\gamma_t] = \phi_t\{\Lambda\{X\}\} = \Re\check{p}_t\pi_t\{\Lambda\{X\}\},$$

where $\pi_t\{X\} = x_t$. This gives (28)–(30) in terms of Ξ_t^j, Γ_t, defining

$$\xi_t^j = \int \mathrm{Tr}_{\mathcal{H}_0}\{\Xi_t^j[g_t](\vartheta)X(\vartheta)\}d\vartheta,\ \gamma_t = \int \mathrm{Tr}_{\mathcal{H}_0}\{\Gamma_t[g_t](\vartheta)X(\vartheta)\}d\vartheta.$$

Example. Let us consider the case $d = 1$ with $L_1 = \frac{1}{2} X_1 = L_1^*$, $X_1 = \sqrt{2\kappa}Q$, $\mathtt{A}_1(t) = \sqrt{\hbar}\mathtt{A}_t$, where $\mathtt{A}_t$ is the canonical annihilation process $[\mathtt{A}_{t'}, \mathtt{A}_t^*] = \min\{t', t\}$, corresponding to the nondemolition observation of the commutative output process

$$dY(t) = \sqrt{2\kappa}Q(t)dt + \sqrt{\hbar}I \otimes d\mathtt{V}_t, \quad Q(t) = U(t)^*QU(t)$$

with $\mathtt{V}_t = 2\Re\mathtt{A}_t$. In that case $\sqrt{\hbar}\check{v}^1(t) = \mathtt{V}_t = \hat{v}_1(t)/\sqrt{\hbar}$ is a standard Wiener process, represented in Fock space, and the equation (19) and (28) are the classical stochastic equations in the linear (for the nonnormalized $\hat{f}_t(\vartheta) = f_t(\vartheta, \mathtt{V})$) and nonlinear (for $\check{g}_t(\vartheta) = f_t(\vartheta, \mathtt{V})/(p_t(\mathtt{V}))$ posterior density operators with respect to the output state $\langle Z(t)\rangle = \langle z(\mathtt{V})p_t(\mathtt{V})\rangle$ on $\hat{z}_t = U_t^* Z(t) U_t = z_t(\mathtt{V})$, $p_t(\mathtt{V}) = \int f_t(\vartheta, \mathtt{V})d\vartheta$.

Suppose, that the quantum reciever is an open oscilator, described by the Hamiltonian $H = \frac{1}{2} A^*A$, where $A = iP + \omega Q$ and Q, P are the canonical coordinate and momentum operators: $[Q, P] = i\hbar I$. Then the quantum Langevian equation (9) for $A(t) = U_t A U_t^*$ is the linear one

$$dA(t) + i\omega A(t)dt = iI \otimes (\kappa d\vartheta_t + \sqrt{\kappa\hbar/2}d\mathtt{U}_t),$$

where $\mathtt{U}_t = 2\Im\mathtt{A}_t^*$ is the standart Wiener process, corresponding to the quantum linear input channel $\dot{\vartheta}_t \mapsto \dot{\mathtt{C}}_t = \sqrt{\kappa/2\hbar}\dot{\vartheta}_t + i\dot{\mathtt{A}}_t$, defining the right hand side of the equation as $iI \otimes \sqrt{2\kappa\hbar}\Re d\mathtt{C}_t$. In the case of Gaussian input process ϑ_t, corresponding to the linear $v(\vartheta) = v\vartheta$ and Gaussian initial state ϕ the optimal estimate of $X(\vartheta_t) = I \otimes \vartheta_t = X(t)$ is given by the linear posterior mean value $\hat{\vartheta}_t = \mathtt{X}_t = \hat{x}_t$ with respect to the output coordinate process $Y(t)$. In the standard Fock representation it is given as the last component of the stochastic row $\hat{\boldsymbol{x}}_t = (\mathtt{Q}_t, \mathtt{P}_t, \mathtt{X}_t)$ of the posterior mean values of $\boldsymbol{X}(t) = (Q(t), P(t), X(t))$, satisfying the Kalman equation

$$d\hat{\boldsymbol{x}}_t + \hat{\boldsymbol{x}}_t \boldsymbol{\Lambda} dt = \boldsymbol{k}_t d\,\widetilde{\mathtt{V}}_t, \quad \boldsymbol{\Lambda} = \begin{pmatrix} 0 & \omega^2 & 0 \\ -1 & 0 & 0 \\ 0 & -\kappa & v \end{pmatrix},$$

where $\boldsymbol{k}_t = (k_t^{11}, k_t^{12}, k_t^{13})$ is the first row of the 3×3–matrix $\boldsymbol{K}_t = (k_t^{jj'})$, defined by the Riccati equation

$$\frac{d}{dt}\boldsymbol{K}_t + \boldsymbol{K}_t\boldsymbol{\Lambda} + \boldsymbol{\Lambda}^\top \boldsymbol{K}_t + \boldsymbol{k}_t^\top \boldsymbol{k}_t = \boldsymbol{\Upsilon}, \quad \boldsymbol{\Upsilon} = \begin{pmatrix} 0 & 0 & 0 \\ 0 & \kappa\hbar/2 & 0 \\ 0 & 0 & \sigma^2 \end{pmatrix},$$

and $\tilde{V}_t = V_t - \sqrt{2\kappa\hbar}Q_t$. The pair $(\hat{x}_t, K_t)$ defines the posterior (normalized) Gaussian state of the quantum system with input signal $x_t = \vartheta_t$ and the mean square error $\langle\tilde{\vartheta}_t^2\rangle$, $\tilde{\vartheta}_t = \vartheta_t - \hat{\vartheta}_t$ is given by the component k_t^{33} of the posterior correlation matrix K_t.

A posterior dynamics of the quantum system under another nondemolition measurement of the received electromagnetic field by a photon counter is considered in [15].

References

[1] Belavkin V.P. Quantum filtering of Markovian signals with quantum white noises. *Radiotechnika i Electronica*, 25:1445, 1980.

[2] Gardiner C.W. and Collet M.J. Input and output in damped quantum systems: quantum statistical differential equations and the master equation. *Phys Rev A*, 31:3761, 1985.

[3] Belavkin V.P. Nondemolition measurement and control in quantum dynamical systems. In Blaquiere A., Diner S., and Lochak G., editors, *Information complexity and control in quantum physics, Proc. of CISM, Udine 1985*, pages 331–336, Springer-Verlag, Wien-New York, 1987.

[4] Belavkin V.P. Nondemolition measurements,nonlinear filtering and dynamic programmic of quantum stochastic processes. In *Modelling and Control of Systems, Proceedings Bellmann Continuous Workshop, Sophia-Antipolis 1988, Lect. Notes in Control and Inform. Sciences, vol.121,*, pages 245–265, Blaquiere A., Springer-Verlag, Berlin-Heidelberg-New York-London-Paris-Tokyo, 1988.

[5] Barchielli A. Input and Output channels in quantum systems and quantum stochastic differential equations. In L. Accardi and von Waldenfels W., editors, *Qüantum Probability and Applications III*, Berlin: Springer–Verlag, 1988. Volume 37.

[6] Belavkin V.P. Nondemolition stochastic calculus in Fock space and nonlinear filtering and control in quantum systems. In R.Gielerak and W. Karwowski, editors, *Stochastic methods in Mathematics and Physics*,

pages 310–324, Proc. XXIV Karpacz winter school, 1988, World Scientific, Singapore-New Jersey-London-Hong Kong, 1988.

[7] Belavkin V.P. A new wave equation for a continuous nondemolition measurement. *Phys. Lett. A*, 140:355, 359, 1989.

[8] Belavkin V.P. and Staszevski P. A quantum particle undergoing continuous observation. *Phys. Lett. A*, 140:359–362, 1989.

[9] Belavkin V.P. A continuous counting observation and posterior quantum dynamics. *J. of Physics A*, 22:L1109–L1114, 1989.

[10] Helstrom C.W. *Quantum Detection and Estimation theory.* Academic Press, New York, 1976.

[11] Hudson R.L. and Parthasarathy K.R. *Quantum Ito's formula and stochastic evolution*, pages 301–323. Volume 93, Comm. Math. Phys., 1984.

[12] Belavkin V.P. and Staszewski P. Conditional entropy in algebraic statistical physics. *Rep. on Math. Phys.*, 20:373, 1984.

[13] Takesaki M. Tomita's theory of modular hilbert algebra and its applications. *Lecture Notes in Mathematics*, 128, 1970.

[14] Holevo A.S. in: AMS Translation Proc. Steclov Math. Inst., 1978. Issue 3.

[15] Belavkin V.P. A stochastic posterior Schrödinger equation for counting nondemolition measurement. *Letters in Math. Phys.*, 20:85, 1990.

ENTROPY PRODUCTION AND NONEQUILIBRIUM STATIONARITY IN QUANTUM DYNAMICAL SYSTEMS

Izumi OJIMA

Reseach Institute for Mathematical Sciences
Kyoto University, Kyoto 606, Japan

ABSTRACT

In statistical physics, Kubo's linear response theory is well known as the most effective method, in the linear-approximation regimes, of calculating transport coefficients which describe dissipative aspects in the macroscopic manifestations of microscopic quantum systems. From the viewpoint of the mutual relationship between the microscopic and macroscopic levels, it is clear that the response theory is essentially concerned with the *information-theoretical* problems. For lack of such key-concepts as *entropy* and/or *entropy production*, however, this theory has long been taken in physics merely as a calculational device, without the deep understanding of the reason for its general validity.

In this report, it is shown that the notion of entropy production as a generating function of various transport coefficients can be defined in a natural way with the aid of relative entropy. We carefully examine the nonlinear responses of quantum dynamical systems against *almost-periodic* perturbations with (countably) *many frequencies* in the general algebraic framework which is applicable to the systems with infinite degrees of freedom without the limitations of standard linear response theory. Then, the positivity of (*time-averaged*) entropy production expressing the dissipativity is proved on a general ground, and also the basis for the validity of linear approximation is clarified on the basis of time-scale change controlled by van Hove limit. By virtue of the methods of dilation/coarse graining, the parallelism becomes clear between the response theory in statistical physics and the theory of quantum measurements as well as of communication channels.

1. INTRODUCTION

In statistical physics, Kubo's linear response theory [1] is well known to provide, in the linear-approximation regimes, the most effective method of calculating transport coefficients, the quantities describing the *dissipative* aspects of the nonequilibrium and irreversible processes. Since it is formulated in the framework of dynamical systems with *reversible* time developments, however, the origin of irreversible dissipative processes described in terms of *positive* transport coefficients has been rather mysterious [2, 3]. From the information-theoretical viewpoint, the essence of the problem is just the information transmission from the microscopic regimes of dynamical fluctuations to

Lecture Notes in Physics, Vol. 378
C. Bendjaballah, O. Hirota, S. Reynaud (Eds.),
Quantum Aspects of Optical Communications

the macroscopic world with controllable external forces perturbing the microscopic systems (*inputs*) which induce state-changing processes as responses of the object systems (*outputs*) to the former. Therefore, it should be clear that the entropy or its related notions play a crucial role here, whereas these notions have not been formulated in an explicit form in the context of response theory.

According to [4, 5], we present here a natural definition of entropy production in terms of relative entropy, which summarizes all the transport coefficients as their generating function. In view of the several restrictions of standard formulation of Kubo theory, we reformulate the problem in the (C^*-)algebraic framework for quantum dynamical systems with infinite degrees of freedom, in which we examine the nonlinear responses under *almost-periodic* perturbations with (countably) *many frequencies.* Then, the positivity of entropy production in its *time-averaged form* is proved in the general context of nonlinear response theory *without* recourse to the linear approximation. This will make it possible to treat the nonequilibrium structures realized outside the thermodynamical branches around equilibria, which is very important for approaching the actual physical world far from a thermal equilibrium with a uniform temperature.

For the natural understanding of the relation between reversible dynamics and irreversible dissipative processes, it is important to clarify the physical meaning of the initial and final time-averaging procedures appearing in our definition of entropy production. This can be done by the dilation/coarse graining methods controlling the relation between subsystems and the total system [5]: The original system is embedded as a subsystem into the composite total system, combined together with the external forces treated as a classical dynamical system with its own time development. Then, the initial-time average can be replaced by the initial ergodic state of this external-force system which is to play a pivotal role in formulating the dissipativity in terms of a *conditional expectation.* Next, the final-time average plays an important role in realizing the stationarity of final nonequilibrium states: It is important to note that the dissipativity associated with positive entropy production in nonequilibrium stationary states implies the coexistence of two conflicting aspects, the *stationarity* of the state on one hand and the constant *increase* in the entropy on the other hand. What reconciles this conflict is just the *discrimination of two levels*, microscopic and macroscopic, which are then *interconnected* by the time-scale change controlled by *van Hove limit.* Through this limiting transition, the macroscopic stationarity is realized in the presence of microscopic fluctuations, which is expressed mathematically as the transition from non-Markovian dynamics with memory effects to Markovian dissipative process of state changes. At the same time, the validity of linear approximation and the origin of irreversibility can be naturally explained in this context. Although the van Hove limit itself is nothing new being a familiar technique in the discussion of irreversible "*downhill processes*" of *return to thermal equilibrium*, it is just in the opposite direction that it works here in the "*uphill process*" from equilibrium *to nonequilibrium.*

In this way, the response theory can be reformulated as the interacting processes of *two physical systems*, a microscopic quantum system as an object system, on one hand, and a macroscopic classical system as a perturbing external-force and/or a probe system, on the other. One of the most important merits of such a reformulation is that it exhibits a remarkable parallelism between the statistical-mechanical response theory and the theory of quantum measurements as well as of the communication channels. In short, all these problems can be viewed as different realizations of a unified framework of *quantum information theory*, according to the differences of applied domains.

In the next section, we briefly recapitulate the result on the entropy production. In Sect.3, our time-inhomogeneous system under the influence of time-dependent external force will be embedded into a larger time-independent system together with the external force driven by its own dynamics. This dilation method clarifies in Sect.4 the essential role of the scale changing procedure of van Hove limit.

2. ENTROPY PRODUCTION IN QUANTUM DYNAMICAL SYSTEM

2.1 Entropy Production in General Framework of Nonlinear Response Theory

Let the observables and dynamics of our object system be described, respectively, by a (C^*-) algebra $\mathcal{A}$ and a (strongly continuous) 1-parameter automorphism group α_t on $\mathcal{A}$. In this general setup, the temperature states corresponding to Gibbs ensembles are characterized by the KMS condition [6, 7]. In view of the *stability* required of the equilibrium, the thermal equilibrium state should be represented by an *ergodic* KMS state [non-ergodic ones will correspond to mixed thermodynamical phases].

In this context, any other states *outside* this category may be called nonequilibrium ones, but, such a negative-form definition cannot convey any positive messages about the dynamic aspects of nonequilibrium structures in the actual world. Instead, we consider here the "adiabatic switching procedure" to generate a nonequilibrium state from a thermal equilibrium state ω_β with inverse temperature $\beta = 1/k_B T$ prepared at "infinite past $t_0 \to -\infty$" by perturbing the system with external force $X(t) \equiv (X_1(t), \cdots, X_n(t))$. If $X(t)$ couples to the variable $\boldsymbol{A} \equiv (A_1, \cdots, A_n), A_i \in \mathcal{A}$, through an interaction Hamiltonian $H_I(t)$ given by

$$H_I(t) = -\boldsymbol{A} \cdot X(t) \equiv -\sum_i A_i X_i(t), \tag{2.1}$$

the time-dependent perturbed dynamics is described by a family of automorphisms $\alpha_{s,t;X}$ on $\mathcal{A}$ characterized by the equation

$$\frac{d}{dt}\alpha_{s,t;X}(B) = \alpha_{s,t;X}(\delta(B) + [iH_I(t), B]) \tag{2.2}$$

with the initial condition $\alpha_{s,t=s} = Id_{\mathcal{A}}$. Here δ is the infinitesimal generator of the unperturbed dynamics α_t: $(d/dt)\alpha_t(B) = \alpha_t \circ \delta(B)$. The solution to this equation can be expressed as

$$\alpha_{s,t;X}(B) = \alpha_{-s}[U(t,s;X)^* \alpha_t(B) U(t,s;X)] \qquad \text{for } B \in \mathcal{A}, \tag{2.3}$$

with the "propagator" $U(t,s;X)$ in the interaction picture given by

$$U(t,s;X) = T\exp\{i \int_s^t d\tau \alpha_\tau(\boldsymbol{A}) \cdot X(\tau)\}. \tag{2.4}$$

In view of the formula [8] for relative entropy [9] between the state $\varphi_t \equiv \omega_\beta \circ \alpha_{t_0,t}$ and the initial state $\varphi_{t=t_0} \equiv \omega_\beta$,

$$S(\varphi_t|\varphi_{t_0} = \omega_\beta) = \beta \int_{t_0}^{t} ds \varphi_s(\delta(A)) \cdot X(s), \tag{2.5}$$

"microscopic" entropy production $P(t, t_0; X)$ [8] and *mean* entropy production $\bar{P}$ [4] are defined, respectively, by

$$P(t, t_0; X) \equiv \frac{d}{dt} S(\varphi_t|\omega_\beta) = \beta \varphi_t(\delta(A)) \cdot X(t) \equiv \beta \langle J \rangle(t) \cdot X(t), \tag{2.6}$$

$$\bar{P} \equiv \lim_{T\to\infty} \frac{1}{T} \int_0^T dt \lim_{T_0\to\infty} \frac{1}{T_0} \int_{-T_0}^{0} dt_0 P(t + t_0, t_0; X)$$
$$= \lim_{T\to\infty} \frac{1}{T} \lim_{T_0\to\infty} \frac{1}{T_0} \int_{-T_0}^{0} dt_0 S(\varphi_{T+t_0}|\varphi_{t_0} = \omega_\beta) \geq 0. \tag{2.7}$$

Here J is the current operator conjugate to the external force $X(t)$ defined by $J \equiv \delta(A)$. While the relative entropy $S(\varphi_t|\omega_\beta)$ is always non-negative ([9]), its time-derivative $P(t, t_0; X)$ can be negative and the initial-time averaging limit is essential for the positivity of $\bar{P}$. The limit $\lim_{T\to\infty}$ is important for attaining the stationarity of the final state. Since the physical meaning of these limits are not manifest in this form, we will clarify in the following their crucial roles in the transition from the regime of reversible dynamics to that of nonequilibrium stationarity according to [5].

2.2 Long Time Averages and Almost Periodicity

In order to make these two limits meaningful, we need to give some suitable characterization to the external force $X(t)$. For instance, *periodic* perturbations (with some damping factor) are usually chosen in Kubo theory. If $X(t)$ is taken as *random noise*, our formulation reduces to the stochastic process approach. In the former case, it is difficult to get rid of thermodynamic branches owing to the stability of equilibrium states [7], and the origin of dissipativity is easily missed in the formal procedure of linear approximation [3]. As for the latter, dynamics of the system is described by an irreversible stochastic process with dissipativity, but the probabilistic law governing the noise remains just an *ad hoc* input.

If we want to seek the dynamical origin of dissipativity and the natural framewrk accommodating the nonequilibrium structures belonging to non-thermodynamic branches, it is crucial to interpolate between these two extreme cases, <periodic regularity> vs. <chaotic randomness>. Although the traditional physics have covered both of these extremes, the intermediate regions have remained to be cultivated. In this respect, the notion of almost periodicity [10, 11] is remarkable as it encompasses the most important aspects of the "complexity" intervening between the above two opposites. Just to fix notations, we start our discussion from the definition of almost periodicity [10] in the special case of an additive group $\boldsymbol{R}$ identified with the time axis:

i) A continuous function $X(t)$ is *almost periodic* in $t \in \boldsymbol{R}$ if it can be approximated uniformly by linear combinations of periodic functions: $X(t) = \sum_k a_k \text{exp} i\omega_k t$ [convergence in the uniform topology with sup-norm]. By Bochner theorem [10], this is equivalent to saying that the orbit $\{X_\lambda; \lambda \in R\}$ under the time flow $X_\lambda(t) \equiv X(t-\lambda)$ is *precompact*.

ii) The latter condition allows one to define the *hull* M_X as the completion of this orbit by $M_X \equiv \overline{\{X_\lambda; \lambda \in R\}}$, which turns out to be a *compact* abelian group having a unique normalized *Haar measure* μ. As a probability measure, μ gives a *state* on the commutative algebra $C(M_X)$. The time translation group R acts on M_X by extending $X \mapsto X_\lambda$ by continuity with each orbit dense in M_X. Thus, the time flows λ_t on M_X and σ_t on $C(M_X)$ given, respectively, by

$$\lambda_t \xi \equiv \xi_t, \qquad \text{for } \xi \in M_X, \tag{2.8}$$

$$(\sigma_t f)(\xi) \equiv f(\lambda_{-t}(\xi)) = f(\xi_{-t}), \quad \text{for } f \in C(M_X), \tag{2.9}$$

define an ergodic classical dynamical system $(C(M_X), \sigma_t, \mu)$. The time dependence of $X(t)$ can be totally absorbed in this time flow σ_t, and the remaining degrees of freedom specifying the function $X(t)$ itself is represented by a time-independent function $\hat{X}(\xi)$ on M_X so that there exists a $\xi^{(0)} \in M_X$ satisfying $X(t) = \hat{X}(\xi^{(0)}_{-t})$. In view of the expansion $X(t) = \sum_k a_k \exp i\omega_k t$, a point ξ and the flow λ_t on the hull M_X can be identified, respectively, with a sequence $\xi = (\xi_k), \xi_k \in \boldsymbol{T}$, on the torus $\boldsymbol{T}$, and with $\lambda_t(\xi) \equiv (e^{-i\omega_k t}\xi_k)$. The function $\hat{X}$ corresponds to the sequence (a_k) of the expansion coefficients of $X(t)$: $\hat{X}(\xi) = \sum_k a_k \xi_k$. Physically, ξ on the hull M_X is interpreted as the *microscopic fluctuations* of the external force $X(t)$ and its coefficients (a_k) as the time-independent "shape" of $X(t)$ determined by the macroscopic setup of the apparatus.

iii) Ergodicity of σ_t: For any L^1 function $G(\xi)$ on M_X, we have

$$\lim_{T-S\to\infty} \frac{1}{T-S} \int_S^T dt G(\xi_{-t}) = \int_{M_X} d\mu(\xi) G(\xi). \tag{2.10}$$

By using these facts, the initial time average $\lim_{T_0\to\infty}(1/T_0)\int_{-T_0}^0 dt_0$ in (2.7) can be shown [4] to exist, being equal to the μ-average over the hull M_X of $\boldsymbol{X}$ if $X(t)$ is almost periodic [5]:

$$\begin{aligned} \bar{P} &\equiv \lim_{T\to\infty} \frac{1}{T} \int_0^T dt \lim_{T_0\to\infty} \frac{1}{T_0} \int_{-T_0}^0 dt_0 P(t+t_0, t_0; \boldsymbol{X}) \\ &= \lim_{T\to\infty} \frac{1}{T} \int_0^T dt \int_{M_X} d\mu(\xi) P(t, 0; \xi). \end{aligned} \tag{2.11}$$

Thus, we can take the limit of "infinite past $t_0 \to -\infty$" even if the limiting state $\lim_{t_0\to-\infty} \omega_\beta \circ \alpha_{t_0,t}$ does not exist [as in the case of oscillating perturbation $X(t)$ without the adiabatic damping factor]. As for the final time averaging limit, $\lim_{T\to\infty}(1/T)\int_0^T dt$, however, it is *not uniquely* determined in general, and we need to choose for convergence some subsequence $(T_n)_{n\in\mathrm{N}}$ tending to $+\infty$, the existence of which is ensured by Markov-Kakutani fixed point theorem [12].

Finally, we note that the group structure does not play an essential role here aside from the uniqueness of the initial-time averaging limit identified with the Haar measure μ. Indeed, the present setting can be extended to any external force $\boldsymbol{X}$ such that the dynamical system $(C(M_X), \sigma_t)$ has a *unique* ergodic measure μ on its hull M_X.

3. ENTROPY PRODUCTION AND CONDITIONAL EXPECTATION

3.1 Embedding of Time-Dependent Dynamics into Time-Independent Composite System

Using the hull M_X of external force $X(t)$, we reformulate the external-force perturbation as the interaction between *two physical systems*, the microscopic object system $(\mathcal{A}, \alpha_t)$ and the classical dynamical system $(C(M_X), \sigma_t)$ on M_X. Through the coupling term (2.1), they are glued together into a composite total system with time-independent dynamics [13].

The algebra $\mathcal{B}$ of this composite system is defined by $\mathcal{B} \equiv C(M_X, \mathcal{A}) = \mathcal{A} \otimes C(M_X)$. It can be identified interchangebly as the (C^*-)algebra of $\mathcal{A}$-valued continuous functions $\xi \mapsto \hat{B}(\xi)$ on M_X or the tensor-product of $\mathcal{A}$ with $C(M_X)$ through the correspondence $(\xi \mapsto f(\xi)A) \leftrightarrow (A \otimes f)$ $[A \in \mathcal{A}, f \in C(M_X)]$. The original object system $\mathcal{A}$ is embedded in $\mathcal{B}$ through the map ι given by

$$\iota : \mathcal{A} \ni B \mapsto B \otimes 1 \equiv \iota(B) \in \mathcal{B}. \tag{3.1}$$

To define a *composite dynamics* on $\mathcal{B}$, we extend the parameter dependence of $\alpha_{s,t;X}$ from $X(t)$ into the variable ξ on M_X. The obtained $\alpha_{s,t;\xi}$ is easily seen to satisfy the following properties:

$$\alpha_{s,t;\xi} \circ \alpha_{t,u;\xi} = \alpha_{s,u;\xi} \qquad \text{[chain rule]}, \tag{3.2}$$

$$\alpha_{s+\lambda,t+\lambda;\xi} = \alpha_{s,t;\xi_{-\lambda}} \qquad \text{[covariance condition]}. \tag{3.3}$$

Then we can define a composite dynamics β_t [13] on $\mathcal{B}$ by

$$[\beta_t(\hat{B})](\xi) \equiv \alpha_{0,t;\xi}(\hat{B}(\xi_{-t})) \quad \text{for} \quad \hat{B} \in \mathcal{B}. \tag{3.4}$$

By virtue of (3.2) and (3.3), β_t gives a time-independent dynamics satisfying the group property $\beta_s \circ \beta_t = \beta_{s+t}$.

The ergodicity (2.10) of μ implies for any state ω of $\mathcal{A}$ and $\hat{B} \in \mathcal{B}$,

$$\begin{aligned}(\omega \otimes \mu)[\beta_t(\hat{B})] &= \int_{M_X} d\mu(\xi)\omega(\beta_t(\hat{B})(\xi)) \\ &= \lim_{T_0 \to \infty} \frac{1}{T_0} \int_{-T_0}^{0} dt_0 \omega(\alpha_{t_0,t+t_0;\xi}(\hat{B}(\xi_{-t-t_0}))),\end{aligned} \tag{3.6}$$

through which the Haar measure μ becomes equivalent to the initial time averaging procedure in (2.7). Namely, the fictitious adiabatic switching procedure with initial-time limit has been replaced by an ergodic *state* μ of the external force system.

On the other hand, the problem of the final time limit is more complicated: Although both the initial states ω_β and μ are ergodic, the product state $\omega_\beta \otimes \mu$ is in general *neither* stationary *nor* ergodic in the composite dynamical system $(\mathcal{B}, \beta_t)$. Therefore, we need to investigate the dynamical behaviour of $\omega_\beta \otimes \mu$ under the dynamics β_t of the *total* system.

3.2 Entropy Production and Stationarity in Composite System

To this end, we introduce the GNS representations $(\pi, \mathcal{H}, \Omega, U_t = \exp itH_\beta)$ and $(\pi_\mu, L^2(M_X, \mu), \mu^{1/2}, V_t = \exp it\Delta)$ associated, respectively, with states ω_β and μ of the dynamical systems $(\mathcal{A}, \alpha_t)$ and $(C(M_X), \sigma_t)$:

$$\omega_\beta(B) = \langle \Omega, \pi(B)\Omega \rangle \qquad \text{for } B \in \mathcal{A}, \tag{3.7}$$

$$\mathcal{H} = \overline{\pi(\mathcal{A})\Omega} \qquad \text{(cyclicity)}, \tag{3.8}$$

$$\pi(\alpha_t(B)) = U_t \pi(B) U_t^*; \quad U_t = \exp iH_\beta t; \quad H_\beta \Omega = 0; \tag{3.9}$$

$$\mu(f) = \int_{M_X} d\mu(\xi) f(\xi) \equiv \langle \mu^{1/2}, \pi_\mu(f)\mu^{1/2} \rangle, \tag{3.10}$$

$$[\pi_\mu(f)\psi](\xi) \equiv f(\xi)\psi(\xi) \quad \text{for } f \in C(M_X), \ \psi \in L^2(M_X, \mu), \tag{3.11}$$

$$(V_t\psi)(\xi) \equiv \psi(\xi_{-t}); \ \ \pi_\mu(\sigma_t f) = V_t \pi_\mu(f) V_t^*; \ \ V_t = \exp it\Delta. \tag{3.12}$$

Then, the system $(\mathcal{B}, \beta_t)$ can be represented in the Hilbert space $\mathcal{H} \otimes L^2(M_X, \mu) = L^2(M_X, \mathcal{H}; \mu)$ of $\mathcal{H}$-valued L^2-functions as

$$[\{(\pi \otimes \pi_\mu)(\hat{B})\}\psi](\xi) = \pi(\hat{B}(\xi))\psi(\xi),$$
$$\text{for } \hat{B} \in \mathcal{B}, \ \ \psi \in L^2(M_X, \mathcal{H}; \mu), \tag{3.13}$$

$$(\pi \otimes \pi_\mu)(\beta_t(\hat{B})) = e^{itH}(\pi \otimes \pi_\mu)(\hat{B})e^{-itH}, \tag{3.14}$$

where the generator H is given by

$$H \equiv H_\beta \otimes \mathbf{1} + \mathbf{1} \otimes \Delta - \sum_i \pi(A_i) \otimes \pi_\mu(\hat{X}_i) \equiv H_\beta \otimes \mathbf{1} + \mathbf{1} \otimes \Delta - \pi \otimes \pi_\mu(A \otimes \hat{X}). \tag{3.15}$$

Note that the state $\omega_\beta \otimes \mu$ is a KMS state on $\mathcal{B}$ for the *decoupled* dynamics $\alpha_t \otimes Id_{C(M_X)} \equiv \hat{\alpha}_t$ with generator $H_\beta \otimes \mathbf{1}$, from which the *coupled* dynamics β_t arises by perturbing the generator by $H - H_\beta \otimes \mathbf{1} = \mathbf{1} \otimes \Delta - \pi \otimes \pi_\mu(A \otimes \hat{X})$. Then, we can apply the formula (2.5) for relative entropy to the composite system $(\mathcal{B}, \beta_t)$ to obtain

$$\begin{aligned}
&S(\omega_\beta \otimes \mu \circ \beta_t | \omega_\beta \otimes \mu) \\
&= \beta \int_0^t ds \, \langle \Omega \otimes \mu^{1/2}, \ e^{iHt} \, [iH_\beta \otimes \mathbf{1}, \ -\mathbf{1} \otimes \Delta + \\
&\qquad + \pi \otimes \pi_\mu(A \otimes \hat{X})] \ e^{-iHt}\Omega \otimes \mu^{1/2} \rangle \\
&= \beta \int_0^t ds \lim_{T_0 \to \infty} \frac{1}{T_0} \int_{-T_0}^0 dt_0 \ \omega_\beta(\alpha_{t_0, s+t_0; \mathrm{X}}(\delta(A))) \cdot X(s) \\
&= \lim_{T_0 \to \infty} \frac{1}{T_0} \int_{-T_0}^0 dt_0 \ S(\varphi_{t+t_0} | \varphi_{t_0} = \omega_\beta).
\end{aligned} \tag{3.16}$$

The last two equalities are due to the ergodicity (3.6). The mean entropy production $\bar{P}$ is now expressed in a very concise form [5]:

$$\bar{P} = \lim_{n \to \infty} \frac{1}{T_n} S(\omega_\beta \otimes \mu \circ \beta_{T_n} | \omega_\beta \otimes \mu) = \beta \hat{\varphi}(\boldsymbol{J} \otimes \hat{X}) \geq 0. \tag{3.17}$$

Here $\hat{\varphi}$ is one of the limiting *stationary* states to which the state $\omega_\beta \otimes \mu$ in equilibrium under the decoupled dynamics $\hat{\alpha}_t$ tends driven by the coupled dynamics β_t:

$$\hat{\varphi} \equiv \lim_{n\to\infty} \frac{1}{T_n} \int_0^{T_n} dt (\omega_\beta \otimes \mu) \circ \beta_t. \tag{3.18}$$

As already noted, the final-time limit in (3.18) involves some subtle points: Although the state ω_β on the object system $(\mathcal{A}, \alpha_t)$ and the external-force system $(C(M_X), \sigma_t)$ are ergodic, the composite system $(\mathcal{B}, \beta_t)$ or the state on it may not be so in general, and the final-time average of the orbit $\omega_\beta \otimes \mu \circ \beta_t$ of $\omega_\beta \otimes \mu$ along β_t may not be uniquely determined. Since the state space on $\mathcal{B}$ is convex and weak*-compact, Markov-Kakutani theorem [12] assures the existence of a fixed point, which can be attained as a limit point of some *subsequence* with $t = T_n$ tending to $+\infty$. However, the resulting limiting state $\hat{\varphi}$ will change depending upon the choice of subsequences. Thus, there may exist many stationary states on $(\mathcal{B}, \beta_t)$ and the state $\hat{\varphi}$ will be decomposed into the direct sum or integral of ergodic stationary states. Further analysis of this problem requires detailed information of the explicit structure of the dynamical system $(\mathcal{A}, \alpha_t)$, but the general setting for the bifurcation problem in nonequilibrium stationarity will be formulated in this framework as such an ergodic decomposition.

The formula (3.17) can naturally be interpreted as the quantized nonlinear verson of Onsager's dissipation function, since it reduces to the product of the fluxes and the forces in the leading approximation. The relation between (3.17) and the Kubo formula will be discussed in the next section from the viewpoint of van Hove limit.

4. STATIONARITY AND DISSIPATIVITY

4.1 Object System as an Open Subsystem of Enlarged Total System

With the aid of the notions of the hull M_X, the Haar measure μ on it and the "time-independent shape" $\hat{X}$ of the time-dependent external force $X(t)$, the problems concerning the initial- and final-time limits have been clearly separated in the expressions (3.17) and (3.18), not only at the technical level of limit proceures but also at that of states and observables: In the original response-theoretical formulation, the physical meaning of the "adiabatic switching-on" process with the limit $t_0 \to -\infty$ has been obscured in the complicated formulae, being kept only at the level of the heuristic arguments. On the contrary, it is explicitly formulated here as the *initial state* μ of the dynamical system $(C(M_X), \sigma_t)$ of external force. The important role of this state μ in generating the dissipativity will be made clear in the following ([5]).

On the other hand, the problem concerning the final state is condensed in the Eq.(3.18). While its stationarity is formulated there in the enlarged system $(\mathcal{B}, \beta_t)$, we also need to know how the *original object system* $\mathcal{A}$ behaves in this total system, in order to clarify what kind of physical system appears in the final long-time limit. For this purpose, it is necessary for us to control freely the shift of description levels between the microscopic total system with the uncontrollable "redundant" variables ξ and the macroscopic observable subsystem of "relevant" variables. This can be achieved by the help of the embedding map ι given in (3.1) in combination with the following map $\hat{\mu}$ projecting out the variables in the total system $\mathcal{B}$ onto its subsystem $\mathcal{A}$:

$$\hat{\mu} : \mathcal{B} \ni \hat{B} \mapsto \hat{\mu}(\hat{B}) \equiv \int_{M_X} d\mu(\xi)\hat{B}(\xi) \in \mathcal{A}. \tag{4.1}$$

It is easily seen that they satisfy the following relations:

$$\omega \otimes \mu = \omega \circ \hat{\mu}, \tag{4.2}$$

$$\hat{\mu} \circ \iota = Id_{\mathcal{A}}. \tag{4.3}$$

In terms of these maps, the state $\hat{\varphi}$ and the dynamics β_t on the total system $\mathcal{B}$ can be pulled back onto the object system $\mathcal{A}$ as follows:

$$\varphi \equiv \iota^*(\hat{\varphi}) \equiv \hat{\varphi} \circ \iota, \tag{4.4}$$

$$\gamma_t \equiv \hat{\mu} \circ \beta_t \circ \iota. \tag{4.5}$$

As a *conditional expectation* characterized by the properties

$$\hat{\mu}(\iota(B_1)B_2\iota(B_3)) = B_1\hat{\mu}(B_2)B_3, \tag{4.6}$$

$$\hat{\mu}(B^*B) \geq \hat{\mu}(B^*)\hat{\mu}(B) \geq 0, \tag{4.7}$$

$\hat{\mu}$ defined by (4.1) is a *completely positive*(CP) map preserving the positivity of the observables in the stronger sense than (4.7),

$$\left[\sum_{k=1}^{N} \hat{\mu}(\hat{B}^*_{ik}\hat{B}_{kj})\right]^N_{i,j=1} \geq 0, \qquad \forall(\hat{B}_{ij})^N_{i,j=1} \in \mathcal{B} \otimes M_N(C) \text{ for } \forall N \in N, \tag{4.8}$$

where $M_N(C)$ is the algebra of complex $N \times N$ matrices and $\mathcal{B} \otimes M_N(C) \simeq M_N(\mathcal{B})$ is the algebra of matrices with each component in $\mathcal{B}$. However, $\hat{\mu}$ is *not* a homomorphism between $\mathcal{B}$ and $\mathcal{A}$ preserving the multiplication structures:

$$\hat{\mu}(AB) \neq \hat{\mu}(A)\hat{\mu}(B). \tag{4.9}$$

Therefore, unless the coupling between the system $\mathcal{A}$ and the external force vanishes, the mapping γ_t defined by (4.5) is not an automorphism of $\mathcal{A}$ but a CP map which can transfer a pure state into a mixed state. In this sense, the system $(\mathcal{A}, \gamma_t)$ defines a *dissipative* dynamics, but, due to the "memory effect", it *cannot* satisfy in general the Markov property:

$$\gamma_s \circ \gamma_t \neq \gamma_{s+t}. \tag{4.10}$$

Therefore, contrary to the stationarity of the state $\hat{\varphi}$ of the composite system $(\mathcal{B}, \beta_t)$, the pull-back state φ on $(\mathcal{A}, \gamma_t)$ *cannot* be ensured straightforwardly to be stationary.

4.2 van Hove Limit as Adiabatic Elimination of High Frequencies and its Relation to Stationarity and Markov Property

To attain the nonequilibrium stationarity with positive entropy production, we note here the role of van Hove limit [14, 15] in reducing the generalized master equation dragging memory effects to the Markovian master equation without memory [16, 17]. Although this problem has been discussed traditionally in the context of "*downill process*" of *return to equilibrium* which is just opposite to our "*uphill proess*" aiming at nonequilibrium stationarity, the formulation of van Hove limit in terms of master equations seems to be quite useful here, at least for the qualtative understanding of the general relationships between scale changes, Markov property and stationarity.

For this purpose, .we introduce a (dimensionless) coupling parameter λ in the interaction Hamiltonian (2.1),

$$H_I(t) = -A \cdot X(t) \to -\lambda A \cdot X(t), \tag{4.11}$$

which causes also a change in the third term of H in (3.15) as

$$H_I = -\pi \otimes \pi_\mu(A \otimes \hat{X}) \to -\lambda\pi \otimes \pi_\mu(A \otimes \hat{X}). \tag{4.12}$$

Then, van Hove limit means the limiting procedure to let the time parameter t tend to infinity with the quantity $\lambda^2 t \equiv \tau$ fixed finite:

$$t \to \infty \qquad \text{with } \lambda^2 t = \tau \text{ fixed.} \tag{4.13}$$

Using the embedding map ι and the conditional expectation $\hat{\mu}$, we define a "projection" operator P_0 picking up the subalgebra $\mathcal{A} \otimes 1$ in $\mathcal{B}$ and its complement P_1 by

$$\iota \circ \hat{\mu} \equiv P_0 = P_0^2, \tag{4.14}$$

$$P_1 \equiv Id_{\mathcal{B}} - P_0 = P_1^2. \tag{4.15}$$

Then we can formulate (at least formally) the generalized master equation governing the non-Markovian dissipative dynamics γ_t on $\mathcal{A}$ with time parameter $t = \tau/\lambda^2$, in the following two forms adapted to discussing the asymptotic behaviours of states φ and observables C, respectively:

$$\begin{aligned} &\varphi \circ \alpha^{-1}_{\tau/\lambda^2} \circ \gamma_{\tau/\lambda^2} \\ &= \varphi - \int_0^\tau du \int_0^{(\tau-u)/\lambda^2} ds \varphi \circ \alpha^{-1}_{u/\lambda^2+s} \circ \hat{\mu} \circ \\ &\quad \circ \operatorname{ad}(A \otimes \hat{X}) \circ P_1 \circ e^{sZ} \circ \operatorname{ad}(A \otimes \hat{X}) \circ \iota \circ \gamma_{u/\lambda^2}, \end{aligned} \tag{4.16}$$

$$\begin{aligned} &\gamma_{\tau/\lambda^2}(\alpha^{-1}_{\tau/\lambda^2}(C)) \\ &= C - \int_0^\tau du \int_0^{(\tau-u)/\lambda^2} ds (\gamma_{u/\lambda^2} \circ \hat{\mu} \circ \\ &\quad \circ \operatorname{ad}(A \otimes \hat{X}) \circ P_1 \circ e^{sZ} \circ \operatorname{ad}(A \otimes \hat{X}) \circ \iota)(\alpha^{-1}_{u/\lambda^2+s}(C)). \end{aligned} \tag{4.17}$$

Here Z denotes the generator of the "renormalized" unperturbed dynamics given by

$$Z = \frac{d}{dt}\alpha_t \otimes \sigma_t \Big|_{t=0} - i\lambda P_1 \operatorname{ad}(A \otimes \hat{X}) P_1 \tag{4.18}$$

on the assumption that $\mu(\hat{X}) = 0$. ad denotes $\operatorname{ad}(A)(B) = [A, B]$.

Thus, we see that the validity of the Markov property in the van Hove limit (4.13) is ensured if the following two points are verified: i) *whether* the upper end $(\tau - u)/\lambda^2$ of the 2nd integrals in the right-hand sides of (4.16) and (4.17) can be replaced by $+\infty$,

$$\int_0^{(\tau-u)/\lambda^2} ds \underset{\lambda\to 0}{\longrightarrow} \int_0^{\infty} ds, \tag{4.19}$$

and ii) (*adiabatic elimination of microscopic rapid motion*) *whether* the factors $\alpha^{-1}_{\tau/\lambda^2}$ and $\alpha^{-1}_{u/\lambda^2+s}$ can be absorbed by some conditions of "quasi-invariance" on the states φ and/or the observables C such as

$$\lim_{\lambda\to 0} \varphi \circ \alpha^{-1}_{\tau/\lambda^2} = \varphi \qquad \text{for } \forall \tau > 0, \tag{4.20}$$

$$\lim_{\lambda\to 0} \alpha^{-1}_{\tau/\lambda^2}(C) = C \qquad \text{for } \forall \tau > 0. \tag{4.21}$$

If these approximations are valid, then the generalized master equation (4.16)-(4.17) will be reduced to the Markovian master equation for $\tilde{\gamma}_\tau \equiv \gamma_{\tau/\lambda^2}$ with $\tau \geq 0$ satisfying

$$\tilde{\gamma}_{\tau+\sigma} = \tilde{\gamma}_\tau \circ \tilde{\gamma}_\sigma, \tag{4.22}$$

and hence the stationarity of the state will be attained for any asymptotic limit state of the form

$$\lim_{T\to\infty} \frac{1}{T} \int_0^T d\tau \varphi \circ \tilde{\gamma}_\tau.$$

In the usual formulation of master equation in the processes of return to equilbrium, the state $\varphi \circ \alpha^{-1}_{\tau/\lambda^2}$ in (4.16) is replaced by the density matrix of the Gibbs state $e^{\beta(F-H)}$, which obscures the relevance of the second problem ii) of the adiabatic elimination of microscopic rapid motion in the infinite future time (τ/λ^2 with $\lambda \sim 0, \tau \geq 0$). Consequently, the natural meaning of van Hove limit has been lost being taken only as a formal recipe to derive a Markovian stochastic dynamics. Taking account properly of the adiabatic elimination, however, we can interpret van Hove limit in a more realistic way as a scale transformation controlling the *change of units between the two different times* $t = t_{\text{micro}}$ and $\tau = t_{\text{macro}}$ of the microscopic and macroscopic levels:

$$\lambda^2 t_{\text{micro}} = t_{\text{macro}}. \tag{4.23}$$

If we take the microscopic time t_{micro} finite in (4.23), the limit $\lambda \to 0$ takes us literally to the situation of *weak coupling* or approximate decoupling between the macroscopic external force X and the microscopic object system $\mathcal{A}$, where we "observe" the purely dynamical motion of the latter one with time t_{micro}. On the contrary, the situation of van Hove limit (4.13) in combination with the adiabatic elimination (4.20)-(4.21) brings into focus the macroscopic level of *state-changing process* with finite time t_{macro}. In spite of the small coupling parameter $\lambda \to 0$, the effects of the coupling term (4.11) accumulates into macroscopically visible state changes through the infinite repetitions of "invisible" microscopic dynamical motions of high frequencies during the *infinite* time interval $t_{\text{micro}} = t_{\text{macro}}/\lambda^2 \to \infty$, as is seen in (4.16)-(4.17).

The physical basis of such an interpretation is that the notion of *time* emerges from the correlations among physical motions which are "fibered" into different levels with certain typical motions in each regime (i.e. *different* "standard clocks" at each size level), and hence that the standard scales of time differ from one level to another according to the changes of the "standard physical motions". Such a scale transformation as (4.23) is just a "calibration" between different "clocks" belonging to different size regimes. In the *idealization limit* of $\lambda \to 0$, simple approximate descriptions emerge as above for either one of microscopic dynamical motions *or* macroscopic state-changing processes, according to the choice between $\{t_{\mathrm{micro}} = \text{finite with } t_{\mathrm{macro}} \to 0\}$ and $\{t_{\mathrm{macro}} = \text{finite with } t_{\mathrm{micro}} \to \infty\}$. This interpretation leads to a natural reformulation of the Kubo formula in direct relation with entropy production in the next subsection.

4.3 Entropy Production and van Hove Limit as Scale Change

If we adopt the above physical viewpoint, the inverse temperature β appearing in the mean entropy production $\bar{P}$ (3.17) should also be transformed in parallel with the time corresponding to the time-scale change (4.23) as shown in [18]:

$$\text{microscopic } \beta \to \text{macroscopic } \lambda^2\beta \equiv \beta_{\mathrm{eff}}. \tag{4.24}$$

Therefore, we obtain the following expression [5] for the mean entropy production $\bar{P}$ in van Hove limit (4.13):

$$\begin{aligned} 0 \le \bar{P} &= \beta\hat{\varphi}_{\lambda\hat{\mathrm{X}}}(\lambda J \otimes \hat{X}) \\ &= \beta_{\mathrm{eff}}\lambda^{-1}\hat{\varphi}_{\lambda\hat{\mathrm{X}}}(J \otimes \hat{X}) \\ &\underset{\lambda\to 0}{\to} \beta_{\mathrm{eff}}\frac{d}{d\lambda}\hat{\varphi}_{\lambda\hat{\mathrm{X}}}(J \otimes \hat{X})\Big|_{\lambda=0} \\ &= \frac{1}{2}\beta_{\mathrm{eff}}\frac{d^2}{d\lambda^2}\hat{\varphi}_{\lambda\hat{\mathrm{X}}}(\lambda\delta(A) \otimes \hat{X})\Big|_{\lambda=0} \\ &= \frac{1}{2}L_{ij}\hat{X}^i\hat{X}^j \ge 0, \end{aligned} \tag{4.25}$$

with L_{ij} defined by

$$L_{ij} \equiv \beta_{\mathrm{eff}}\frac{\partial^2}{\partial\hat{X}^i\partial\hat{X}^j}\hat{\varphi}_{\hat{\mathrm{X}}}(\delta(A) \otimes \hat{X})\Big|_{\hat{\mathrm{X}}=0}. \tag{4.26}$$

Here the $\lambda\hat{X}$ dependence of the state $\hat{\varphi}$ is made explicit as $\hat{\varphi}_{\lambda\hat{\mathrm{X}}}$ and we have used the simple facts that

$$\hat{\varphi}_{\lambda\hat{\mathrm{X}}}(J \otimes \hat{X})\big|_{\lambda=0} = (\omega_\beta \otimes \mu)(\delta(A) \otimes \hat{X})) = 0$$

and that the external force $\hat{X}$ is proportional to the coupling constant λ.

In order to rewrite the formula (4.25) into a more convenient form, we utilize an integral equation for the state $\omega_\beta \otimes \mu \circ \beta_t$,

$$\omega_\beta \otimes \mu \circ \beta_t(\hat{B}) = \omega_\beta \otimes \mu(\hat{B}) - \int_0^t ds\omega_\beta \otimes \mu([i\lambda A \otimes \hat{X}, \beta_s(\hat{B})]), \tag{4.27}$$

In view of the remark in [3] on the relation between kinetic-theoretical approaches and linear-response theoretical ones toward the understanding of dissipativity, it would be interesting to consider here the problem of ordering consistency between *stochastization* and *linearization*: It is usually said (see [2], [3]) that in the former case first comes the stochastization procedure followed by linearization, and that it is just in the opposite order in the latter. Here in the above discussion of master equation as one of the kinetic-theoretical approaches, the *stochasization* is due to the conditional expectation map $\hat{\mu}$ which induces a non-Markovian dissipative dynamics $\gamma_t = \hat{\mu} \circ \beta_t \circ \iota$ governed by the generalized master equation (4.16) and (4.17). To attain a genuine kinetic equation in the form of a Markovian master equation, the adiabatic elimination mechanism of van Hove limit to focus upon the state-changing processes is indispensable as a kind of "*linearization*" procedure. However, we can see, from such an expression as

$$\begin{aligned}\varphi \circ \tilde{\gamma}_\tau &= \varphi \circ \lim_{\lambda \to 0} \hat{\mu} \circ \beta_{\tau/\lambda^2} \circ \iota \\ &= \lim_{\lambda \to 0} (\varphi \otimes \mu) \circ ([\alpha_{\tau/\lambda^2} \otimes \sigma_{\tau/\lambda^2}]^{-1} \circ \beta_{\tau/\lambda^2}) \circ \iota,\end{aligned}$$

that the "difference in the ordering" of the "stochastization" $\hat{\mu}$ and the "linearization" $\lambda \to 0$ is at most in their *appearances* of the same procedures viewed from the "Heisenberg picture" for *observable* dynamics and from the "Schödinger one" for *state-changing* dynamics which are *contragrediently* related through the duality between algebras and states.

Finally, we stress the intrinsic dynamical natures of the above two "procedures". First, the *probability* measure μ on $C(M_X)$ occupying the pivotal position responsible for the *stochastization* should not be taken as an *ad hoc* device, but its origin should be traced to the ergodicity of the *dynamics* of external force system which couples to the object system. Secondly, the validity of van Hove limit as a *level-transition* mechanism requires the consistency between macroscopic *stochasiticity* and the long-time asymptotic behaviours of the microscopic *dynamics* in such a form as the *mixing property* (4.29), which is of renormalization-group-theoretical nature in essence. Namely, the total dynamical system $(\mathcal{B}, \beta_t)$ should choose $\lambda = 0$ as its *infrared-stable* fixed point for the consistency of our discussion. If $\lambda = \infty$ is infrared-stable on the contrary, the roles of t_{micro} and t_{macro} should be just interchanged, and hence we would encounter the situation where the singular coupling limit [19]* is relevant.

In [5], the problem of the inhomogeneous substructures inherent in nonequilibrium stationarity is discussed from the viewpoint of *multiple-reservoir system* [20] and of *groupoid dynamical systems* [21, 22]. For shortage of the space, we omit it here in spite of its relevance to the consistent treatment of space-time dependent nonequilibrium phenomena such as inhomogeneous temperature distributions, thermal diffusions, and also to the quantum-field-theoretical extension of the present framework involving the continuous version of multiple-reservoir systems which may be closely to the problem of continuous measurements.

---

* See also the references cited in [17]

In view of the remark in [3] on the relation between kinetic-theoretical approaches and linear-response theoretical ones toward the understanding of dissipativity, it would be interesting to consider here the problem of ordering consistency between *stochastization* and *linearization*: It is usually said (see [2], [3]) that in the former case first comes the stochastization procedure followed by linearization, and that it is just in the opposite order in the latter. Here in the above discussion of master equation as one of the kinetic-theoretical approaches, the *stochasization* is due to the conditional expectation map $\hat{\mu}$ which induces a non-Markovian dissipative dynamics $\gamma_t = \hat{\mu} \circ \beta_t \circ \iota$ governed by the generalized master equation (4.16) and (4.17). To attain a genuine kinetic equation in the form of a Markovian master equation, the adiabatic elimination mechanism of van Hove limit to focus upon the state-changing processes is indispensable as a kind of "*linearization*" procedure. However, we can see, from such an expression as

$$\begin{aligned}\varphi \circ \tilde{\gamma}_\tau &= \varphi \circ \lim_{\lambda \to 0} \hat{\mu} \circ \beta_{\tau/\lambda^2} \circ \iota \\ &= \lim_{\lambda \to 0} (\varphi \otimes \mu) \circ ([\alpha_{\tau/\lambda^2} \otimes \sigma_{\tau/\lambda^2}]^{-1} \circ \beta_{\tau/\lambda^2}) \circ \iota,\end{aligned}$$

that the "difference in the ordering" of the "stochastization" $\hat{\mu}$ and the "linearization" $\lambda \to 0$ is at most in their *appearances* of the same procedures viewed from the "Heisenberg picture" for *observable* dynamics and from the "Schödinger one" for *state-changing* dynamics which are *contragrediently* related through the duality between algebras and states.

Finally, we stress the intrinsic dynamical natures of the above two "procedures". First, the *probability* measure μ on $C(M_X)$ occupying the pivotal position responsible for the *stochastization* should not be taken as an *ad hoc* device, but its origin should be traced to the ergodicity of the *dynamics* of external force system which couples to the object system. Secondly, the validity of van Hove limit as a *level-transition* mechanism requires the consistency between macroscopic *stochasiticity* and the long-time asymptotic behaviours of the microscopic *dynamics* in such a form as the *mixing property* (4.29), which is of renormalization-group-theoretical nature in essence. Namely, the total dynamical system $(\mathcal{B}, \beta_t)$ should choose $\lambda = 0$ as its *infrared-stable* fixed point for the consistency of our discussion. If $\lambda = \infty$ is infrared-stable on the contrary, the roles of t_{micro} and t_{macro} should be just interchanged, and hence we would encounter the situation where the singular coupling limit [19]* is relevant.

In [5], the problem of the inhomogeneous substructures inherent in nonequilibrium stationarity is discussed from the viewpoint of *multiple-reservoir system* [20] and of *groupoid dynamical systems* [21, 22]. For shortage of the space, we omit it here in spite of its relevance to the consistent treatment of space-time dependent nonequilibrium phenomena such as inhomogeneous temperature distributions, thermal diffusions, and also to the quantum-field-theoretical extension of the present framework involving the continuous version of multiple-reservoir systems which may be closely to the problem of continuous measurements.

* See also the references cited in [17]

REFERENCES

1. R. Kubo, J. Phys. Soc. Japan **12** (1957), 570.
2. N. G. van Kampen, Physica Norvegica **5** (1971), 279.
3. R. Kubo, M. Toda, and N. Hashitsume, *Statistical Physics*, vol. 2 (Springer-Verlag, Berlin, 1985).
4. I. Ojima, H. Hasegawa, and M. Ichiyanagi, J. Stat. Phys. **50** (1988), 633.
5. I. Ojima, J. Stat. Phys. **56** (1989), 203.
6. R. Haag, N. M. Hugenholtz and M. Winnink, Comm. Math. Phys. **5** (1967), 215.
7. O. Bratteli and D. W. Robinson, *Operator Algebras and Quantum Statistical Mechanics*, vol. 2 (Springer-Verlag, Berlin, 1981).
8. M. Ichiyanagi, J. Phys. Soc. Japan **55** (1986), 2093.
9. H. Araki, Publ. RIMS, Kyoto Univ. **11** (1976), 809 ; **13** (1977), 173; A. Uhlmann, Comm. Math. Phys. **54** (1977), 21.
10. K. Yosida, *Functional Analysis* 6th ed. (Springer-Verlag, Berlin, 1980).
11. J. Avron and B. Simon, Comm. Math. Phys. **82** (1981), 101; J. Bellissard, R. Lima and D. Testard, in *Mathematics + Physics: Lectures on Recent Results*, vol.1, L. Streit, ed. (World Scientific, Singapore, 1985), and references cited therein.
12. N. Dunford and J. T. Schwartz, *Linear Operators*, Vol.I (Wiley-Interscience, New York, 1958); N. Bourbaki, *Éléments de Mathématique, Espaces Vectoriels Topologiques*, 2nd ed. (Hermann, Paris, 1966).
13. J. Bellissard, private communication (1987); J. S. Howland, Math. Ann. **207**(1974), 315; K. Yajima, J. Math. Soc. Japan **29** (1977), 729.
14. L. van Hove, Physica **21** (1955), 517.
15. L. van Hove, Physica **23** (1957), 441.
16. E. B. Davies, Comm. Math. Phys. **39** (1974), 91; Math. Ann. **219** (1976), 147; Ann. Inst. Henri Poincaré **11** (1975), 265.
17. V. Gorini, A. Frigerio, M. Verri, A. Kossakowski and E. C. G. Sudarshan, Rep. Math. Phys. **13** (1978), 149, and references cited therein.
18. I. Ojima, in *Quantum Field Theory*, F. Mancini, ed. (Elsevier, 1986), p.443,; in *Proceedings 2nd International Symposium on Foundations of Quantum Mechanics* (Tokyo, 1986), M. Namiki, *et al.*, eds.(Physical Society of Japan, Tokyo, 1987), p.91.
19. K. Hepp and E. Lieb, Helv. Phys. Acta **46** (1973), 573.
20. J. Lebowitz, Phys. Rev. **114** (1959), 1192; H. Hasegawa, T. Nakagomi, M. Mabuchi, and K. Kondo, J. Stat. Phys. **23** (1980), 281.
21. A. Connes, in *Lecture Notes in Mathematics*, No.725 (Springer-Verlag, Berlin, 1979), p.19; J. Renault, *Lecture Notes in Mathematics*, No.793 (Springer-Verlag, Berlin, 1980); J. Bellissard and D. Testard, preprint CPT-81/P.1311 (1981).
22. T. Masuda, Publ. RIMS, Kyoto Univ. **20** (1984), 929, 959.

Detection Theory in Quantum Optics and Quantum Stochastic Calculus

Alberto Barchielli

Dipartimento di Fisica dell'Università di Milano,
Via Celoria 16, I–20133 Milano, Italy,
and Istituto Nazionale di Fisica Nucleare, Sezione di Milano

1 Quantum Stochastic Calculus

Quantum stochastic calculus (QSC) [1–3] is a noncommutative analogue of Itô's stochastic calculus. It was developed originally as a mathematical theory of quantum noise in open systems, but soon after it was applied also to measurement theory in quantum mechanics [4–6] and to quantum optics [7–13]. The main ingredients of QSC are certain Bose fields which play the role of quantum analogues of independent Wiener processes. The starting point for the applications of QSC in quantum optics is to take these Bose fields as an approximation of the electromagnetic field. The explicit introduction of QSC in quantum optics was made in [7], but the use of the related δ-correlated noise is older [14].

One of the possible applications of QSC to quantum optics concerns the development of a quantum photodetection theory (direct, heterodyne and homodyne detection) [11] [15–18]. This is the topic we want to discuss in this paper. However we need first to recall something about QSC; we consider the simplest version of it, namely that one based on the boson Fock space [1,2].

Let us introduce d Bose fields $a_j(t)$, $a_j^\dagger(t)$, satisfying the canonical commutation rules (CCR's)

$$[a_j(t), a_i(s)] = 0\,, \qquad [a_j(t), a_i^\dagger(s)] = \delta_{ji}\,\delta(t-s)\,. \tag{1}$$

We consider the Fock representation; this means that the Hilbert space on which the field operators act is the symmetric Fock space over the "one-particle space" $\mathbb{C}^d \otimes L^2(\mathbb{R})$. This space, which we denote by Γ, is spanned

Lecture Notes in Physics, Vol. 378
C. Bendjaballah, O. Hirota, S. Reynaud (Eds.),
Quantum Aspects of Optical Communications

by the exponential vectors $\psi(f)$, whose components in the $0, 1, \ldots, n, \ldots$ particle spaces are

$$\psi(f) = \left(1, f, \ldots, (n!)^{-1/2} f \otimes \ldots \otimes f, \ldots\right), \qquad f \in \mathbb{C}^d \otimes L^2(\mathbb{R}). \quad (2)$$

Apart from the normalization, the exponential vectors are the usual coherent vectors for the fields $a_j(t)$; the vector $\psi(0)$ is the Fock vacuum.

Then we define the annihilation, creation and gauge (or number) processes by

$$A_j(t) := \int_0^t a_j(s)\,\mathrm{d}s\,, \quad A_j^\dagger(t) = \int_0^t a_j^\dagger(s)\,\mathrm{d}s\,, \quad \Lambda_{ij}(t) := \int_0^t a_i^\dagger(s) a_j(s)\,\mathrm{d}s\,. \quad (3)$$

The rigorous definition of these processes in terms of their action on the exponential vectors is given in [1], §2 and 4. In particular, the exponential vectors are eigenstates of $A_j(t)$ and $a_j(t)$.

A QSC of Itô type, based on the integrators $\mathrm{d}A_j(t)$, $\mathrm{d}A_j^\dagger(t)$, $\mathrm{d}\Lambda_{ij}(t)$ and $\mathrm{d}t$, has been developed by Hudson and Parthasarathy [1]. The couples $A_j(t)$, $A_j^\dagger(t)$ are noncommutative analogues of independent Wiener processes and the operators $\Lambda_{ij}(t)$ are the main ingredients in the construction of quantum analogues of Poisson processes. Let $\mathcal{H}$ be a separable Hilbert space associated to some system S. By QSC, one can give a rigorous meaning to equations for operators in $\mathcal{H} \otimes \Gamma$ of the type

$$\mathrm{d}M(t) = \sum_j \left[\textstyle\sum_i D_{ji}(t)\,\mathrm{d}\Lambda_{ji}(t) + E_j(t)\,\mathrm{d}A_j(t) + F_j(t)\,\mathrm{d}A_j^\dagger(t)\right] + G(t)\,\mathrm{d}t\,, \quad (4)$$

where the differentials are intended to "point into the future" (Itô prescription), i.e. $\mathrm{d}A_j(t) = A_j(t+\mathrm{d}t) - A_j(t)$, $\ldots$, and $M(t)$, $D_{ji}(t)$, $E_j(t)$, $F_j(t)$, $G(t)$ are *adapted* operator-valued processes; roughly speaking, this means that they depend on the fields $a_j(s)$ and $a_j^\dagger(s)$ only for times s less than t (see [1], Def. 3.1). By the CCR's (1) adapted processes commute with the fundamental differentials (see [1], Theor. 4.5), i.e. $[M(t), \mathrm{d}\Lambda_{ij}(t)] = 0$, $[M(t), \mathrm{d}A_j(t)] = 0, \ldots$. If $N(t)$ is another process satisfying an equation analogous to (4), then the differential of the product $M(t)N(t)$ is given by $\mathrm{d}\big(M(t)N(t)\big) = \big(\mathrm{d}M(t)\big)N(t) + M(t)\big(\mathrm{d}N(t)\big) + \big(\mathrm{d}M(t)\big)\big(\mathrm{d}N(t)\big)$, where the "Itô correction" $\big(\mathrm{d}M(t)\big)\big(\mathrm{d}N(t)\big)$ has to be computed by means of the "Itô table"

$$\begin{gathered}
\mathrm{d}A_j(t)\,\mathrm{d}A_i^\dagger(t) = \delta_{ji}\,\mathrm{d}t\,, \qquad \mathrm{d}A_j(t)\,\mathrm{d}\Lambda_{ki}(t) = \delta_{jk}\,\mathrm{d}A_i(t)\,, \\
\mathrm{d}\Lambda_{ji}(t)\,\mathrm{d}A_k^\dagger(t) = \delta_{ik}\,\mathrm{d}A_j^\dagger(t)\,, \qquad \mathrm{d}\Lambda_{ji}(t)\mathrm{d}\Lambda_{lk}(t) = \delta_{il}\,\mathrm{d}\Lambda_{jk}(t)\,, \\
\mathrm{d}A_i^\dagger(t)\,\mathrm{d}A_j(t) = \mathrm{d}\Lambda_{ki}(t)\,\mathrm{d}A_j(t) = \mathrm{d}A_k^\dagger(t)\,\mathrm{d}\Lambda_{ji}(t) = 0\,;
\end{gathered} \quad (5)$$

all the products of $\mathrm{d}A_j(t)$, $\mathrm{d}A_j^\dagger(t)$ or $\mathrm{d}\Lambda_{ij}(t)$ with $\mathrm{d}t$ vanish (see [1], (7.1)).

The Hudson and Parthasarathy theory allows us, in particular, to give meaning to "stochastic Schrödinger equations" of the type

$$dU_t = \left\{ \sum_j \left[-R_j^\dagger \, dA_j(t) + R_j \, dA_j^\dagger(t) - \tfrac{1}{2} R_j^\dagger R_j \, dt \right] - iH \, dt \right\} U_t \,, \qquad U_0 = \mathbb{1} \,, \tag{6}$$

where $R_j, H \in \mathcal{B}(\mathcal{H})$ (bounded operators on $\mathcal{H}$) and $H = H^\dagger$. Here and in the following we identify R_j with $R_j \otimes \mathbb{1}$, $A_j(t)$ with $\mathbb{1} \otimes A_j(t)$ and so on. The solution U_t of this equation exists and is unique; $\{U_t,\ t \geq 0\}$ is an adapted process and, for any $t \geq 0$, U_t is a *unitary* operator on $\mathcal{H} \otimes \Gamma$ ([1], §7).

The Bose fields $a_j(t)$ can be considered as an approximation of the electromagnetic field; in this case the index j stays for polarization, direction of propagation (discretized), and so on. Then, (6) gives the dynamics of the composed system (S plus electromagnetic field) in the so called broadband approximation [7]. U_t is the evolution operator in the interaction picture with respect to the free dynamics of the fields. Rigorous results showing how the dynamics (6) can be obtained in the so called weak coupling limit are given in [19].

Let now ϱ be the initial state of S (a statistical operator on $\mathcal{H}$) and let the initial state of the fields be the coherent vector

$$\Psi(f) := \frac{|\psi(f)\rangle\langle\psi(f)|}{\|\psi(f)\|^2} \,. \tag{7}$$

If Tr_Γ is the partial trace over Fock space, the quantity

$$\varrho(t) := \mathrm{Tr}_\Gamma \left\{ U_t \big(\varrho \otimes \Psi(f) \big) U_t^\dagger \right\} \tag{8}$$

represents the reduced statistical operator of system S at time t. Let us differentiate this equation. By the fact that we are using the Itô prescription and the fact that U_t is an adapted process, the mean values of $dA_j(t)$, $dA_j^\dagger(t)$ and $d\Lambda_{ij}(t)$ are the same as in the free case and are given by $f_j(t)\, dt$, $f_j^*(t)\, dt$ and $f_i^*(t) f_j(t)\, dt$ respectively. Then, by the rules of QSC we obtain that $\varrho(t)$ satisfies the *quantum master equation*

$$\frac{d}{dt} \varrho(t) = \mathcal{L}_t \, \varrho(t) \,, \tag{9}$$

$$\mathcal{L}_t \, \varrho := -i \Big[H + i \textstyle\sum_j \big(f_j^*(t) R_j - f_j(t) R_j^\dagger \big), \varrho \Big] + \dfrac{1}{2} \displaystyle\sum_j \big([R_j, \varrho R_j^\dagger] + [R_j \varrho, R_j^\dagger] \big) \,. \tag{10}$$

By using infinitely many independent fields, the most general master equation (with bounded Liouvillian) can be obtained [2]. At least at a formal level, the same results hold in the physically more interesting case of H

and R_j unbounded system operators. It is also possible to show that time-ordered multi-time quantum correlation functions for system operators can be re-expressed by means of the reduced dynamics (quantum regression theorem; see [7], §IV.D). The whole procedure can be generalized in order to take into account "thermal" and "squeezed" fields [3,7] (this corresponds to the use of certain non-Fock representations).

Let us consider now a situation in which we cannot act directly on our system S, but any action is mediated by some quantum input and output channel. We can think of an atom driven by a laser (input) and emitting fluorescence light (output) or of the light entering (input) and leaving (output) an optical cavity. In these examples the role of input and output channels is played by the electromagnetic field and, again, we can think of approximating it by the Bose fields $a_j(t)$ introduced before.

We call $A_j(t)$, $A_j^\dagger(t)$, $\Lambda_{ij}(t)$ (or, equivalently, $a_j(t)$ and $a_j^\dagger(t)$) the "input fields"; they represent the fields before the interaction with system S. In this order of ideas, the initial state (7) for the fields describes a coherent input signal (e.g. a laser field). Conversely, the output channel will be represented by the fields after the interaction with system S and what we have to do is to show how to describe mathematically these fields.

Let U_t be given by (6) and let us introduce the "output fields" by [6]

$$A_j^{\mathrm{out}}(t) := U_t^\dagger\, A_j(t)\, U_t\,, \qquad \Lambda_{ij}^{\mathrm{out}}(t) := U_t^\dagger\, \Lambda_{ij}(t)\, U_t\,. \tag{11}$$

The quantities $A_j^{\mathrm{out}}(t)^\dagger$ are given by analogous equations.

By (6) one has that the quantity $U_{s,t} := U_t\, U_s^\dagger$, $s \leq t$, depends only on the fields $a_j(r)$, $a_j^\dagger(r)$ with times r in between s and t. By the commutation rules (1) we have $[U_{s,t}, A_j(\tau)] = 0$ if $\tau \leq s$. This gives

$$A_j^{\mathrm{out}}(t) = U_t^\dagger\, U_{t,T}^\dagger\, A_j(t)\, U_{t,T}\, U_t = U_T^\dagger\, A_j(t)\, U_T\,, \qquad \forall T \geq t\,, \tag{12}$$

and similar equations for $A_j^{\mathrm{out}}(t)^\dagger$ and $\Lambda_{ij}^{\mathrm{out}}(t)$. In particular, by taking T sufficiently large, we see that the output processes are obtained from the input processes by a single unitary transformation and, therefore, the commutation rules are preserved. In particular their formal derivatives $a_j^{\mathrm{out}}(t) := \mathrm{d}A_j^{\mathrm{out}}(t)/\mathrm{d}t$, $a_j^{\mathrm{out}}(t)^\dagger := \mathrm{d}A_j^{\mathrm{out}}(t)^\dagger/\mathrm{d}t$ satisfy the CCR's (1). Equation (12) shows indeed that the operators $a_j^{\mathrm{out}}(t)$ represent the fields after the interaction with system S.

By (11) and the rules of QSC, one obtains that the output processes satisfy the quantum stochastic differential equations (QSDE's) [6]

$$\begin{aligned}
&\mathrm{d}A_j^{\mathrm{out}}(t) = \mathrm{d}A_j(t) + U_t^\dagger\, R_j\, U_t\, \mathrm{d}t\,,\\
&\mathrm{d}\Lambda_{ij}^{\mathrm{out}}(t) = \mathrm{d}\Lambda_{ij}(t) + U_t^\dagger\, R_j\, U_t\, \mathrm{d}A_i^\dagger(t) + U_t^\dagger\, R_i^\dagger\, U_t\, \mathrm{d}A_j(t) + U_t^\dagger\, R_i^\dagger R_j\, U_t\, \mathrm{d}t\,.
\end{aligned} \tag{13}$$

From these equations one sees explicitly that the output fields carry information on system S: the quantities R_j are the system operators appearing in the system-field interaction.

2 Direct Detection

Let us consider again the situation of a system S with input and output channels. Information about system S can be "extracted" by some measurement on the output fields. In this section we consider a standard photon counting, realized by a photoelectron counter (*direct detection*). For simplicity, we assume the detector to be sensitive only to the photons of field 1. The cases of a counter sensitive to more fields or of many counters present no conceptual complication. The output current from the photoelectron counter will be given by

$$I(t) = \int_0^t \mathrm{d}s\, F(t-s)\, \frac{\mathrm{d}N_s}{\mathrm{d}s}\,, \tag{14}$$

where $F(t-s)$ is the detector response function and $\mathrm{d}N_s/\mathrm{d}s$ is the rate of arrival of photons. N_t is the number of photons up to time t carried by the output field $a_1^{\mathrm{out}}(t)$. By using the same symbol for an observable and the associated selfadjoint operator, we can write

$$N_t := \Lambda_{11}^{\mathrm{out}}(t) \equiv U_t^\dagger \int_0^t \mathrm{d}s\, a_1^\dagger(s) a_1(s)\, U_t\,. \tag{15}$$

The operators $\Lambda_{11}(t)$, $t \geq 0$, are a family of commuting selfadjoint operators and, by the results of the previous section, the same holds for N_t, $t \geq 0$. By (14), we can say that our photoelectron counter realizes a measurement of the compatible observables N_t, $t \geq 0$ [16–18].

The easiest way of treating this measurement is to introduce the "characteristic operator" [4–6] [11,12] (the analogue of the characteristic functional of a classical stochastic process)

$$V_t[k] := \exp\left[\mathrm{i}\int_0^t k(s)\,\mathrm{d}N_s\right] = U_t^\dagger \exp\left[\mathrm{i}\int_0^t k(s)\,\mathrm{d}\Lambda_{11}(s)\right] U_t\,, \tag{16}$$

where k is a real test function; the last step in (16) results from the analogue of (12). By the rules of QSC, one shows that this operator satisfies the QSDE [6] [11,12]

$$\mathrm{d}V_t[k] = V_t[k]\left[\mathrm{e}^{\mathrm{i}k(t)} - 1\right]\mathrm{d}N_t\,, \qquad \mathrm{d}N_t \equiv \mathrm{d}\Lambda_{11}^{\mathrm{out}}(t)\,, \qquad V_0[k] = \mathbb{1}\,. \tag{17}$$

Let now ϱ be the initial state of system S and $\Psi(f)$ be the initial state of the fields. Then the quantity

$$\Phi_t[k] := \mathrm{Tr}_{\mathcal{H}\otimes\Gamma}\left\{V_t[k]\,\varrho\otimes\Psi(f)\right\} \tag{18}$$

is by construction the characteristic functional (Fourier transform of the probability distribution) of a point process, describing the counting of photons in the output field. The characteristic functional for the stochastic process describing the behaviour of the output current $I(s)$ can be obtained from (18) by letting the final time t go to infinity and by performing the substitution $k(s) \to \int_s^{+\infty} \mathrm{d}\tau\, k(\tau)F(\tau - s)$. The whole statistics of the photodetection process is contained in (18).

In [20] there is an interesting observation, which allows to relate the approach presented here to that one of Kelley and Kleiner [21]. Let us forget for a moment the system S (the source of the output fields) and take $U_t = \mathbb{1}$. By using the normal ordering operation : :, we can write

$$V_t[k] \equiv \exp\left[\mathrm{i}\int_0^t k(s)\, a_1^\dagger(s)a_1(s)\,\mathrm{d}s\right] =: \exp\left\{\int_0^t \left[\mathrm{e}^{\mathrm{i}k(s)} - 1\right] a_1^\dagger(s)a_1(s)\,\mathrm{d}s\right\}: \tag{19}$$

This equation can be proven by taking the matrix elements of both sides with respect to two coherent vectors. Then, one shows that these matrix elements satisfy the same differential equation with the same initial conditions and, therefore, they are identical [18].

By taking $k(s) = w$ (a constant) if $s \in (t_1, t_2)$, $k(s) = 0$ otherwise, $0 \le t_1 < t_2 \le t$, equations (18) and (19) give

$$\Phi_t[k] \to \Phi(w; t_1, t_2) = \sum_{m=0}^{\infty} \mathrm{e}^{\mathrm{i}wm}\, P_m(t_1, t_2), \tag{20}$$

$$P_m(t_1, t_2) = \left\langle : \frac{1}{m!}\left[\int_{t_1}^{t_2} a_1^\dagger(s)a_1(s)\mathrm{d}s\right]^m \exp\left[-\int_{t_1}^{t_2} a_1^\dagger(s)a_1(s)\mathrm{d}s\right] : \right\rangle, \tag{21}$$

where the symbol $\langle\ \rangle$ means the quantum expectation with respect to the state of the fields. Equation (21) is the Kelley–Kleiner formula for the probability of m photocounts in the time interval (t_1, t_2) and $\Phi(w; t_1, t_2)$ is its Fourier transform. Equation (21) refers to a counter of unit efficiency; in our approach non-unit efficiency is taken into account by the response function F. Similar considerations hold in the case we take into account system S: we have only to reinterpret the symbol : : as normal ordering of the output fields.

In the first section we have seen how a quantum master equation can be obtained by eliminating the degrees of freedom of the fields. Something similar can be done for $\Phi_t[k]$ [4–6] [11,12]: we can re-express it in terms of operators of system S alone (the source of the output fields). We define an operator $\mathcal{G}_t[k]$ on the trace class on $\mathcal{H}$ (the space spanned by the statistical operators) by

$$\mathrm{Tr}_{\mathcal{H}}\left\{X\,\mathcal{G}_t[k]\varrho\right\} = \mathrm{Tr}_{\mathcal{H}\otimes\Gamma}\left\{U_t^\dagger X U_t V_t[k]\,\varrho\otimes\Psi(f)\right\}, \qquad \forall X\in\mathcal{B}(\mathcal{H}), \quad (22)$$

so that the characteristic functional (18) can be written as

$$\Phi_t[k] = \mathrm{Tr}_{\mathcal{H}}\left\{\mathcal{G}_t[k]\varrho\right\}. \quad (23)$$

A differential equation for $\mathcal{G}_t[k]$ can be obtained from (22). The differential of the right hand side of (22) can be computed by using the rules of QSC; the final result is some kind of master equation for $\mathcal{G}_t[k]$:

$$\frac{\mathrm{d}}{\mathrm{d}t}\mathcal{G}_t[k] = \left\{\mathcal{L}_t + \mathcal{K}_t\big(k(t)\big)\right\}\mathcal{G}_t[k], \qquad \mathcal{G}_0[k] = \mathbb{1}, \quad (24)$$

$$\mathcal{K}_t(k)\varrho = \left(\mathrm{e}^{\mathrm{i}k} - 1\right)\left[R_1 + f_1(t)\right]\varrho\left[R_1^\dagger + f_1^*(t)\right], \quad (25)$$

and $\mathcal{L}_t$ is again given by (10).

By defining

$$\tilde{\mathcal{L}}_t\varrho := -\mathrm{i}\left[H + \mathrm{i}\sum_{j=2}^d\left(f_j^*(t)R_j - f_j(t)R_j^\dagger\right), \varrho\right] - \tfrac{1}{2}\left\{R_1^\dagger R_1, \varrho\right\} -$$

$$- f_1(t)R_1^\dagger\varrho - f_1^*(t)\varrho R_1 - |f_1(t)|^2\,\varrho + \frac{1}{2}\sum_{j=2}^d\left(\left[R_j, \varrho R_j^\dagger\right] + \left[R_j\varrho, R_j^\dagger\right]\right), \quad (26)$$

$$\mathcal{J}(t)\varrho := \left[R_1 + f_1(t)\right]\varrho\left[R_1^\dagger + f_1^*(t)\right], \quad (27)$$

(where $\{a,b\} = ab + ba$) we can write

$$\mathcal{L}_t + \mathcal{K}_t\big(k(t)\big) = \tilde{\mathcal{L}}_t + \mathrm{e}^{\mathrm{i}k(t)}\mathcal{J}(t). \quad (28)$$

Then, the solution of (24) can be given as an expansion in $\mathcal{J}(t)$ (Dyson series)

$$\mathcal{G}_t[k] = \Upsilon(t,0) + \sum_{m=1}^\infty \int_0^t \mathrm{d}t_m \int_0^{t_m}\mathrm{d}t_{m-1}\cdots\int_0^{t_2}\mathrm{d}t_1\,\exp\left[\mathrm{i}\sum_{r=1}^m k(t_r)\right]$$
$$\times\,\Upsilon(t,t_m)\,\mathcal{J}(t_m)\,\Upsilon(t_m,t_{m-1})\,\mathcal{J}(t_{m-1})\cdots\Upsilon(t_2,t_1)\,\mathcal{J}(t_1)\,\Upsilon(t_1,0), \quad (29)$$

where $\Upsilon(t,s)$, $t\geq s$, is defined by the equation

$$\frac{\mathrm{d}}{\mathrm{d}t}\Upsilon(t,s) = \tilde{\mathcal{L}}_t\,\Upsilon(t,s), \qquad \Upsilon(s,s) = \mathbb{1}. \quad (30)$$

By inserting (29) into (23) we obtain the following expression for the characteristic functional

$$\Phi[k] = P_t(0|\varrho) + \sum_{m=1}^\infty\int_0^t\mathrm{d}t_m\int_0^{t_m}\mathrm{d}t_{m-1}\cdots$$
$$\cdots\int_0^{t_2}\mathrm{d}t_1\,\exp\left[\mathrm{i}\sum_{r=1}^m k(t_r)\right]\,p_t(t_m,t_{m-1},\ldots,t_1|\varrho), \quad (31)$$

$$P_t(0|\varrho) := \mathrm{Tr}_{\mathcal{H}} \left\{ \Upsilon(t,0)\varrho \right\} , \tag{32}$$

$$p_t\big(t_m, t_{m-1}, \ldots, t_1|\varrho\big) := \mathrm{Tr}_{\mathcal{H}} \Big\{ \Upsilon(t,t_m)\, \mathcal{J}(t_m)\, \Upsilon(t_m,t_{m-1})\, \mathcal{J}(t_{m-1}) \cdots \cdots \Upsilon(t_2,t_1)\, \mathcal{J}(t_1)\, \Upsilon(t_1,0)\, \varrho \Big\} . \tag{33}$$

The meaning of the various quantities is apparent from the structure (31) of $\Phi_t[k]$ (recall that a characteristic functional is the Fourier transform of a probability distribution). The quantity (32) is the probability of having no count in the time interval $(0,t]$, when the initial state of the system is ϱ. The quantity (33) is the probability density of a count at time t_1, a count at time t_2, ..., and no other count in the interval $(0,t]$; these quantities are called exclusive probability densities and they determine all the probabilities for counts. For instance, the probability of exactly n counts in the interval $(0,t]$ is $P_t(n|\varrho) = \int_0^t \mathrm{d}t_n \int_0^{t_n} \mathrm{d}t_{n-1} \cdots \int_0^{t_2} \mathrm{d}t_1\, p_t\big(t_n, t_{n-1}, \ldots, t_1|\varrho\big)$. In principle the quantities (32) and (33) determine the whole statistics of the photons arriving at the counter and, therefore, the whole statistics of the output current (14).

Let us stress that the quantity $\Upsilon(t,s)$ is not a dynamics: it is not trace preserving. Instead, the reduced dynamics of system S is $\mathcal{U}(t,s)$, $t \geq s$, defined by the equation

$$\frac{\mathrm{d}}{\mathrm{d}t}\mathcal{U}(t,s) = \mathcal{L}_t \mathcal{U}(t,s) , \qquad \mathcal{U}(s,s) = 1\!\!1 . \tag{34}$$

Note that $\mathcal{G}_t(0)\varrho = \mathcal{U}(t,0)\varrho = \varrho(t)$ (cf. equations (24), (25), (9), (10)).

The kind of counting processes described here were introduced by Davies [22,23] (see also [24,25]). Essentially, he gave directly the prescriptions for computing the quantities (32)–(33), while characteristic operators and QSC have been introduced in this context in [26] [4,5] [16–18]. The results presented in this section constitute an alternative approach to Davies' counting processes; moreover, they show that the dissipative dynamics $\mathcal{U}(t,s)$, appearing in Davies' theory, comes naturally from a unitary one (U_t) for the total system (S + fields) and they relate the counting process with the notion of output fields discussed in the previous section.

3 Optical Heterodyne and Homodyne Detection

Also a different kind of measurement, the so called *heterodyne detection* [27–29], can be described by using QSC [16–18]. By inserting a beam splitter (a partially transparent mirror of transmissivity η, $0 < \eta < 1$) before the photoelectron counter, the output field from S is made to beat with an intense laser field (local oscillator). The laser field is described by a Bose field $a_0(t)$, independent from the fields $a_j(t)$, $j = 1, \ldots, d$, involved in the

The characteristic functional of the process is again given by (18). As in the previous section, we can eliminate the remaining fields and we obtain equations (22)–(24) and (10), while (25) becomes

$$\mathcal{K}_t(k)\,\varrho = -\tfrac{1}{2}\,|f_0(t)|^2\,k^2\,\varrho + \mathrm{i}k\left\{f_0(t)\varrho\big[R_1^\dagger + f_1^*(t)\big] + f_0^*(t)\big[R_1 + f_1(t)\big]\varrho\right\}. \tag{40}$$

Now $\mathcal{G}_t[k]$ corresponds to a "Gaussian" (or diffusive) continuous measurement, first introduced in [26] (see also [4–6]).

If also $f_1(t)$ is a known signal, we can subtract from the output current the term $\int_0^t \mathrm{d}s\, F(t-s)\,[f_0(s)f_1^*(s) + f_0^*(s)f_1(s)]$. With this final change, the expression of $\mathcal{K}_t(k)$ becomes

$$\mathcal{K}_t(k)\,\varrho = -\tfrac{1}{2}k^2\,|f_0(t)|^2\,\varrho + \mathrm{i}k\left[f_0(t)\,\varrho\,R_1^\dagger + f_0^*(t)R_1\,\varrho\right]. \tag{41}$$

Again equations (23)–(24), (10) and (41) contain in principle the whole statistics of our random variable $Z(t)$ and, by (36), of $I(t)$.

Let us now go back to the situation described in the previous section. The "true" direct detection corresponds to the case $f_1(t) = 0$. In the case $f_1(t) \neq 0$, the laser light either stimulates the system either reaches the counter. In this case, the detector "measures" the interference between the laser light in field 1 and the light re-emitted by the system in the same channel. This measurement configuration is called optical *homodyne detection.*

We want to see what happens when the laser light is very intense. The laser signal $f_1(t)$ is known and we can subtract from the output current the known deterministic term $\int_0^t \mathrm{d}s\, F(t-s)\,|f_1(t)|^2$. This amounts to multiply the characteristic operator $\mathcal{G}_t[k]$ by $\exp\left[-\mathrm{i}\int_0^t \mathrm{d}s\, k(s)\,|f_1(s)|^2\right]$. Then, we consider a very intense laser field $\big(f_1(t) \to \lambda f_1(t)\big)$ and rescale the output current by a factor $1/\lambda$. Correspondingly, we have to replace $k(s)$ by $k(s)/\lambda$ in the characteristic operator. Under these transformations and for λ large, the equation for $\mathcal{G}_t[k]$ is (24) with [18]

$$\begin{aligned}\mathcal{L}_t\varrho := {}& -\mathrm{i}\left[H + \mathrm{i}\textstyle\sum_{j=2}^d \left(f_j^*(t)R_j - f_j(t)R_j^\dagger\right), \varrho\right] + \\ & + \frac{1}{2}\sum_{j=1}^d \left(\left[R_j, \varrho R_j^\dagger\right] + \left[R_j\varrho, R_j^\dagger\right]\right) + \lambda\left[f_1^*(t)R_1 - f_1(t)R_1^\dagger, \varrho\right],\end{aligned} \tag{42}$$

$$\mathcal{K}_t(k)\,\varrho := -\tfrac{1}{2}k^2\,|f_1(t)|^2\,\varrho + \mathrm{i}k\left(f_1^*(t)R_1\,\varrho + f_1(t)\,\varrho\,R_1^\dagger\right). \tag{43}$$

interaction with S. The initial state of this new field is taken to be again a coherent vector. At the two input ports of the beam splitter they arrive the fields $a_1^{\mathrm{out}}(t)$ and $a_0(t)$. Let us call $b_1(t)$ and $b_2(t)$ the fields leaving the two output ports; they are given by

$$b_1(t) = \sqrt{\eta}\, a_1^{\mathrm{out}}(t) + \sqrt{1-\eta}\, a_0(t)\,, \qquad b_2(t) = \sqrt{\eta}\, a_0(t) - \sqrt{1-\eta}\, a_1^{\mathrm{out}}(t)\,. \tag{35}$$

The phases of all the fields can always be redefined in order to have no additional phase shifts in (35). The coefficients are such that CCR's hold among the new fields $b_1(t)$, $b_2(t)$, $b_1^\dagger(t)$, $b_2^\dagger(t)$.

Let us consider now the so called *balanced heterodyne detection* [30–32] [20]. Two identical photoelectron counters are used for detecting the photons in both the fields $b_1(t)$ and $b_2(t)$. The two output currents will be $I_j(t) = \int_0^t \mathrm{d}s\, F(t{-}s) b_j^\dagger(s) b_j(s)$, but we measure only their linear combination

$$I(t) = \sqrt{\tfrac{\eta}{1-\eta}}\, I_1(t) - \sqrt{\tfrac{1-\eta}{\eta}}\, I_2(t) \equiv \int_0^t \mathrm{d}s\, F(t-s) \frac{\mathrm{d}Z(s)}{\mathrm{d}s}\,. \tag{36}$$

The new observables $Z(t)$ are associated with the commuting operators

$$Z(t) = \int_0^t \mathrm{d}s\, \left[\sqrt{\tfrac{\eta}{1-\eta}}\, b_1^\dagger(s) b_1(s) - \sqrt{\tfrac{1-\eta}{\eta}}\, b_2^\dagger(s) b_2(s)\right] \tag{37}$$

and the characteristic operator is now $\overline{V}_t[k] = \exp\left\{\mathrm{i} \int_0^t k(s)\, \mathrm{d}Z(s)\right\}$ (cf. (16)).

As in the previous section, we can eliminate the degrees of freedom of the fields. We make this in two steps: first we eliminate only the field $a_0(t)$ and then the remaining fields. Let us denote by $V_t[k] \equiv \langle \overline{V}_t[k] \rangle_{\Gamma_0}$ the quantum expectation of the characteristic operator with respect to the initial state $\Psi(f_0)$ of $a_0(t)$; by the rules of QSC, one can compute the QSDE satisfied by this quantity. We do not write this equation in general, but we consider only the case of a very intense laser field $\big(f_0(t) \to \lambda f_0(t)$, λ large$\big)$. By scaling the output current $\big(I(t) \to I(t)/\lambda$, $k(t) \to k(t)/\lambda\big)$, in the limit $\lambda \to +\infty$ we obtain the QSDE

$$\mathrm{d}V_t[k] = V_t[k] \left\{\mathrm{i}k(t) \left[f_0(t)\, \mathrm{d}A_1^{\mathrm{out}}(t)^\dagger + f_0^*(t)\, \mathrm{d}A_1^{\mathrm{out}}(t)\right] - \tfrac{1}{2} k^2(t)\, |f_0(t)|^2\, \mathrm{d}t\right\}. \tag{38}$$

With the initial condition $V_0[k] = \mathbb{1}$, the solution of this equation is

$$V_t[k] = \exp\left\{\mathrm{i} \int_0^t k(s) \left[f_0(t)\, \mathrm{d}A_1^{\mathrm{out}}(s)^\dagger + f_0^*(s)\, \mathrm{d}A_1^{\mathrm{out}}(s)\right]\right\}. \tag{39}$$

This equation shows that heterodyne detection corresponds to a measurement of the compatible observables $f_0(t)\, a_1^{\mathrm{out}}(t)^\dagger + f_0^*(t)\, a_1^{\mathrm{out}}(t)$, $t \geq 0$. Typically, $f_0(t) \propto \exp(-\mathrm{i}\Omega_1 t)$ so that the measured observable is a quadrature of the field.

References

1. R.L. Hudson, K.R. Parthasarathy: Commun. Math. Phys. **93** 301 (1984)
2. R.L. Hudson, K.R. Parthasarathy: Acta Appl. Math. **2** 353 (1984)
3. R.L. Hudson, J.M. Lindsay: J. Funct. Anal. **61** 202 (1985)
4. A. Barchielli, G. Lupieri: J. Math. Phys. **26** 2222 (1985)
5. A. Barchielli, G. Lupieri: in *Quantum Probability and Applications II*, Lecture Notes in Mathematics, Vol. 1136, ed. by L. Accardi, W. von Waldenfels (Berlin, Springer, 1985) pp. 57–66
6. A. Barchielli: Phys. Rev. A **34** 1642 (1986)
7. C.W. Gardiner, M.J. Collet: Phys. Rev. A **31** 3761 (1985)
8. C.W. Gardiner: Phys. Rev. Lett. **56** 1917 (1986)
9. P. Alsing, G.J. Milburn, D.F. Walls: Phys. Rev. A **37** 2970 (1988)
10. M.A. Marte, H. Ritsch, D.F. Walls: Phys. Rev. A **38** 3577 (1988)
11. A. Barchielli: J. Phys. A: Math. Gen. **20** 6341 (1987)
12. A. Barchielli: in *Quantum Probability and Applications III*, Lecture Notes in Mathematics, Vol. 1303, ed. by L. Accardi, W. von Waldenfels (Berlin, Springer, 1988) pp. 37–51
13. P. Robinson, H. Maassen: *Quantum stochastic calculus and the dynamical Stark effect.* Report 9025, May 1990, Department of Mathematics, Catholic University, Toernooiveld, 6525 ED Nijmegen, The Netherlands
14. M. Lax: Phys. Rev. **145** 110 (1966)
15. G.J. Milburn: Phys. Rev. A **36** 5271 (1987)
16. A. Barchielli: *Quantum stochastic calculus: Applications to quantum optics.* To appear in *Stochastic Processes – Geometry and Physics*, ed. by S. Albeverio *et al.* (World Scientific, Singapore)
17. A. Barchielli: *Applications of quantum stochastic calculus to quantum optics.* To appear in *Quantum Probability and Applications VI*, ed. by L. Accardi, W. von Waldenfels (Kluwer)
18. A. Barchielli: *Direct and heterodyne detection and other applications of quantum stochastic calculus to quantum optics.* To appear in Quantum Opt.
19. L. Accardi, A. Frigerio, Lu Yun Gang: in *Quantum Probability and Applications IV*, Lecture Notes in Mathematics, Vol. 1396, ed. by L. Accardi, W. von Waldenfels (Berlin, Springer, 1989) pp. 20–58
20. M.J. Collet, R. Loudon, C.W. Gardiner: J. Mod. Optics **34** 881 (1987)
21. P.L. Kelley, W.H. Kleiner: Phys. Rev. A **30** 844 (1964)
22. E.B. Davies: Commun. Math. Phys. **15** 277 (1969); **19** 83 (1970); **22** 51 (1971)
23. E.B. Davies: *Quantum Theory of Open Systems* (London, Academic, 1976)
24. M.D. Srinivas, E.B. Davies: Opt. Acta **28** 981 (1981); Opt. Acta **29** 235 (1982)
25. A.S. Holevo: Sov. Math. **26** 1 (1982) [Izv. Vuz. Mat. **26** 3 (1982)]
26. A. Barchielli, L. Lanz, G.M. Prosperi: Found. Phys. **13** 779 (1983)
27. J.H. Shapiro, H.P. Yuen, J.A. Machado Mata: IEEE Trans. Inf. Theory **IT-25** 179 (1979)
28. H.P. Yuen, J.H. Shapiro: IEEE Trans. Inf. Theory **IT-26** 78 (1980)
29. J.H. Shapiro, S.S. Wagner: IEEE J. Quantum Electr. **QE-20** 803 (1984)

Session IV

– Quantum Detection and Photon Channels –

Chairman : M. Nakagawa

Calculating Photocount Distributions by Saddlepoint Methods

Carl W. Helstrom
University of California, San Diego
La Jolla, California, 92093, U.S.A.

Abstract

Probability generating functions of the number of photoelectrons emitted by a detector during a given interval can often be relatively easily calculated from a physical model of the emissive processes, but inverting them analytically to obtain the distributions of the number of so-called "photocounts" is usually complicated or impossible. Techniques for computing those distributions can be based on numerical evaluation of the inversion integral along a suitably chosen contour passing through a saddlepoint of the integrand and lying as close as possible to a path of steepest descent of that integrand. The method can be extended to include the effects of additive Gaussian noise, intersymbol interference, and postdetector filtering.

Let $p(m) = \Pr(\mathbf{n} = m)$ be the probability that some photoelectric device emits $\mathbf{n} = m$ electrons during an interval $(0, T)$, $0 \le m < \infty$. The probability generating function for this distribution of probabilities is defined as

$$h(z) = \sum_{m=0}^{\infty} p(m) z^m . \qquad (1)$$

It has no singularities within or on the unit circle, for

$$|h(z)| \le \sum_{m=0}^{\infty} p(m) |z|^m \le \sum_{m=0}^{\infty} p(m) = 1, \quad |z| \le 1 . \qquad (2)$$

The probabilities can be recovered from $h(z)$ by the inversion integral

$$p(m) = \int_C h(z) z^{-m-1} \frac{dz}{2\pi i} , \qquad (3)$$

where C is a closed curve containing the origin, but no singularities of $h(z)$.

Lecture Notes in Physics, Vol. 378
C. Bendjaballah, O. Hirota, S. Reynaud (Eds.),
Quantum Aspects of Optical Communications

The performance of a communication system utilizing a photodetector is characterized by probabilities of error that involve the cumulative distribution of the number $\mathbf{n}$ of output photoelectrons. We define this as

$$q_-(k) = \Pr(\mathbf{n} < k) = \sum_{m=0}^{k-1} p(m), \tag{4}$$

and from (3) it can be expressed as

$$q_-(k) = \int_C h(z) \sum_{m=0}^{k-1} z^{-(m+1)} \frac{dz}{2\pi i} = \int_C h(z) \frac{1-z^{-k}}{z-1} \frac{dz}{2\pi i}. \tag{5}$$

With C_- a circle centered at the origin and lying inside the unit circle, and with C_+ one lying outside the unit circle, but enclosing no singularities of the probability generating function $h(z)$,

$$\int_C \frac{h(z)}{z-1} \frac{dz}{2\pi i} = \begin{cases} 0, & C = C_-, \\ 1, & C = C_+, \end{cases} \tag{6}$$

and we find for the cumulative distribution

$$q_-(k) = \int_{C_-} \frac{h(z) z^{-k}}{1-z} \frac{dz}{2\pi i}, \tag{7}$$

and for its complement

$$q_+(k) = \Pr(\mathbf{n} \geq k) = 1 - q_-(k) = \int_{C_+} \frac{h(z) z^{-k}}{z-1} \frac{dz}{2\pi i}. \tag{8}$$

For many photodetectors one can derive the probability generating function $h(z)$ from the physics of the device, but inverting it to calculate the probabilities $p(m)$ and then summing to determine the "tail" probabilities $q_-(k)$ and $q_+(k)$ may be difficult or tedious, particularly when the numbers k are large and lie far from the expected value $E(\mathbf{n})$, as is the case when the probability of error is small. One might, for instance, substitute $z = \exp(i\theta)$ into (3) and use the Fast Fourier Transform (FFT) algorithm to evaluate the probabilities $p(m)$, finally summing these probabilities to calculate $q_-(k)$ or $q_+(k)$; but with k in either tail of the distribution, one would need to sample the Fourier integrand at a great many points, and the FFT would require large storage and a long computation time. When photomultipliers are being treated, the numbers k and $E(\mathbf{n})$ may lie in the thousands. Even when closed-form expressions for the probabilities are known, the summations may then be prone to overflow and round-off errors. Some more efficient method for calculating the cumulative distribution and its complement must be sought.

The integrand of (7) as a function of real values of z in $0 < \operatorname{Re} z < 1$ can be shown to be convex $\cup$ and hence to possess a single minimum at $z = x_0^-$ in that region. Likewise, with

x_1 the first singularity of $h(z)$ in $1 < \mathrm{Re}\, z < \infty$, the integrand of (8) is convex $\cup$ in $1 < \mathrm{Re}\, z < x_1$ and possesses a single minimum x_0^+ in that region. An analytic function possessing a minimum along a certain line in the complex plane has a maximum there in a direction perpendicular to that line, and such a point x_0^- or x_0^+ is therefore called a "saddlepoint" or "col" of the integrand.

We write each integrand as

$$\frac{h(z)\,z^{-k}}{\pm(z-1)} = e^{\Phi(z)}, \ \Phi(z) = \ln h(z) - k \ln z - \ln[(\pm(z-1)] , \tag{9}$$

and we call $\Phi(z)$ the "phase" of the integrand. At the saddlepoint,

$$\Phi'(z) = \frac{d\Phi}{dz} = 0, \quad z = x_0^-, x_0^+ . \tag{10}$$

In the neighborhood of the saddlepoint,

$$\Phi(z) \approx \Phi(z_0) + \tfrac{1}{2}\Phi''(z_0)(z - z_0)^2 + 0(|z - z_0|^3), \ z_0 = x_0^-, x_0^+ . \tag{11}$$

If we take the contour C_- to pass through the inner saddlepoint x_0^-, and C_+ to pass through the outer one x_0^+, the integrands of (7) and (8) approximate Gaussian functions in the neighborhood of their saddlepoints, and by extending the range of integration to $\pm\infty$ we obtain the "saddlepoint approximations"

$$q_-(k) \approx [2\pi\Phi''(x_0^-)]^{-\frac{1}{2}} \exp \Phi(x_0^-), \quad 0 < x_0^- < 1, \tag{12}$$

$$q_+(k) \approx [2\pi\Phi''(x_0^+)]^{-\frac{1}{2}} \exp \Phi(x_0^+), \quad 1 < x_0^+ < x_1. \tag{13}$$

This type of approximation has been known for a long time. In an early application to Hankel functions [1] Debye attributed it to Riemann [2]. It is much used in studying wave phenomena, as in the propagation of radio waves through the ionosphere [3]. Daniels [4] first applied it to evaluating both probabilities such as $p(m)$ and probability density functions of continuous random variables, but not to approximating their cumulative distributions. The first application to calculating the latter seems to have been by Mazo and Salz [5], who were studying the distributions of quadratic forms in Gaussian random variables. By taking further terms of the expansion in (11) one can derive an asymptotic series involving derivatives of the phase of higher order, of which (12) and (13) are the leading terms [4], [6]. These approximations are most accurate in the tails of the distribution, far from the expected value of the random variable. In the neighborhood of the expected value one often uses Edgeworth's series [7]. A uniform asymptotic expansion that combines the advantages of both was developed by S. O. Rice [8], [9]. Its coefficients involve combinations of derivatives of the moment generating function of the random variable at the saddlepoint of the Laplace inversion integral for its probability density function. It could be applied to distributions of discrete random variables by formulating it instead in terms of probability generating functions.

For complicated probability generating functions, however, calculating derivatives of high order for use in such asymptotic expansions may be analytically intractable beyond a certain point, and one may have to resort to numerical differentiation with its attendant imprecision. Rather than going to all that trouble, one may as well integrate (7) and (8) numerically. In order to minimize the number of samples required by the numerical quadrature algorithm, one would like the amplitude of the integrand to drop off to zero as rapidly as possible, and it does so when one integrates along the so-called "path of steepest descent" of the integrand, which passes through the saddlepoint x_0^- or x_0^+ [10]. Along that path the phase $\Phi(z)$ remains real, and $\operatorname{Im}\Phi(z) \equiv 0$. One would deform the original contour C_- or C_+ into such a path and integrate along that instead.

Computing the path of steepest descent would, however, much protract the numerical integration, and one seeks a simpler path that approximates it. In [11] and [12], which deal with distributions of continuous random variables, a straight line inclined at a suitable angle to the y-axis was chosen. Often the path of integration can more efficiently be taken as a parabola passing through the saddlepoint and having at that point the same curvature $c = \Phi'''(x_0)/[3\Phi''(x_0)]$, $x_0 = x_0^-$, x_0^+, as the path of steepest descent [13]. Along this "osculatory parabola" $x = x_0 + \frac{1}{2} c y^2$, and the integrals for the tail probabilities can be written

$$q_\pm(k) = \frac{1}{\pi} \operatorname{Re} \int_0^\infty e^{\Phi(z)}(1 - icy)\, dy, \quad z = x_0 + \tfrac{1}{2} c y^2 + iy \,. \tag{14}$$

One must truncate the numerical integration at some finite point z' on the osculatory parabola, introducing a "truncation error" that is in general the smaller, the farther z' lies from the origin. We shall derive an upper bound to this truncation error ε for the class of probability generating functions having the form

$$h(z) = \prod_{m=1}^{\infty} [1 - n_m(z-1)]^{-1} \exp\left[\sum_{m=1}^{\infty} \frac{s_m(z-1)}{1 - n_m{}'(z-1)}\right], \quad n_m \ge 0, \; n_m{}' \ge 0, \; s_m \ge 0 \,. \tag{15}$$

A probability generating function of this kind arises, for example, when the light falling on a photoelectrically emissive surface is a combination of a coherent component $\operatorname{Re} S(t) \exp i\Omega t$ and incoherent background light having a narrowband spectral density $W(\omega - \Omega)$ in the neighborhood of the angular carrier frequency Ω [14], [15]. The coherent component might be the output of an ideal pulsed laser; it ejects an expected number

$$\bar{n}_s = \tfrac{1}{2}\eta \int_0^T |S(t)|^2 \, dt \tag{16}$$

of photoelectrons during an interval $(0, T)$, where $\eta = \eta'/\hbar\Omega$; η' is the quantum efficiency of the detector and $\hbar$ is Planck's constant. The temporal coherence function of the light is proportional to Re $\phi(\tau) \exp i\Omega\tau$ with

$$\phi(\tau) = \int_{-\infty}^{\infty} W(\omega) e^{i\omega\tau} \frac{d\omega}{2\pi} . \tag{17}$$

We normalize it and the spectral density so that $\phi(0) = 1$. It defines an orthonormal set of eigenfunctions $f_n(t)$ with eigenvalues g_n through the integral equation

$$g_n f_n(t) = \frac{1}{T} \int_0^T \phi(t-u) f_n(u)\, du\,, \quad \sum_{n=1}^{\infty} g_n = 1\,. \tag{18}$$

Then in (15) $n_m = n_m{}' = n_0 g_m$ and $s_m = \frac{1}{2}\eta\, |S_m|^2$, where the

$$S_m = \int_0^T f_m{}^*(t)\, S(t)\, dt \tag{19}$$

are the Fourier coefficients of the coherent signal, and n_0 is the expected number of electrons ejected by the incoherent component of the incident light.

The Poisson and Laguerre distributions also possess probability generating functions of the form (15). For the Poisson $n_m \equiv n_m{}' \equiv 0$, $s_1 = \mu$, $s_m \equiv 0$, $m > 1$. For the generalized Laguerre distribution [16] some integral number M of the n_m's are equal and the rest are zero. Still another probability generating function of this class represents the number of electrons at the output of an avalanche photodiode in which no holes are generated internally and in which the number of primary electrons has a Poisson distribution with expected value μ [17],

$$h(z) = \exp\left[\frac{S(z-1)}{1-n(z-1)}\right], \tag{20}$$

where $S = G\mu$ with G the gain, and $n = G-1$; $n_m \equiv 0$, $n_1{}' = n$, $n_m{}' \equiv 0$, $m > 1$.

In order to estimate the error introduced by truncating the numerical integration of (14) at a finite point z', the originally circular contours C_- or C_+ of integration in (7) and (8) are deformed into closed curves composed of the parabola P in (14), a straight vertical line L from $z' = x' + iy'$ to $x' + i\infty$, an arc A of infinite radius passing counterclockwise from $x' + i\infty$ to $x' - i\infty$, and a straight vertical line from $x' - i\infty$ to $z'^* = x' - iy'$. This can be done without crossing any singularities of the integrand. The integral along the arc A can easily be shown to vanish, and the absolute value of the truncation error is

$$\varepsilon = \frac{1}{\pi} \left| \int_{x'+iy'}^{x'+i\infty} \frac{h(z) z^{-k}}{z-1}\, dz \right| \le \frac{1}{\pi} \int_{y'}^{\infty} \frac{|h(z)|\,|z|^{-k}}{|z-1|}\, dy, \quad z = x' + iy \in L\,,$$

with

$$|h(z)| = \prod_m |1-n_m(z-1)|^{-1} \exp\left[\sum_{m=1}^{\infty} \mathrm{Re}\frac{s_m(z-1)}{1-n_m'(z-1)}\right].$$

On L, $|1-n_m(z-1)|^{-1} \le |1-n_m(z'-1)|^{-1}$, $|z-1|^{-1} \le |z'-1|^{-1}$, and

$$\exp\left[\mathrm{Re}\frac{s_m(z-1)}{1-n_m'(z-1)}\right] \le \delta_m = \begin{cases} \exp \mathrm{Re}\left[\dfrac{s_m(z'-1)}{1-n_m'(z'-1)}\right], & x' \le 1+n_m'^{-1} \\ \exp(-s_m/n_m'), & x' > 1+n_m'^{-1} \end{cases}$$

Then

$$\varepsilon \le \frac{1}{\pi|z'-1|} \prod_m \{|1-n_m(z'-1)|^{-1}\delta_m\} \int_{y'}^{\infty} |z|^{-k}\,dy,$$

and by the same technique as in Appendix A of [13] we find that the truncation error is bounded by

$$\varepsilon \le \frac{|z'|^2}{(k-2)y'} |I(z')| \prod_{m=1}^{M} \exp\left[-\mathrm{Re}\frac{s_m/n_m'}{1-n_m'(z'-1)}\right], \quad k>2, \tag{21}$$

$$1 < 1+n_m'^{-1} < x', \quad 1 \le m \le M,$$

where

$$I(z) = \frac{z^{-k}h(z)}{\pi[\pm(z-1)]} = \frac{1}{\pi} e^{\Phi(z)}.$$

For $k=2$ the factor before $|I(z')|$ in (21) can be replaced by $|z'|^2/y'$. For $k=1$, $q_-(k) = p(0) = h(0)$, and no contour integration is required.

When all $n_m' \equiv 0$, as for the Poisson distribution, the contour of integration can instead be completed by a circular arc of radius $|z'|$ running from z' to z'^*, and one easily finds the simpler bound

$$\varepsilon \le 2\pi|z'|\,|I(z')| \tag{22}$$

on the truncation error. These inequalities provide simply programmed rules for stopping the numerical integration in order to attain a prescribed absolute or relative truncation error. They hold whether the osculatory parabola opens to the left or to the right, and indeed, the curve P along which one integrates need not be a parabola, but needs only to recede continuously from the origin and to lie symmetrically about the Re z-axis.

For reasons discussed by Schwartz [18] and Rice [19] the most suitable quadrature algorithm for evaluating the integrals in (7) and (8) is the trapezoidal rule. Roughly speaking, halving the step-size and thereby doubling the number of samples of the integrand doubles the number of reliable significant figures in the result. Numerical examples are given in [14] and [20]. The number of samples required for a prescribed accuracy tends to increase slightly as the number k approaches the expected value $E(\mathbf{n})$. With proportionally increasing values of both k and $E(\mathbf{n})$, the required number of samples tends to decrease, but slowly.

Applications of this method to calculating the distribution of the number of electrons at the output of an avalanche photodiode and at the output of a single-stage multiplier are to be found in [21] and [22], respectively, and photomultipliers with many stages are treated in [20].

In optical receivers the decision about the presence or absence of an incident light pulse is generally based not on a number of photocounts, but on the electric current at the output of the photodetector, amplified and sampled at an appropriate time. The amplifier adds Gaussian thermal noise, and in a communication system the output may contain components due to intersymbol interference. The voltage $V(t)$ at the output of the amplifier can be represented as the sum of shot noise and a thermal component $n(t)$,

$$V(t) = w(t) + n(t). \tag{23}$$

At any time t, the term $n(t)$ is a Gaussian random variable with zero expected value and variance σ^2.

In an on-off binary optical communication receiver the output voltage $V(t_s)$ at a sampling time t_s is compared with a decision level V, and if $V(t_s) > V$, the receiver decides that the current information bit is a '1', otherwise that it is a '0'. In order to determine the probability of error of such a communication system, we need the cumulative distribution $q_-(V) = \Pr(V(t_s) \le V)$ of the output voltage, or its complement $q_+(V) = \Pr(V(t_s) > V)$. In terms of the moment-generating function $H(s) = E(\exp sV)$ of the voltage $V = V(t_s)$, we can write these probabilities as Laplace inversion integrals,

$$q_-(V) = \int_{C_-} \frac{H(s)}{(-s)}\, e^{-sV} \frac{ds}{2\pi i}, \quad q_+(V) = \int_{C_+} \frac{H(s)}{s}\, e^{-sV} \frac{ds}{2\pi i}, \tag{24}$$

where C_- and C_+ are straight vertical contours lying in the convergence strip of the Laplace transform

$$H(s) = \int_{-\infty}^{\infty} p(V)\, e^{sV}\, dV \tag{25}$$

of the probability density function $p(V)$ of the output voltage; C_- lies to the left, C_+ to the right of the origin of the s-plane. The equations (24) are obtained by integrating the Laplace inversion integral for $p(V)$. When the output voltage has the form in (23), the moment-generating function is

$$H(s) = H_0(s) \exp(\tfrac{1}{2}\sigma^2 s^2), \tag{26}$$

where $H_0(s) = E(\exp sw)$ is the moment-generating function of the shot-noise component $w(t_s)$.

The integrands of (24), like those of (7) and (8), possess saddlepoints lying on the real s-axis within the convergence strip of the transform (25). Denoting those integrands by $\exp \Phi(s)$, we can calculate saddlepoint approximations of the same form as in (12) and (13). Like those, they are most accurate in the tails of the distribution of V, far from the expected value $E(V)$. Precise values of the probabilities $q_-(V)$ and $q_+(V)$ can be computed by numerical quadrature of (24), and this is again most efficient when the paths C_- and C_+ pass through the saddlepoints of their integrands and when these paths are deformed into contours approximating the path of steepest descent of each integrand.

When the numerical integration is carried along a straight vertical contour, $s = s_0 + iy$, and stopped at a point $s' = s_0 + iy'$, the truncation error ε is bounded by

$$\varepsilon = \frac{1}{\pi}\left|\int_{y'}^{\infty} s^{-1} H_0(s) \exp(\tfrac{1}{2}\sigma^2 s^2 - sV)\, dy\right|$$

$$\leq \pi^{-1} |s_0|^{-1} |H_0(s_0)| \exp(\tfrac{1}{2}\sigma^2 s_0^2 - s_0 V) \int_{y'}^{\infty} \exp(-\tfrac{1}{2}\sigma^2 y^2)\, dy$$

$$\leq (\pi\sigma^2 y')^{-1} \exp \Phi(s_0) \exp(-\tfrac{1}{2}\sigma^2 y'^2),$$

where $\exp \Phi(s)$ is the integrand in (24). The first inequality follows from the obvious facts that $|H_0(s_0 + iy)| \leq H_0(s_0)$ and that $|s_0 + iy|^{-1} \leq |s_0|^{-1}$.

In [23] this method was applied to a receiver in which the output voltage $V(t_s)$ was proportional to the sum of Gaussian thermal noise and the number of electrons at the output of an avalanche photodiode, integrated over a fixed interval. The contour of integration was the path of steepest descent itself, computed step by step along with the numerical integration. A recent paper [24] extended this work to include the effects of intersymbol interference, the incident light intensity consisting of Gaussian pulses. The impulse response of the post-detector amplifier was taken as either a Gaussian or an exponential pulse. A suitable approximation to the path of steepest descent was determined by trial and error in advance.

References

[1] Debye, P., "Näherungsformeln für die Zylinderfunktionen für grosse Werte des Arguments und unbeschränkt veränderliche Werte des Index," Math. Annalen, vol. 67, 535-558 (1909).

[2] Riemann, B., "Sullo svolgimento del quoziente di due serie ipergeometriche in frazione continua infinita," (Oct. 1863). In *Collected Works of Bernhard Riemann,* H. Weber, Ed., 424-430, New York: Dover Publications Inc. (1953).

[3] Budden, K. G., *The Propagation of Radio Waves,* Cambridge: Cambridge University Press (1985).

[4] Daniels, H. G., "Saddle point approximations in statistics," Ann. Math. Stat., vol. 25, 631-650 (1954).

[5] Mazo, J. E., Salz, J., "Probability of error for quadratic detectors," Bell System Tech. J., vol. 44, 2165-2186 (Nov. 1965).

[6] Helstrom, C. W., "Approximate evaluation of detection probabilities in radar and optical communications," IEEE Trans. on Aerospace & Electronic Systems, vol. AES-14, 630-640 (July, 1978).

[7] Fry, Thornton, C., *Probability and Its Engineering Uses.* 2nd ed., 257-264, Princeton, N.J.: D. Van Nostrand Company (1965).

[8] Rice, S. O., "Uniform asymptotic expansions for saddle point integrals -- Application to a probability distribution occurring in noise theory," Bell Sys. Tech. J., vol. 47, 1971-2013 (Nov. 1968).

[9] Lugannani, R., Rice, S. O., "Saddlepoint approximation for the distribution of the sum of independent random variables," Adv. Appl. Prob., vol. 12, 475-490 (1980).

[10] Carrier, G. F., Krook, M., Pearson, C. E., *Functions of a Complex Variable,* 257-264, New York: McGraw-Hill Book Co. (1966).

[11] Rice, S. O., "Distribution of quadratic forms in normal random variables -- Evaluation by numerical integration," SIAM J. Sci. Stat. Comput., vol. 1, 438-448 (Dec. 1980).

[12] Lugannani, R., Rice, S. O., "Distribution of the ratio of quadratic forms in normal variables -- Numerical methods," SIAM J. Sci. Stat. Comput., vol. 5, 476-488 (June, 1984).

[13] Helstrom, C. W., Ritcey, J. A., "Evaluating radar detection probabilities by steepest descent integration," IEEE Trans. on Aerospace & Electronic Systems, vol. AES-20, 624-634 (Sept. 1984).

[14] Helstrom, C. W., "Computation of photoelectron counting distributions by numerical contour integration," J. Opt. Soc. Am., vol. A2, 674-682 (May, 1985).

[15] Helstrom, C. W., "Computation of photoelectron counting distributions by numerical contour integration. II. Light with rational spectral density," J. Opt. Soc. Am., vol. A4, 1245-1255 (July, 1987).

[16] Gagliardi, R. M., Karp, S., *Optical Communications,* 67-70, New York: J. Wiley & Sons (1976).

[17] Personick, S. D., "New results on avalanche multiplication statistics with applications to optical detection," Bell Sys. Tech. J., vol. 50, 167-189 (Jan. 1971).

[18] Schwartz, C., "Numerical integration of analytic functions," J. Computational Physics, vol. 4, 19-29, 1969.

[19] Rice, S. O., "Efficient evaluation of integrals of analytic functions by the trapezoidal rule," Bell System Tech. J., vol. 52, 707-722 (May-June, 1973).

[20] Helstrom, C. W., "Output distributions of electrons in a photomultiplier," J. Appl. Phys., vol. 55, 2786-2792 (Apr. 1, 1984).

[21] Helstrom, C. W., "Computation of output electron distributions in avalanche photodiodes," IEEE Trans. on Electron Devices, vol. ED-31, 955-958 (July, 1984).

[22] Helstrom, C. W., Rice, S. O., "Computation of counting distributions arising from a single-stage multiplicative process," J. Computational Physics, vol. 54, 289-324 (May, 1984).

[23] Helstrom, C. W., "Computing the performance of optical receivers with avalanche diode detectors," IEEE Trans. on Communications, vol. COM-36, 61-66 (Jan. 1988).

[24] Helstrom, C. W., Ho, C.-L., "Analysis of avalanche diode receivers by saddlepoint integration," submitted to IEEE Trans. on Communications.

Rigorous Derivation of Error Probability in Coherent Optical Communication

Masanori OHYA and Hiroki SUYARI

Department of Information Sciences
Science University of Tokyo
Noda City, Chiba 278, Japan

Abstract

The optical communication theory has been developed in the fields of both mathematics and engineering, and it has become very important recently. In this paper we mathematically treat quantum control communication processes from information theoretical points of view, and we derive error probabilities for a concrete quantum channel in each digital modulation and detection. In paticular, several error probabilities for a coherent state and a squeezed state are rigorously calculated. It is shown that a proper squeezed state provides us a lower error probability, namely, better efficiency for optical communication.

§1. Introduction

The quantum information theory on optical communication has provided new ideas in the fields of quantum entropy theory [1,2] and quantum control theory [3]. It is recently considered that the squeezed state [4,5,6] may be useful for communication of information. Especially we can have many new treatments for the applications of a squeezed state to the optical communication system [7,8]. In this paper we rigorously derive some probabilities and discuss the efficiency of communication processes by them.

It is known from Shannon's work [9] that the concept of channel is essential and necessary to mathematically construct the communication theory. Therefore in this paper we briefly review the channel in quantum systems and we discuss the new derivation for the attenuation process. Under the above preparations, we apply the quantum mechanical channel to the rigorous derivation of the error probability in each digital modulation. Moreover we calculate the error probabilities for BPSK-homodyne detection and we compare them for each input state.

At first, in Sec. 2, we review the quantum mechanical channel [1,2] and show the new expression for the attenuation process [10]. In Sec. 3, we derive the error probability for each modulation and detection by using the above channel and the P.O.V.M. (Positive Operator Valued Measure) [11]. Especially, we compute each error probability for a coherent state and a squeezed state taken as an input state. Here we show the derivation of error probability in only case of coherent detection. In Sec. 4, we discuss the results obtained by computing the error probabilities, and we consider the efficiency for each modulation and detection.

Lecture Notes in Physics, Vol. 378
C. Bendjaballah, O. Hirota, S. Reynaud (Eds.),
Quantum Aspects of Optical Communications

§2. Quantum Mechanical Channel

2.1 A Construction of Quantum Mechanical Channel

When an input state is transmitted to a output system it will be influenced by noise and loss. We show the quantum communication channel by taking account of the effects of noise and loss existing in the course of the information transmission.

Let $\mathcal{H}_1$, $\mathcal{H}_2$, $\mathcal{K}_1$, $\mathcal{K}_2$ be the Hilbert spaces describing an input space, an output space, a noise space and a loss space, respectively. We denote the set of all states (density operators) on the Hilbert spaces $\mathcal{H}_k$ and $\mathcal{K}_k$ by $\mathfrak{S}(\mathcal{H}_k)$ and $\mathfrak{S}(\mathcal{K}_k)$, respectively. A quantum mechanical channel Λ^* is a mapping from $\mathfrak{S}(\mathcal{H}_1)$ to $\mathfrak{S}(\mathcal{H}_2)$ such that the dual map Λ of Λ^* is completely positive from the set $B(\mathcal{H}_2)$ of all bound linear operators on $\mathcal{H}_2$ to $B(\mathcal{H}_1)$, that is, $\sum_{i,j} A_i^* \Lambda(B_i^* B_j) A_j \geqq 0$ for any $A_i \in B(\mathcal{H}_1)$ and $B_j \in B(\mathcal{H}_2)$.

A quantum mechanical channel Λ^* is given by [1,2]

$$\Lambda^*(\rho) \equiv \mathrm{tr}_{\mathcal{K}_2} \Pi^*(\rho \otimes \sigma) \tag{2.1}$$

for any $\rho \in \mathfrak{S}(\mathcal{H}_1)$, where σ is a state described by a noise and Π^* is a channel mapping from $\mathfrak{S}(\mathcal{H}_1 \otimes \mathcal{K}_1)$ to $\mathfrak{S}(\mathcal{H}_2 \otimes \mathcal{K}_2)$. Moreover, $\mathrm{tr}_{\mathcal{K}_2}$ is the partial trace : $< \Phi_1, \mathrm{tr}_{\mathcal{K}_2} Q \Phi_2 > \equiv \sum_n < \Phi_1 \otimes \Psi_n, Q\Phi_2 \otimes \Psi_n >$ for any $Q \in B(\mathcal{H}_2 \otimes \mathcal{K}_2)$, any $\Phi_1, \Phi_2 \in \mathcal{H}_2$, and any $CONS\{\Psi_n\}$ of $\mathcal{K}_2$.

In this paper, we take the noise state σ the vaccum state in the space $\mathcal{K}_1$, namely, we only consider the zero point fluctuation as the noise disturbing the system.

2.2 Attenuation Process

In any communication processes the effect of loss is inevitable. Therefore we regard the quantum mechanical channel Λ^* as an attenuation process. This attenuation process is shown [10] to be expressed by an operator V as follows:

$$\Lambda^*(\rho) \equiv \mathrm{tr}_{\mathcal{K}_2} V(\rho \otimes \sigma) V^* \tag{2.2}$$

with

$$V \mid \theta > = \mid \alpha\theta > \otimes \mid \beta\theta > \tag{2.3}$$

where $\mid \theta >$ is a coherent vector [12] and we take $\alpha \geqq 0$, $\beta \geqq 0$ with $\alpha^2 + \beta^2 = 1$ without loss of generality.

§3. Derivation of Error Probability for a Quantum Channel

Positive operator valued measure is a useful tool to describe the quantum measurement process. And we apply the quantum attenuation channel and each P.O.V.M. expressing to the derivation of error probability for a coherent input state and a squeezed input state.

Here we assume the signals "0" and "1" are transmitted with equal probability 1/2. Namely, in general, an error probability P_e for digital modulation is given by

$$P_e = \frac{P_{e0} + P_{e1}}{2} \tag{3.1}$$

where P_{e0} and P_{e1} are the error probabilities of the signal "0" and the signal "1", respectively. And the indexes "$_0$" and "$_1$" represent each signal "0" and "1", respectively.

3.1 Error Probability for Direct Detection

Direct detection is the measurement of the photons in a transmitted state, so that the P.O.V.M. for direct detection is given by

$$E_{DD}(n) = \mid n >< n \mid \tag{3.2}$$

where $\mid n >$ is the n-th number photon vector in $\mathcal{H}_2$.

Especially in digital modulation - direct detection we usually encode the information by OOK. Namely, ρ_0 is a vaccum state and ρ_1 is another state in the input space $\mathcal{H}_1$. Therefore we have only to consider the case that the signal "1" is recognized as the signal "0" by a certain mistake because the noise state in $\mathcal{K}_1$ is assumed to be vaccum. (That's, we consider only Z-type channel).

In this case the error probability of each signal is given by

$$\begin{aligned} P_{e0}^{OOK-DD} &= 0 \\ P_{e1}^{OOK-DD} &= \mathrm{tr}_{\mathcal{H}_2}\Lambda^* \rho_1 \, E_{DD}(0) \end{aligned}$$

Therefore, from Eq.(3.1), we can compute the error probability P_e^{OOK-DD} by

$$P_e^{OOK-DD} = \frac{1}{2}\,\mathrm{tr}_{\mathcal{H}_2}\Lambda^* \rho_1 \mid 0 >< 0 \mid \tag{3.3}$$

3.1.1 In case that the input state is a coherent state

By substituting a coherent state $\rho_1 = \mid \theta >< \theta \mid$ in Eq.(3.3), the error probability $P_{e(CO)}^{OOK-DD}$ is given by

$$\begin{aligned} P_{e(CO)}^{OOK-DD} &= \frac{1}{2}\,\mathrm{tr}_{\mathcal{H}_2}(\mathrm{tr}_{\mathcal{K}_2} V \mid \theta >< \theta \mid V^*) \mid 0 >< 0 \mid \\ &= \frac{1}{2}\,\mathrm{tr}_{\mathcal{H}_2}(\mathrm{tr}_{\mathcal{K}_2} \mid \alpha\theta >< \alpha\theta \mid \otimes \mid \beta\theta >< \beta\theta \mid) \mid 0 >< 0 \mid \\ &= \frac{1}{2}\,\mathrm{tr}_{\mathcal{H}_2} \mid \alpha\theta >< \alpha\theta \mid\mid 0 >< 0 \mid \\ &= \frac{1}{2} \mid < 0 \mid \alpha\theta > \mid^2 = \frac{1}{2}\exp(- \mid \alpha\theta \mid^2) = \frac{1}{2}\exp(-\eta \mid \theta \mid^2) \end{aligned} \tag{3.4}$$

where $\eta = \alpha^2$ and η is constant describing the transmission efficiency for the channel.

3.1.2 In case that the input state is a squeezed state

A squeezed state can be expressed by a unitary operator $U(z)\,(z \in \mathbf{C})$ such that

$$\rho_1 = U(z) \mid \theta >< \theta \mid U(z)^*$$

where $| \theta >$ is a certain coherent state. Therefore the error probability $P_{e(SQ)}^{OOK-DD}$ is given by

$$\begin{aligned}
P_{e(SQ)}^{OOK-DD} &= \frac{1}{2} \operatorname{tr}_{\mathcal{H}_2}(\operatorname{tr}_{\mathcal{K}_2} V U(z) \mid \theta >< \theta \mid U(z)^* V^*) \mid 0 >< 0 \mid \\
&= \frac{1}{2} \operatorname{tr}_{\mathcal{H}_1} U(z) \mid \theta >< \theta \mid U(z)^* \, (V^*(\mid 0 >< 0 \mid \otimes I) V) \\
&= \frac{1}{2} < U(z)\theta , V^*(\mid 0 >< 0 \mid \otimes I) V U(z)\theta > \\
&= \frac{1}{2\pi^2} \int\!\!\int d^2v d^2w < U(z)\theta, w >< \alpha w, 0 >< \beta w, \beta v > \\
&\quad \times < 0, \alpha v >< v, U(z)\theta >
\end{aligned}$$

This can be computed by the following Gaussian type integration :

$$\frac{1}{\pi} \int d^2 w \exp\{- \mid w \mid^2 + aw + b\overline{w} + cw^2 + d\overline{w}^2\} = \frac{1}{\sqrt{1-4cd}} \exp\left\{ \frac{a^2d + ab + b^2c}{1-4cd} \right\}. \tag{3.5}$$

The result is

$$P_{e(SQ)}^{OOK-DD} = \frac{1}{2} \sqrt{\tau} \exp\left[\{(1-\eta)\tau - 1\} \mid \theta \mid^2 + \{1 - (1-\eta)^2 \tau\} \left\{ \frac{\overline{\nu}\theta^2}{2\mu} + \frac{\nu\overline{\theta}^2}{2\overline{\mu}} \right\} \right] \tag{3.6}$$

where $\tau = \{ \mid \mu \mid^2 - (1-\eta)^2 \mid \nu \mid^2 \}^{-1}$, μ and ν are complex numbers which satisfy $\mid \mu \mid^2 - \mid \nu \mid^2 = 1$.

3.2 Error Probability for Homodyne Detection

Homodyne detection is a measurement of the real part of the complex amlitude of a transmitted state. Therefore the P.O.V.M. E_{HO} for homodyne detection is given by

$$E_{HO}(\Delta^{HO}) = \int_{\Delta^{HO}} \mid \theta_x >< \theta_x \mid d\theta_x \tag{3.7}$$

where $\mid \theta_x >$ is the eigenvector of the operator $a_x = (a + a^*)/2$, a is the annihilation operator of photon, and Δ^{HO} is the set of real variables θ_x.

Therefore the infinitesimal nonnegative definite Hermitian operator $dE_{HO}(\theta_x)$ is given by

$$dE_{HO}(\theta_x) = \mid \theta_x >< \theta_x \mid d\theta_x \tag{3.8}$$

The probability density function $p^{HO}(\theta_x)$ of the outcome is

$$\begin{aligned}
p^{HO}(\theta_x)\, d\theta_x &= \operatorname{tr}_{\mathcal{H}_2} \Lambda^* \rho \, dE_{HO}(\theta_x) \\
&= \operatorname{tr}_{\mathcal{H}_2} \Lambda^* \rho \mid \theta_x >< \theta_x \mid d\theta_x ,
\end{aligned}$$

so that the probability density function $p^{HO}(\theta_x)$ is

$$p^{HO}(\theta_x) = \operatorname{tr}_{\mathcal{H}_2} \Lambda^* \rho \mid \theta_x >< \theta_x \mid \tag{3.9}$$

Using (3.9), we derive error probabilities for OOK and BPSK.

3.2.1 In case that the input state is a coherent state

At first, we derive the probability density function $p_{CO}^{HO}(\theta_x)$ for a coherent input state.

$$
\begin{aligned}
p_{CO}^{HO}(\theta_x) &= \mathrm{tr}_{\mathcal{H}_2}\Lambda^*(\mid \theta >< \theta \mid) \mid \theta_x >< \theta_x \mid \\
&= \mid < \theta_x \mid \alpha\theta > \mid^2 \\
&= \sqrt{\frac{2}{\pi}}\exp\left(-2\left(\theta_x - \alpha Re(\theta)\right)^2\right)
\end{aligned}
\tag{3.10}
$$

This probability density function $p_{CO}^{HO}(\theta_x)$ is a Gaussian type. Then m_{CO} and ${\sigma_{CO}}^2$, the average and the variance for this distribution $p_{CO}^{HO}(\theta_x)$, are calculated as

$$
m_{CO} = \alpha Re(\theta), \quad {\sigma_{CO}}^2 = \frac{1}{4}. \tag{3.11}
$$

<u>(i) OOK</u>

In OOK, ρ_0 is a vaccum state and ρ_1 is another state in the input system $\mathcal{H}_1$.

The probability density functions $p_{0(CO)}^{HO}(\theta_x)$ and $p_{1(CO)}^{HO}(\theta_x)$ for the signal "0" and "1" are respectively obtained by (3.10) as

$$
p_{0(CO)}^{HO}(\theta_x) = \sqrt{\frac{2}{\pi}}\exp\left(-2\theta_x^2\right) \tag{3.12}
$$

$$
p_{1(CO)}^{HO}(\theta_x) = \sqrt{\frac{2}{\pi}}\exp\left(-2(\theta_x - \alpha Re(\theta_1))^2\right) \tag{3.13}
$$

Therefore every error probability of OOK for each signal is identical. That is,

$$
P_{e0(CO)}^{OOK-HO} = P_{e1(CO)}^{OOK-HO} = \int_{\alpha Re(\theta_1)/2}^{\infty} p_{0(CO)}^{HO}(\theta_x)\, d\theta_x \tag{3.14}
$$

Hence the error probability $P_{e(CO)}^{OOK-HO}$ is given by

$$
P_{e(CO)}^{OOK-HO} = \frac{1}{2}\,\mathrm{erfc}\left(\frac{\sqrt{\eta}\, Re(\theta_1)}{\sqrt{2}}\right), \tag{3.15}
$$

where erfc(x) is the complementary error function given by

$$
\mathrm{erfc}(x) = \frac{2}{\sqrt{\pi}}\int_x^{\infty}\exp\left(-t^2\right) dt \tag{3.16}
$$

(ii) BPSK

In BPSK, ρ_0 is a state with phase 0 and ρ_1 is a state with phase π. Therefore the probability density functions $p_{0(CO)}^{HO}(\theta_x)$ and $p_{1(CO)}^{HO}(\theta_x)$ for the signal "0" and "1" are respectively obtained from Eq.(3.10)

$$p_{0(CO)}^{HO}(\theta_x) = \sqrt{\frac{2}{\pi}} \exp\left(-2\left(\theta_x - \alpha \mid \theta \mid\right)^2\right) \tag{3.17}$$

$$p_{1(CO)}^{HO}(\theta_x) = \sqrt{\frac{2}{\pi}} \exp\left(-2\left(\theta_x + \alpha \mid \theta \mid\right)^2\right) \tag{3.18}$$

where $\mid \theta \mid$ is the amplitude in an input state ρ_0 or ρ_1.

The error probability of each signal for BPSK is equal, that is, so

$$P_{e0(CO)}^{BPSK-HO} = P_{e1(CO)}^{BPSK-HO} = \int_0^\infty p_{1(CO)}^{HO}(\theta_x)\, d\theta_x \tag{3.19}$$

Then the error probability $P_{e(CO)}^{BPSK-HO}$ is given by

$$P_{e(CO)}^{BPSK-HO} = \frac{1}{2} \operatorname{erfc}\left(\sqrt{2\eta} \mid \theta \mid\right) \tag{3.20}$$

3.2.2 In case that the input state is a squeezed state

At first, from Eq.(3.8) and Eq.(3.9), we derive probability density function p_{SQ}^{HO} for a squeezed input state.

$$\begin{aligned}
p_{SQ}^{HO}(\theta_x) &= \operatorname{tr}_{\mathcal{H}_2}\left(\operatorname{tr}_{\mathcal{K}_2}\left(VU(z) \mid \theta >< \theta \mid U(z)^* V^*\right) \mid \theta_x >< \theta_x \mid\right. \\
&= \frac{1}{2} \operatorname{tr}_{\mathcal{H}_1} U(z) \mid \theta >< \theta \mid U(z)^* \left(V^*(\mid \theta_x >< \theta_x \mid \otimes I) V\right) \\
&= \frac{1}{\pi^2} \int\!\!\int d^2v d^2w < U(z)\theta, w >< \alpha w, \theta_x >< \beta w, \beta v > \\
&\quad \times < \theta_x, \alpha v >< v, U(z)\theta > \\
&= \frac{1}{\sqrt{2\pi\{\frac{1}{4}\eta \mid \mu - \nu \mid^2 + \frac{1}{4}(1-\eta)\}}} \exp\left(-\frac{(\theta_x - \alpha Re((\overline{\mu} - \overline{\nu})\theta))^2}{2\{\frac{1}{4}\eta \mid \mu - \nu \mid^2 + \frac{1}{4}(1-\eta)\}}\right)
\end{aligned} \tag{3.21}$$

This probability density function $p_{SQ}^{HO}(\theta_x)$ is again a Gaussian type. Then m_{SQ} and ${\sigma_{SQ}}^2$, the average and the variance of this distribution (3.21), are calculated as

$$m_{SQ} = \alpha Re((\overline{\mu} - \overline{\nu})\theta), \quad {\sigma_{SQ}}^2 = \frac{1}{4}\eta \mid \mu - \nu \mid^2 + \frac{1}{4}(1-\eta) \tag{3.22}$$

(i) OOK

Under the similar discussion as the section 3.2.1(i), the error probability is given by

$$P_{e(SQ)}^{OOK-HO} = \frac{1}{2}\,\mathrm{erfc}\left(\frac{\sqrt{\eta}Re((\overline{\mu}-\overline{\nu})\theta_1)}{\sqrt{2\eta \mid \mu-\nu \mid^2 + 2(1-\eta)}}\right) \tag{3.23}$$

(ii) BPSK

Under the similar discussion as the section 3.2.1(ii), the error probability is given by

$$P_{e(SQ)}^{BPSK-HO} = \frac{1}{2}\,\mathrm{erfc}\left(\frac{\sqrt{2\eta} \mid \theta \mid Re(\overline{\mu}-\overline{\nu})}{\sqrt{\eta \mid \mu-\nu \mid^2 + (1-\eta)}}\right) \tag{3.24}$$

3.3 Error Probability for Heterodyne Detection

Heterodyne detection is a simultaneous measurement of the real and imaginary part of the complex amlitude in a transmitted state. Therefore the heterodyne detection may not depend on the effect of squeezing. In this section we derive error probabilities for a coherent input state only.

Let E_{HE} be the P.O.V.M. for heterodyne detection.

$$E_{HE}(\Delta^{HE}) = \int_{\Delta^{HE}} \mid \theta >< \theta \mid \frac{d^2\theta}{\pi} \tag{3.25}$$

where $\mid \theta >$ is a coherent vector, and Δ^{HE} is the set of complex variables θ.

Therefore the infinitesimal nonnegative definite Hermitian operator $dE_{HE}(\theta)$ is given by

$$dE_{HE}(\theta) = \mid \theta >< \theta \mid \frac{d^2\theta}{\pi} \tag{3.26}$$

The joint probability density function $p^{HE}(\theta_x, \theta_y)$ of the outcomes becomes

$$\begin{aligned} p^{HE}(\theta_x, \theta_y)\, d^2\theta &= \mathrm{tr}_{\mathcal{H}_2} \Lambda^* \rho\, dE_{HE}(\theta) \\ &= \mathrm{tr}_{\mathcal{H}_2} \Lambda^* \rho \mid \theta >< \theta \mid \frac{d^2\theta}{\pi} \end{aligned}$$

so that the joint probability density function $p^{HE}(\theta_x, \theta_y)$ is :

$$p^{HE}(\theta_x, \theta_y) = \frac{1}{\pi}\mathrm{tr}_{\mathcal{H}_2} \Lambda^* \rho \mid \theta >< \theta \mid \tag{3.27}$$

Now ρ is a coherent state. Then $p^{HE}(\theta_x, \theta_y)$ is concretely derived as follows:

$$\begin{aligned} p^{HE}(\theta_x, \theta_y) &= \frac{1}{\pi}\mathrm{tr}_{\mathcal{H}_2} \Lambda^* \rho \mid \theta >< \theta \mid \\ &= \frac{1}{\pi} \mid < \theta \mid \alpha\theta_S > \mid^2 \\ &= \frac{1}{\pi}\exp(- \mid \theta - \alpha\theta_S \mid^2) \end{aligned} \tag{3.28}$$

where the index "s" represents the signal "0" or "1". The coherent detection demodulate the part "cos wt" from the transmitted signal. We let $p^{HE}_{cos}(\theta_x)$ the marginal probability density function of $p^{HE}(\theta_x, \theta_y)$, and from Eq.(3.28) $p^{HE}_{cos}(\theta_x)$ is given by

$$\begin{aligned} p^{HE}_{cos}(\theta_x) &= \int p^{HE}(\theta_x, \theta_y)\, d\theta_y \\ &= \frac{1}{\sqrt{\pi}} \exp\left(-\left(\theta_x - \alpha Re(\theta_S)\right)^2\right) \end{aligned} \tag{3.29}$$

Under the above preparation we derive error probabilities for OOK, FSK and BPSK in the similar way as the derivation in the section 3.2.

(i) OOK

From Eq.(3.29) the probability density functions $p^{HE}_{0(CO)}(\theta_x)$ and $p^{HE}_{1(CO)}(\theta_x)$ for the signal "0" and "1" are respectively given by

$$p^{HE}_{0(CO)}(\theta_x) = \frac{1}{\sqrt{\pi}} \exp\left(-\theta_x^2\right) \tag{3.30}$$

$$p^{HE}_{1(CO)}(\theta_x) = \frac{1}{\sqrt{\pi}} \exp\left(-(\theta_x - \alpha Re(\theta_1))^2\right) \tag{3.31}$$

As is analogized from the case OOK - homodyne, the error probability $P^{OOK-HE}_{e(CO)}$ is given by

$$P^{OOK-HE}_{e(CO)} = \frac{1}{2} \operatorname{erfc}\left(\frac{\sqrt{\eta} Re(\theta_1)}{2}\right) \tag{3.32}$$

(ii) FSK

In FSK, ρ_0 is a state with frequency w_0 and ρ_1 is a state with w_1. The transmitted state $\Lambda^* \rho_0$ or $\Lambda^* \rho_1$ is separated by an IF dual filter and demodulated by coherent detectors. Here we can consider the only case that the signal "0" is transmitted without loss of genenality. We let A and B the above two coherent detectors, and we let θ_A and θ_B each outcome of A and B, respectively. From Eq.(3.32) the probability density functions $p^{HE}_{A(CO)}(\theta_A)$ and $p^{HE}_{B(CO)}(\theta_B)$ for the outcomes of A and B are respectively given by

$$p^{HE}_{A(CO)}(\theta_A) = \frac{1}{\sqrt{\pi}} \exp\left(-(\theta_A - \alpha Re(\theta_0))^2\right) \tag{3.33}$$

$$p^{HE}_{B(CO)}(\theta_B) = \frac{1}{\sqrt{\pi}} \exp\left(-\theta_B^2\right) \tag{3.34}$$

The error probability $P^{FSK-HE}_{e(CO)}$ is given by

$$\begin{aligned} P^{FSK-HE}_{e(CO)} &= \operatorname{Prob}(v_A - v_B < 0) \\ &= \int_{-\infty}^{0} \frac{1}{\sqrt{2\pi}} \exp\left(-\frac{(v_{A-B} - \alpha Re(\theta_0))}{2}\right) dv_{A-B} \\ &= \frac{1}{2} \operatorname{erfc}\left(\sqrt{\eta} Re(\theta_0)\right) \end{aligned} \tag{3.35}$$

where $v_{A-B} = v_A - v_B$.

(iii) BPSK

From Eq.(3.29) the probability density functions $p^{HE}_{0(CO)}(\theta_x)$ and $p^{HE}_{1(CO)}(\theta_x)$ for the signal "0" and "1" are respectively given by

$$p^{HE}_{0(CO)}(\theta_x) = \sqrt{\frac{1}{\pi}} \exp(-(\theta_x - \alpha \mid \theta \mid)^2) \tag{3.36}$$

$$p^{HE}_{1(CO)}(\theta_x) = \sqrt{\frac{1}{\pi}} \exp(-(\theta_x + \alpha \mid \theta \mid)^2) \tag{3.37}$$

where $\mid \theta \mid$ is the amplitude in an input state ρ_0 or ρ_1.

By a analogy of the case BPSK - homodyne, the error probability $P^{BPSK-HE}_{e(CO)}$ is given by

$$P^{BPSK-HE}_{e(CO)} = \frac{1}{2} \operatorname{erfc}(\sqrt{\eta} \mid \theta \mid) \tag{3.38}$$

§4. Comparison of Error Probabilities for BPSK-Homodyne Detection

In this section we compare the error probabilities for BPSK-homodyne dection in the case that the input state is a coherent state and a squeezed state. When we calculate the error probability for a squeezed input state we assume that the parameters μ and ν are real for simplicity. That is, we consider the following four cases :

$$\begin{array}{llcl} \text{(i)} & \Delta a_x : \Delta a_y = 1 : 4 & \text{that is} & \mu = 1.25, \ \nu = 0.75 \\ \text{(ii)} & \Delta a_x : \Delta a_y = 4 : 1 & \text{that is} & \mu = 1.25, \ \nu = -0.75 \\ \text{(iii)} & \Delta a_x : \Delta a_y = 1 : 16 & \text{that is} & \mu = 2.125, \nu = 1.875 \\ \text{(iv)} & \Delta a_x : \Delta a_y = 16 : 1 & \text{that is} & \mu = 2.125, \nu = -1.875 \end{array}$$

From (3.20), (3.24) and $\mid \theta \mid = \sqrt{20}$, we get the graphes shown in Fig.1. Here we call in the above four cases $P_{(1:4)}$, $P_{(4:1)}$, $P_{(1:16)}$ and $P_{(16:1)}$, respectively.

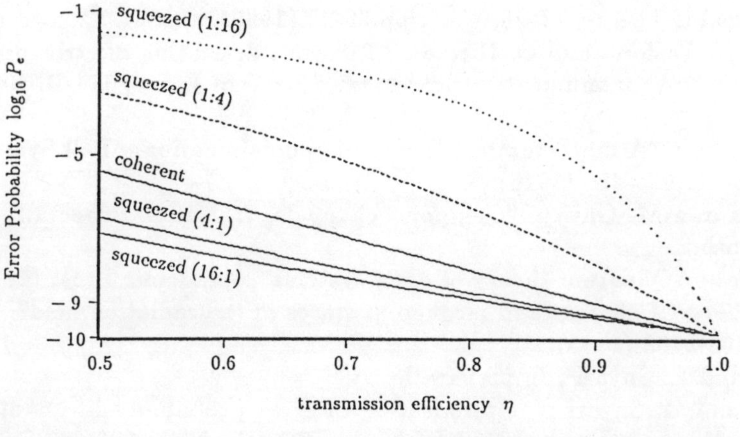

Fig. 1 Error probability for BPSK-homodyne detection

The efficiency of squeezed input states $P_{(1:4)}$ and $P_{(1:16)}$ are worse than that of a coherent state because of the effect of loss in the course of attenuation. On the other hand, the efficiency of squeezed input states $P_{(4:1)}$ and $P_{(16:1)}$, which have the opposite squeezing to $P_{(1:4)}$ and $P_{(1:16)}$, are better than that of a coherent state. Moreover, in this interesting case, the graphs of $P_{(4:1)}$ and $P_{(16:1)}$ means that the larger squeezing is, the better the efficincy is. This result might carry a deep physical reason for the optical communication.

§5. Conclusion

In this paper, using mathematical formulation of quantum mechanical channel attenuation processes, we rigorously derive error probabilities in digital modulations for a coherent input state and a squeezed input state. The results obtained in this paper might give us some clues why and in which sense a squeezed state is useful for optical communication, about which we are still working in terms of SNR (signal to noise ratio) and the optimization problem [13], along the line of [14].

References

[1] M. Ohya : "On compound state and mutual information in quantum information theory", IEEE.Trans.Inf.Theory, 29, pp.770-774 (1983).

[2] M. Ohya : "Some aspects of quantum information theory and their applications to irreversible processes", Rep. on Math. Phys., 27, pp.19-47 (1989).

[3] C.W. Helstrom : "Quantum detection and estimation theory", Academic Press (1976).

[4] D. Stoler : "Equivalence class of minimum uncertainty packets", Phys. Rev., D1, pp.3217-3219 (1970) and D4, pp.1925-1926 (1971).

[5] J.N. Hollenhorst : "Quantum limits on resonant-mass gravitational-radiation detectors", Phys. Rev. D19, pp.1669-1679 (1979).

[6] H.P. Yuen : "Two-photon coherent states of the radiation field", Phys. Rev., A13, pp.2226-2243 (1976).

[7] H.P. Yuen and J.H. Shapiro : "Optical communication with two-photon coherent states - Part III : Quantum measurements realizable with photoemissive detectors", IEEE.Trans.Inf.Theory, IT-26, No.1, pp.78-92 (1980).

[8] M. Ohya, H. Yoshimi and O. Hirota : "Rigorous derivation of error probability in quantum control communication processes", IEICE of Japan, J71-B, No.4, 533-539 (1988).

[9] C.E. Shannon : "A mathematical theory of communication", Bell System Tech.J., 27, pp.379-423,623-656 (1948).

[10] L. Accardi and M. Ohya : "Compoud channels, transition expectations and liftings", preprint.

[11] E.B. Davies : "Quantum theory of open systems", Academic Press (1976).

[12] R.J. Glauber : "Coherent and incoherent states of the radiation field", Phys. Rev. 131, pp.2766-2788 (1963).

[13] M. Ohya and H. Suyari : in preparation.

[14] M. Ohya and H. Suyari : "Optimization of error probability in quantum control communication processes", IEICE of Japan, J73-B-I, No.3, 200-207 (1990).

COMPARISON OF PPAM AND PPM FORMATS IN PHOTON-COUNTING CHANNELS

A. M. S. Gameiro and J.R.F. da Rocha
Department of Electronics and Telecommunications, University of Aveiro, 3800 Aveiro, Portugal

Abstract

In a recent paper, a hybrid format called PPAM has been introduced. It differs from PPM by adding an extra symbol represented by no light at all time slots. In this communication the new format is analysed in the context of a photon-counting Poisson channel. Information theoretic concepts of capacity and cutoff-rate as well as the probability of error for uncoded data will be quantified. We conclude that to get advantage of the extra symbol, alphabet sizes should have a moderate value. Otherwise, PPAM becomes very similar to PPM.

Introduction.

Optical communications with direct-detection photon-counting receivers have been extensively studied in recent years. Modulation formats such as OOK, PPM and OPPM, have been characterized by their capacity, cutoff-rate and probability of error [1-6]. More recently Kanaya et al [7] in a paper studying some features of optical channels using photon number quantum states, introduced an hybrid format which they called PPAM. It differs from the Pierce's M-ary PPM channel by adding to the alphabet an extra symbol corresponding to zero photons at all time-slots.

In this communication we analyse the PPAM format for channels following the Poissonian law associated with the coherent quantum state. We quantify the capacity, cutoff-rate and probability of error for uncoded data and compare them against PPM.

Channel model

The PPAM channel in the noiseless case is described by the following matrix of transition probabilities:

$$\mathbf{P} = \begin{bmatrix} 1 & 0 \ldots 0 \\ 1\text{-}\alpha & \alpha \ldots 0 \\ & \vdots \\ 1\text{-}\alpha & 0 \ldots \alpha \end{bmatrix} \qquad (1)$$

Lecture Notes in Physics, Vol. 378
C. Bendjaballah, O. Hirota, S. Reynaud (Eds.),
Quantum Aspects of Optical Communications

where α is the probability that a symbol other than zero is received without error.

For the noiseless Poisson channel, the number n of photons in the slot where the light source is on, has the following Poisson distribution:

$$\text{Prob}(n/\text{light on}) = \frac{Q^n e^{-Q}}{n!} \tag{2}$$

where Q is the average number of photons. This gives $\alpha = 1 - e^{-Q}$.

Capacity

The expression for the capacity was given in [7] as a function of α and consequently substituting $1-e^{-Q}$ we get for our case:

$$C = \log\left(1 + M(1-e^{-Q})\exp\left(-\frac{Q}{e^{Q}-1}\right)\right) \text{ nats/symbol} \qquad \text{for } M=1,2 \tag{3}$$

and for $M \geq 3$:

$$C = \begin{cases} \log\left(1 + M(1-e^{-Q})\exp\left(-\frac{Q}{e^{Q}-1}\right)\right) & Q^* < Q \leq \infty \\ (1-e^{-Q})\log M & 0 \leq Q \leq Q^* \end{cases} \quad \text{nats/symbol} \tag{4}$$

where Q^* is the unique nonzero solution of the equation:

$$(1-e^{-Q})\log M = Q \tag{5}$$

For M=1 we fall in the OOK case and the expression for the capacity is identical to that derived in [2].

The capacity is achieved with the following a priori probability for zeros:

$$p_0(M,Q) = \begin{cases} \dfrac{1 - Me^{-Q}\exp\left(-\frac{Q}{e^{Q}-1}\right)}{1 + M(1-e^{-Q})\exp\left(-\frac{Q}{e^{Q}-1}\right)} & Q^* < Q < \infty \\ 0 & 0 \leq Q \leq Q^* \end{cases} \tag{6}$$

The other symbols are equiprobable and have probabilities $p_j(M,Q) = (1-p_0(M,Q))/M$, which means that for Q below threshold Q*, to achieve the capacity the PPAM format must degenerate into the PPM format. As $Q \to \infty$, $p_0(M,Q)$ tends to $1/(M+1)$ and the capacity, equation (4), tends to the maximum entropy $\log(M+1)$. The typical behavior of the zero a priori probability is shown in figure 1. The probability is zero below

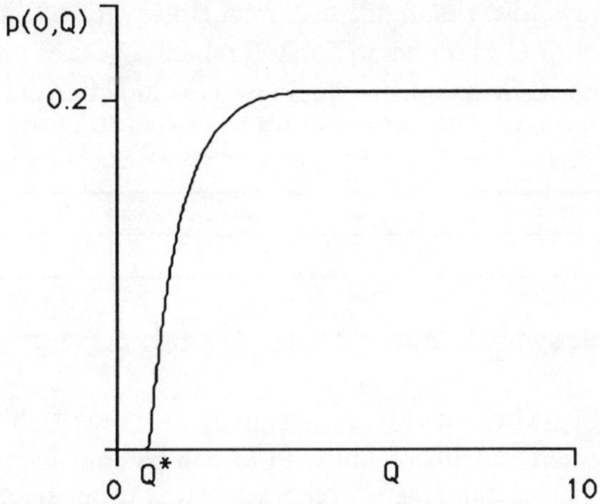

Fig.1 Behavior of $p_0(M,Q)$ as function of Q for M=4.

threshold and then increases, reaching a maximum for Q_{max} given by equation (7), decreasing slowly to 1/(M+1) for higher values of Q.

$$Me^{-Q}\exp\left(-\frac{Q}{e^{Q}-1}\right)=\frac{Qe^{Q}-e^{Q}+1}{(e^{Q}-1)^{2}} \qquad (7)$$

The solution of (7) tends to (M+1) as M tends to infinity. This can be seen in Table I, where the overshoot is the relative error between the value of $p_0(M,Q)$ for Q_{max} and 1/(M+1) and $Q_{1\%}$ is the value of Q beyond which the capacity is achieved with $p_0(M,Q)$ within 1% of the equiprobable distribution. From Table I one concludes also that the a priori probabilities are close to the equiprobable distribution for small values of Q.

Table I

M	Q^*	Q_{max}	Overshoot	$Q_{1\%}$
3	0.191	3.5	1.83%	5.11
4	0.693	4.77	0.62%	3.31
5	1.041	5.89	0.22%	4.08
7	1.520	7.98	0.029%	5.07
15	2.482	16.	1.10^{-7}	6.66
31	3.308	32.	$1.2.10^{-14}$	7.72

On the other hand if we start with an equiprobable distribution the average mutual information is given by:

$$I(X;Y) = \log(M+1) - \frac{1}{M+1}\left(MQe^{-Q} + (1+Me^{-Q})\log(1+Me^{-Q})\right) \quad (8)$$

which is obviously always lower than the capacity. However, I(X;Y) is very close to C even for moderate values of Q as shown in Table II where $Q_{c1\%}$ is the value of Q beyond which I(X;Y) differs from C by less than 1%. So we conclude that the PPAM

Table II

M	2	3	4	7	15
$Q_{c1\%}$	0	0.72	1.17	1.78	2.28

format with equiprobable symbols, does not entail a serious degradation.

The capacity of the PPAM channel is always greater (or equal) than that of the M-ary PPM, which is an expected conclusion since PPM can be considered as a restriction of PPAM. As Q increases the capacity tends to log(M+1) that is the maximum entropy of the source alphabet. It is important to note that we are comparing a format with an alphabet of size M (M-ary PPM) with an alphabet of size M+1 (M-PPAM), that is we implicitly assume the same width for the individual time -slot. This means that in an MT interval we can transmit 1 of M symbols with PPM and 1 of M+1 symbols with PPAM. A better basis for comparison is the capacity per time slot which for PPM is given by $CT=(1-\exp(-Q))\log(M)/M$ and for PPAM by (3) and (4) divided by M.

In a situation of average power restriction exists we must in the case of PPM maximize:

$$CT = (1-\exp(-P_{av}MT))\frac{\log(M)}{M} \quad (9)$$

For the case of PPAM and Q above the threshold the average power is given by:

$$P_{av}T = \frac{Q}{M}(1-p_0(M,Q)) \quad (10)$$

which means that the optimization involves a search for the optimum source symbol probability distribution.

In TableIII we present the capacity per time-slot for three cases: PPM optimized relatively to M, PPAM optimized in relation both to M and a priori symbol probability distribution, and PPAM with an equiprobable symbols (PPAM-Eq). From this table one concludes that for for high values of $P_{av}T$, the best performance is achieved with the optimized PPAM format. However, the difference between PPAM and PPAM-Eq is only slight. For $P_{av}T<0.217$, the optimized PPAM degenerates into the PPM format.

Cutoff-rate

Although the capacity gives us the maximum rate that a channel can support with an arbitrarily small symbol error, it does not give insight into the complexity of the corresponding coding scheme. For this reason the cutoff-rate criterion is sometimes preferred by some authors[4,9], since according to Massey [4], it provides a measure of code complexity.

Table III

$P_{av}T$	PPM	PPAM	PPAM-Eq
1	$M_{opt}=3$ CT=0.348 nats/slot	$M_{opt}=1$ CT=0.517	$M_{opt}=1$ CT=0.482
0.8	$M_{opt}=3$ CT=0.333	$M_{opt}=1$ CT=0.462	$M_{opt}=2$ CT=0.443
0.6	$M_{opt}=4$ CT=0.315	$M_{opt}=2$ CT=0.393	$M_{opt}=2$ CT=0.387
0.4	$M_{opt}=5$ CT=0.278	$M_{opt}=3$ CT=0.313	$M_{opt}=3$ CT=0.318
0.3	$M_{opt}=5$ CT=0.250	$M_{opt}=4$ CT=0.264	$M_{opt}=4$ CT=0.275
0.2	$M_{opt}=7$ CT=0.209		$M_{opt}=6$ CT=0.221
0.1	$M_{opt}=10$ CT=0.146		$M_{opt}=9$ CT=0.149
0001	$M_{opt}=55$ CT=0.031		$M_{opt}=54$ CT=0.031

The cutoff-rate for a constant discrete memoryless channel (CDMC) is given by [4]:

$$R_0 = -\min_{Q} \log\Big(\sum_{y}\Big(\sum_{x} Q(x)\sqrt{P(y/x)}\Big)^2\Big) \qquad (11)$$

Due to the nature of the channel it is easy to guess that the minimum in (11) will be achieved with the symbol zero occuring with probability $p_0(M,Q)$ and the other symbols with probability $(1-p_0(M,Q))/M$. Using the Kuhn-Tucker theorem about convex functions [10], we find:

$$R_0 = \log(1+M\,\mathrm{tgh}(\tfrac{Q}{4})) \qquad \text{for } M=1,2 \qquad (12)$$

and for $M\geq3$

$$R_0 = \begin{cases} \log\left(1+M\,\mathrm{tgh}(\frac{Q}{4})\right) & Q>2\log(M-1) \\ \log\left(\dfrac{M}{(M-1)e^{-Q}+1}\right) & Q\leq 2\log(M-1) \end{cases} \qquad (13)$$

The cutoff-rate is achieved with the following probability for the symbol zero:

$$p_0(M,Q) = \begin{cases} \dfrac{\dfrac{1-e^{-Q}}{M} + e^{-Q} - e^{-Q/2}}{(1-e^{-Q/2})^2 + \dfrac{1-e^{-Q}}{M}} & Q > 2\log(M-1) \\ 0 & Q \le 2\log(M-1) \end{cases} \qquad (14)$$

Similarly to the capacity case, as Q->∞ the distribution tends to an equiprobable one. If in (11) we assume an equiprobable distribution for the source symbols, that is we do not maximize R_0 over **Q,** we obtain a lower bound for the cutoff-rate given by:

$$R'_0 = 2\log(M-1) - \log(M^2 e^{-Q} + M(1+2e^{-Q/2} - e^{-Q}) + 1) \qquad (15)$$

From the above one concludes that R'_0 converges to R_0 as Q->∞,a behaviour similar to the average mutual information. Although the convergence is not as fast as in the case of I(X;Y), R'_0 becomes close to R_0 even for moderate values of Q, as shown in Table IV where $Q_{r1\%}$ is the values of Q beyond which R'_0 differes from R_0 by less than 1%.

Table IV

M	2	3	4	7	15
$Q_{r1\%}$	2.83	3.86	4.38	5.15	5.95

The cutoff-rate/photon (R_{0ph}) increases with M, as can be seen in figure 2. However, for M=1 and 2, R_{0ph} is limited to 0.5 nats/photon. The same conclusion applies to M≥3, when all symbols have a non-null probability (that is for Q>2log(M-1)). Above threshold (Q>2log(M-1)), PPAM is superior to PPM , and both R_{0ph}s take the same value of 0.5 nats/photon at threshold. Below threshold , as PPAM degenerates into PPM we get the same limitation of 1nat/photon for both formats.

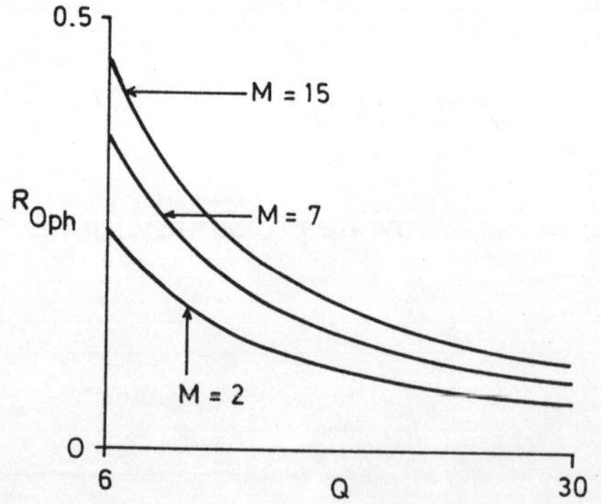

Fig.2 PPAM Cutoff-rate per photon for PPAM with M=2,7,15

Optimization of alphabet size under an average power bound, leads to the results shown in tableV.

Table V

$P_{av}T$	PPM	PPAM	PPAM-Eq
1	M_{opt}=3 R_0T=0.335 nats/slot	M_{opt}=2 R_0T=0.392	M_{opt}=2 R_0T=0.393
0.8	M_{opt}=4 R_0T=0.318	M_{opt}=3 R_0T=0.345	M_{opt}=2 R_0T=0.359
0.6	M_{opt}=4 R_0T=0.286	M_{opt}=4 R_0T=0.485	M_{opt}=3 R_0T=0.308
0.4	M_{opt}=6 R_0T=0.236		M_{opt}=5 R_0T=0.240
0.2	M_{opt}=10 R_0T=0.151		M_{opt}=8 R_0T=0.148
0.1	M_{opt}=18 R_0T=0.086		M_{opt}=16 R_0T=0.085
0001	M_{opt}=161 R_0T=0.0098		M_{opt}=152 R_0T=0.0098

For products $P_{av}T<0.555$, the optimized PPAM format degenerates into PPM which is a more serious limitation than the one related to capacity (table III). So we can conclude that the joint optimization of source symbol probabilities and alphabet size is only effective for moderately high values of average power. Moreover, for these values an equiprobable distribution gives almost the same results as the optimized one.

Probability of error

Let us consider now uncoded data with the noisy channel. In this case even in the absence of a transmitted pulse, there will be an average number (Q_0) of noise photons which we assume Poisson distributed. The matrix of transition probabilities takes on the following form:

$$\mathbf{P} = \begin{bmatrix} \alpha & \frac{1-\alpha}{M} \cdots & \frac{1-\alpha}{M} \\ \gamma & \beta \quad \cdots & \frac{1-\beta-\gamma}{M-1} \\ & \vdots & \\ \gamma & \frac{1-\beta-\gamma}{M-1} \quad \cdots & \beta \end{bmatrix} \tag{16}$$

Assuming equiprobable symbols we get for the probability of error:

$$P_e = \frac{1}{M+1}[1-\alpha+M(1-\beta)] \tag{17}$$

We will consider maximum a posteriori (MAP) detection which for equiprobable input symbols reduces to the maximum likelihood (ML) detector. The receiver computes the (M+1) conditional probabilities Pr(**N**/j) and chooses the symbol that gives the maximum (**N** is the M-dimensional vector of counts over each time-slot and j is the transmitted symbol). As the counts over disjoint intervals are independent we have the following rule:

$$\text{Chosen symbol} = \begin{cases} 0 & \text{if } n_j \leq [\eta] = \mathrm{int}\left(\dfrac{Q_1 - Q_0}{\log(Q_1) - \log(Q_0)}\right) \quad \text{for } j=1..M \\ j & \text{if } n_j > [\eta] \text{ and } n_j > n_i \quad \text{for } i \neq j \end{cases}$$

Due to the discrete nature of the data there will be a non-zero probability that several counts be equal and exceed the threshold and in this case one of these symbols will be chose at random. With this detection rule the probability of a symbol error is given by:

$$P_e = 1 - \frac{1}{M+1}\left\{ \left(F_0([\eta])\right)^M + \sum_{k=[\eta]+1}^{\infty} \Lambda(k)\left(\left(F_0(k)\right)^M - \left(F_0(k-1)\right)^M\right) \right\} \qquad (18)$$

where $[\eta]$ is the largest integer contained in the optimum threshold, $F_0(k)$ is the accumulated distribution of $p_0(k)$ and $\Lambda(k)$ is the likelihood ratio $p_1(k)/p_0(k)$, $p_1(k)$ and $p_0(k)$ being the probability distributions when a pulse is present and absent respectively.

In figure 3 we plot the probability of error as a function of the average number of photons when a pulse is transmitted, Q_1, taking Q_0 and M as parameters.

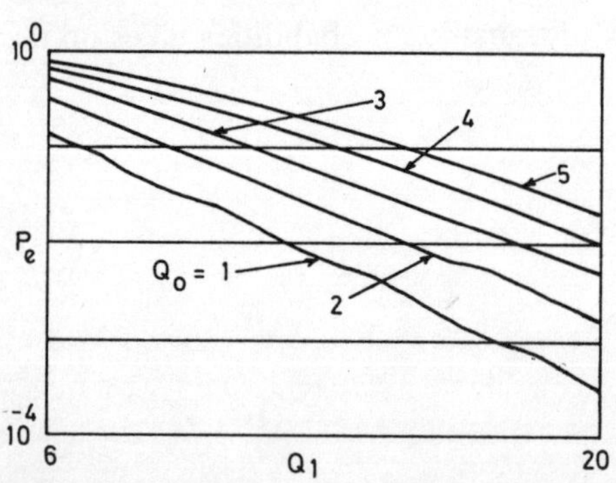

Fig.3 Probability of a symbol error for M=7, Q0=1,2,3,4,5.

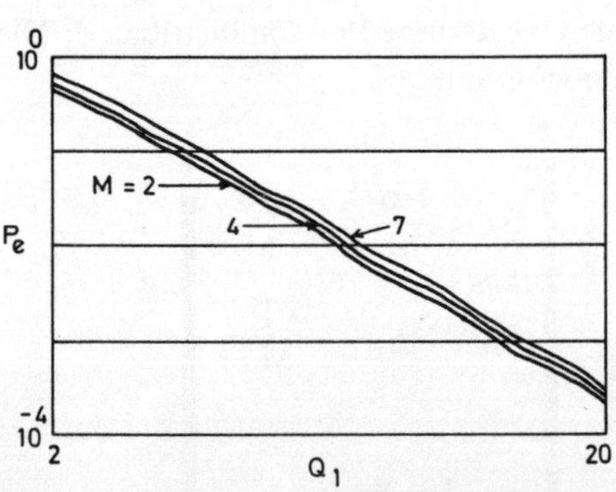

Fig.4 Probability of a symbol error for M=2,4,7 and Q0=1.

The probability of a symbol error inreases with M for Q_0 fixed as can be seen in figure 4. The probability of symbol error for M-PPAM can be related to the the probability of a bit error by the well known formula:

$$P_b = \frac{1}{2}\frac{M+1}{M} P_e \qquad (19)$$

provided M+1 is a power of 2. Calculations based on equation (19) show that the probability of a bit error still increases with M but less significantly than the probability of a symbol error.

Comparison of the basis of similar Q_1, gives a better performance for both M-PPM and (M+1)-PPM, at least for the values of Q_0 shown in figure 5. However, we must recall that with the same Q_1 and Q_0 , PPAM requires less power, since the average number of photons per symbol, Q_{av}, is given by $Q_1+(M-1)Q_0$ for M-PPM and by $Q_1+(Q_1-Q_0)/M$ for M-PPAM. If the comparisons are made on the basis of Q_{av}, M-PPM is still superior to M-PPAM in most cases, but for the same alphabet length and for small values of Q_{av} M-PPAM approaches M-PPM and can even show better performance, as shown in figure 6.

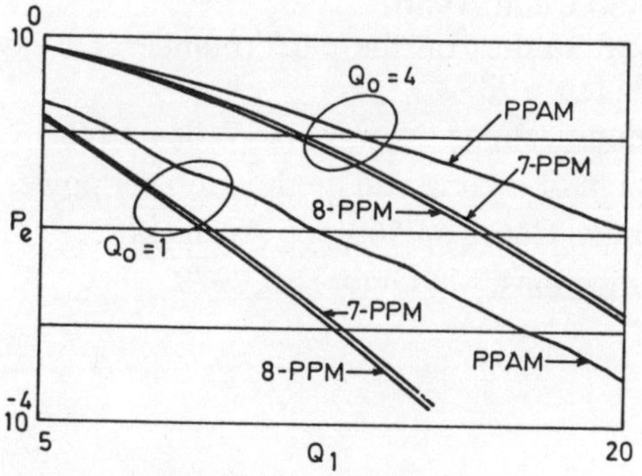

Fig.5 Comparison of 7-PPAM with 7 and 8-PPM for Q_0=1 and 4.

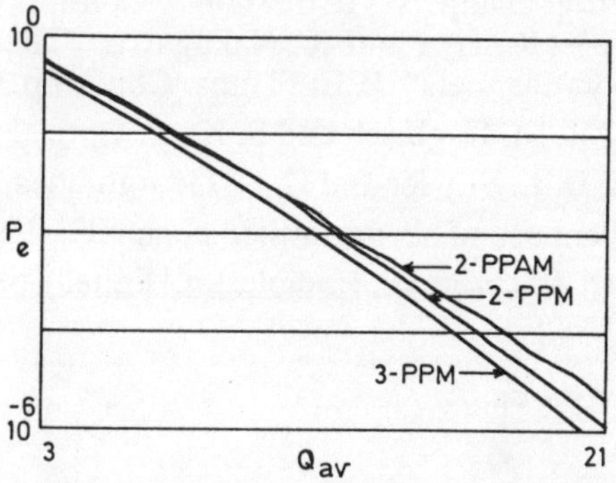

Fig.6 Comparison of 2-PPAM with 2 and 3-PPM as a function of average power per symbol, for Q_0=1.

Conclusions

We have analyzed the PPAM format over Poisson channels. The information theoretic concepts of capacity and cutoff-rate for the noiseless case have been used to get insight into the limits of the format. Comparing with the more common PPM format we have conclude that there is no significative improvement and even for low values of Q, the PPAM degenerates into the PPM case. The cases that could be of interest are those of moderately high power and small alphabet size (M=2,3), where PPAM show a better capacity/photon.

Calculations of the probability of error for the noisy case and ML detection, show that without coding PPAM does not offer advantages over PPM. Moreover, the optimum receiver requires a threshold which is function of Q_1 and Q_0, that is the receiver must know the signal and noise powers.

References

[1] J. R. Pierce, "Optical channels: Practical limits with photon counting", IEEE Trans. Com., pp.1819-1821, Dec. 1978.

[2] J. R. Pierce, E. C. Posner and E. R. Rodemich, "The capacity of the photocounting channel",IEEE Trans. Inform. Theory, pp.61, Jan. 1981.

[3] R. J. Mc Eliece, "Practical codes for photon communications", IEEE Trans. Inform. Theory, pp.393-397, July 1981.

[4] J. L. Massey, "Capacity, cutoff-rate and coding for photon communications", IEEE Trans. Com., pp.1615-1621, Nov. 1981.

[5] I. Bar-David and G. Kaplan, "Information rates of photon-limited overlapping pulse-position-modulation channels",IEEE Trans. Inform. Theory, pp.455-464, May1984.

[6] M. Charbit and C. Bendjallah, "Probability of error and capacity of PPM photon counting channels", IEEE Trans. Com., pp.600-605, Jun. 1986.

[7] F. Kanaya and K. Nakagawa, "Theoretical studies on the performance of lossy photon-channels", IEEE Trans. Com., pp.39-45, Jan. 1989.

[8] R. M. Gagliardi and S. Karp, Optical Communications, New York: Wiley 1976.

[9] D. L. Snyder and C. N. Georghiades, "Design of coding and modulation for power-efficient use of aband-limited channel", IEEE Trans. Com., pp.560-565 April 1983.

[10] A. J. Viterbi, Priciples of Digital Communications, Mc Graw Hill,1979.

Optical communication with coherent squeezed state. Realization of received quantum state control

Osamu Hirota

Department of Information and Communication Science
Tamagawa University, Tokyo 194, Japan

1 Introduction

In general, the reliability of optical communication systems is limited by noises in optical and electrical regions. Recent progress in communication technology has verified that systems operated under the standard quantum limitation are available. By the standard quantum limitation we mean a system whose reliability is bounded by the Poisson shot noise or the vacuum noise which are inherent in coherent state.

The quantum communication theory [1,2], however, has predicted the possibility of going beyond the standard quantum limitation by controlling quantum natures of light. Especially, an application of a squeezed state is hopful. The generation of a squeezed states has recently been performed by several groups (see for example [3]). From a detection theory point of view, certain advantages of such a light source have already been indicated [4,5,6].

Unfortunately, the advantage of the new kind of quantum state is reduced by the vacuum noise caused by energy loss in transmission channels because we generally assume to have about 50 dB considered as channel loss. In order to cope with this problem, we have proposed a concept of Received quantum state control and given some mathematical predictions for systems with ultimate properties [7,8].

In this paper, the feasibility of such a system is considered and some physical realizations are proposed.

Lecture Notes in Physics, Vol. 378
C. Bendjaballah, O. Hirota, S. Reynaud (Eds.),
Quantum Aspects of Optical Communications

2 Description of the received quantum state control

Let us first present the concept of the received quantum state control. In this system, transmitter states conveying signals are assumed to be coherent state, because they can keep the same quantum nature, even if there is transmission loss. Then states of signals are transformed to quantum states, which will give a satisfying performance to the decision process before quantum measurements. A general model of the received quantum state control system is given in Fig.1. Let us assume hereafter, for simplification, that quantum controlled state and quantum measurement process are squeezed state and a balanced homodyne receiver, respectively. Here it is essential to classify the set of squeezed states for our discussion.

Definition 1

The squeezed state, transformed from a coherent state, is called scaling squeezed state, and it is generated by

$$\mid \alpha_{sq}; r, \phi >= S(\zeta)D(\alpha) \mid 0 > \tag{1}$$

where $\alpha_{sq} = \alpha cosh(\zeta) - \alpha^* sinh(\zeta)$, and

$$S(\zeta) = \exp[\zeta^* a^2 - \zeta a^{\dagger 2}] \tag{2}$$

$$D(\alpha) = \exp[\alpha a^\dagger - \alpha^* a] \tag{3}$$

Definition 2

The coherent squeezed state is defined by [9]

$$\mid \alpha; r, \phi >= D(\alpha)S(\zeta) \mid 0 > \tag{4}$$

The new operator should be defined to describe the generation process of the coherent squeezed state from the coherent state with non zero amplitude [7, Eq(7 a, 8)].

Definition 3

The generalized squeezed operator is defined by

$$U_g(\tilde{\alpha} \mid \alpha) = D_g(\tilde{\alpha})S(\zeta) \tag{5}$$

where $S(\zeta)$ is given by Eq.(2), and

$$D_g(\tilde{\alpha}) = \exp(\tilde{\alpha} a^\dagger - \tilde{\alpha}^* a) \exp -i\Im(\tilde{\alpha}\alpha) \tag{6}$$

$$\tilde{\alpha} = \alpha - \alpha_{sq} \tag{7}$$

and the coherent state is generated by

$$| \alpha; r, \phi >= U_g(\tilde{\alpha} \mid \alpha) \mid \alpha > \qquad (8)$$

The outline of this operation is that the scaling squeezed state with amplitude α_{sq} transformed from the coherent state with amplitude α and the amplitude of the squeezed state α_{sq} is permuted again with the amplitude of the initial coherent state. Such a squeezing process is called a generalized squeezer because it is,to some extent, a generalization of the usual squeezed vacuum state. Although the result of the operations of Eqs.4 and 5 give the same coherent squeezed state, the meaning of latter one differs from another in the information theory point of view, namely, input output theory. In addition, this transform does not remove the uncertainty principle, because the transform acts as unitary to each incident state. But the difference from the conventional unitary operator is that this should be interpreted as conditional unitarity. The mathematical properties of this operator are given in [10]. Let us denote the physical interpretation of this operator.The operator requires a physical phenomenon to break the correlation between amplitude and noises in the squeezed mode of the physical process for the squeezing. This property is essential at the realization of this transform. However, we do not require in general the process $(\alpha - \alpha_{sq})$ of this operation, because (α_{sq}) is only parameter in the black box which is generalized squeezer. The purpose of this operator is the description of the change from a coherent state to a coherent squeezed state.
Thus we have two kinds of squeezed states. In other words, these have perfectly different properties in the communication process, which we will give below:

Property 1

The scaling squeezed state cannot have the signal to noise ratio larger than that of a coherent state, because there is a perfect correlation between amplitude and noises [9].

Property 2

The coherent squeezed state cannot generate from coherent state with non zero amplitude without increasing the signal to noise ratio.

These two properties predict the existence of the process which increases signal to noise ratio. Now, let us consider the communication system with the received quantum state control. So far, we proposed two kinds of the received quantum state control systems [7,8]. One of them is the system using the coherent squeezed state, and the other one is the system with some control schemes of the scaling squeezed state.
First, we discuss the first type. It is well known that the quantum limitation of the conventional coherent communication system is

$$SNR(signal.to.noise.ratio) =< 4n_R > \qquad (9)$$

where $< 4n_R >$ is the received average photon number. On the other hand, the received quantum state control constructed by the operator U_g can provide [7,8]

$$SNR = 4 < n_R > \exp(2r) \tag{10}$$

Thus an important advantage in the signal to noise ratio is obtained even though there is a large energy loss in transmission channels. Secondly, the scaling squeezed state also enables us to obtain some new quantum limitations. This, however, requires several conditions in the system constructions. In this system, the transmitter states are also the coherent state. The performance of the receivers with a generator of a squeezed vacuum state is switched by the incident lights with coherent states. That is, the quantum state control should be constructed by quantum state switch. Although there are many variations, these systems can be mathematically described by

$$\rho_i = \rho_{G1}\delta(\rho_{R1} - \rho_R) + \rho_{G0}(1 - \delta(\rho_{R1} - \rho_R)) \tag{11}$$

where ρ_{G1} and ρ_{G0} are density operators of the output of the quantum state controller, and the received state is described by ρ_R. More ρ_R can be equal to ρ_{R1} or ρ_{R0}. As a result,the operator $\delta(\rho_{R1} - \rho_R)$ acts as a unity when the received state is equal to ρ_{R1} and otherwise it is 0. This operation means that the performance of the receiver is controlled by signals. It is easy to verify that it enables us to get some new quantum limitations, if the incident quantum states are transformed into those of differnt types for symbols 1, and 0, respectively. This type of system does not, in general,require an improvement of signal to noise ratio at the receving process. That is, these may be realized by certain constructions using the scaling squeezed state. Unfortunately, there is no general theory for these systems, in spite of many physical examples. Fig.2 shows a typical example of the error rate property of such a system, in which the controlled states are assumed to be a scaling squeezed state and a vacuum state for symbols 1 and 0, respectively, and when a homodyne receiver is employed as the quantum measurement.
These new systems are attracting noteworthy attention in view of the information theory. The physical processes for incident signals are realized by unitary systems in both cases, but the property in the information theory point of view becomes non unitary, because the inner product is reduced in comparison to that of the transmitter states. Let us call their properties, information theoretical non unitary. In general, these processes are described by conditional unitary operator.

Definition 4

The conditional unitary operator is defined by

$$U^{\dagger}(. \mid A)U(. \mid A) = I \tag{12}$$

$$U^{\dagger}(. \mid A)U(. \mid B) \neq I \tag{13}$$

The operators of U_g and δ belong to the conditional unitary operator. More details will be discussed in the subsequent papers.

3 Realization of received quantum state control systems

3.1 Coherent squeezed state

The generation of the coherent squeezed state is one of special topics in quantum optics [11]. Recently, it was verified, based on the input -output formalism, that the coherent squeezed state can be generated by superposition of a coherent state and a squeezed vacuum state in a nonlinear medium [12]. This physical process is called nonlinear beam splitter. It was pointed out that the nonlinear beam splitter can be realized by a single mode optical bistability. This physical process can be understood by Eq.4, but it should be realized that, in fact, they are the result of the transform defined by Eq.5. In fact, the received quantum state control described by Eq.5 can be realized by using the process as shown in Fig.3. Furthermore, this process is the result of the system breaking the correlation between amplitude and noise of incident light as mentioned in section 2, which is a special feature of U_g. The nonlinear beam splitter produces the coherent squeezed state with the amplitude of the coherent state from the beam combine of the incident coherent state and the squeezed vacuum state generated in receiver with the efficiency unity, while the conventional beam splitter cannot produce that with the amplitude of the input coherent state. Such a system provides a great advantage to the conventional fiber and space optical communications, as discussed in section 2, but many problems remain for these realizations in the field of quantum optics, especially, the sensitivity of the nonlinear beam splitter. But the optical bistability is attractive in the perspective of realizing practical optical devices. More physical phenomenons as bright squeezing [13], may be also applicable to our system.

3.2 Quantum switch

A realization of the quantum state switch system was first proposed by Kagami, Bun, and Koyano [14]. Here we analyse a new system [15], which consists of an optical bistability as a nonlinear superposer and homodyne receiver as a measurement process. The optical bistability consists of two driving mechanisms as shown in Fig.4. The continuous driving light E_d passes through the cavity mirror and is injected. The incident light E_p comes from the communication channel which is directly coupled with the excited atomic beam. This scheme is equivalent to the one proposed by Knight as a bright squeezer system. If this system works as the bright squeezer, then it corresponds to the nonlinear beam splitter. That is, the signal–to–noise ratio of the incident light is improved by the bright squeezer process. Here we would like to show that this system can be used as the quantum state switch system even if it does not work as the nonlinear beam splitter. It works as the nonlinear superposer which is a conditional unitary process. Let us assume that the transmitter uses the on-off keying of the laser with a coherent state, and the signals are injected as E_p. When the incident light is a coherent state with non zero amplitude, the output is a coherent squeezed state if it is a bright squeezer. The mechanism of the bright squeezer is that the coherent amplitude in the cavity squeezing mode is generated from the stimulated dipole emission caused by the direct interaction between the atoms and the direct driving light. However, the proof of the bright squeezer is not sufficient. The difficulty is the degree of the suppression of the quantum noise of the incident light. Of course, it does not become the bright squeezer, if there is no suppression of the incident quantum noise. Let us assume that there is not such a suppression. In this case, even if the squeezing mode has a coherent amplitude, the output noise is not less than that of the conventional vacuum noise. That is, the quantum noise of the incident light is preserved in the squeezing mode. This means that the assumption of the noise suppression in the analysis of Knight is relaxed. Such a situation is not difficult to produce. As a result, the output is regarded as the superposition of the squeezed vacuum and the coherent state with non zero amplitude. On the other hand, if the incident light is the vacuum state, then the direct interaction between the atoms and the driving light E_p disappears, because this is an energy interaction process. Thus the received quantum states, coherent state with amplitude and vacuum state, are changed into the state as the superposition of the squeezed vacuum state and the coherent state with amplitude or the pure squeezed vacuum state. The output of the homodyne receiver is Gaussian with different variances and means for each signal. This system can overcome the standard quantum limitation when the double thresholds decision rule and large squeezing gains are used.

4 Conclusions

Our results show that the quantum communication theory and the related theories including squeezed state can provide excellent possibilities for practical communication systems which are being developed in the optical fiber and space optical communications. By merely adding the received quantum state controller to the conventional systems, they enable us to obtain the ultimate communication system.

Acknowledgements

This work was prepared in the Laboratoire des Signaux et Systèmes, École Supérieure d'Électricité. The author is grateful to Drs. C. Bendjaballah (CNRS) and S. Reynaud (École Normale Supérieure) for their helpful discussions.

References

1. C.W. Helstrom : *Quantum Detection and Estimation Theory* (Academic Press, New York 1976)
2. A.S. Holevo : *Probabilistic and statistical aspects of quantum theory* (North-Holland, Amsterdam 1982)
3. A. Heidmann, S. Reynaud : Jour. Mod. Opt. **34**, 923 (1987)

4. H. Yuen, J. Shapiro : IEEE Trans. Inf. Th. **IT-29**, 47 (1983)
5. B.E.A. Saleh, M.C. Teich : Phys. Rev. Lett. **58**, 2656 (1987)
6. C. Bendjaballah, M. Charbit, G. Oliver : *Coherence and Quantum Optics V*, edited by L. Mandel and E. Wolf (Plenum Press, New York 1984)
7. O. Hirota : Optics Communications **67**, 204 (1988)
8. O. Hirota, H. Tsushima : Trans. IEICE Japan **E-72,5**, 460 (1989)
9. C. Caves: Phys. Rev. **D-23**,1669 (1981)
10. O. Hirota : *Seminar of the Institute of Basic Physics Research*, (Kyoto University, March 1990)
11. S. Reynaud, C. Fabre, E. Giacobino : Jour.Opt.Soc.Am. **B-4**,1520 (1987)
12. M.J. Collett, D.F. Walls : *Quantum Optics V*, edited by J.D. Farvey and D.F. Walls (Springer Verlag 1989)
13. F.A.M. de Oliveira, P.L. Knight : Phys. Rev. **A-39** 3417 (1989)
14. O. Hirota : *Squeezed Light*, (Morikita Publ. Co., Tokyo 1990)
15. S. Reynaud : *private communication*
16. V. Braginsky : Science **209**, 547 (1980)

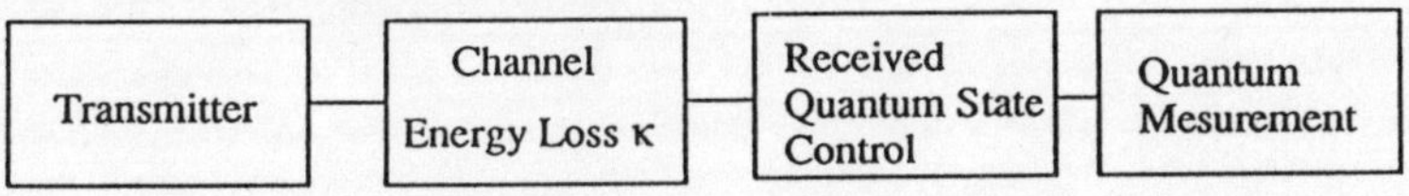

Figl. Schematic Diagram of Received Quantum State Control System

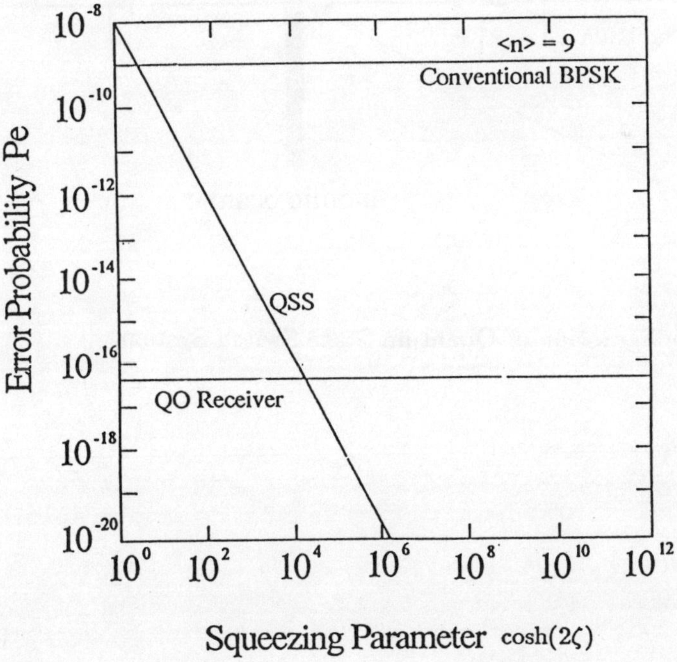

Fig2. Error performance of Quantum State Switch
QO is Quantum Optimum, QSS is Quantum State Switch.

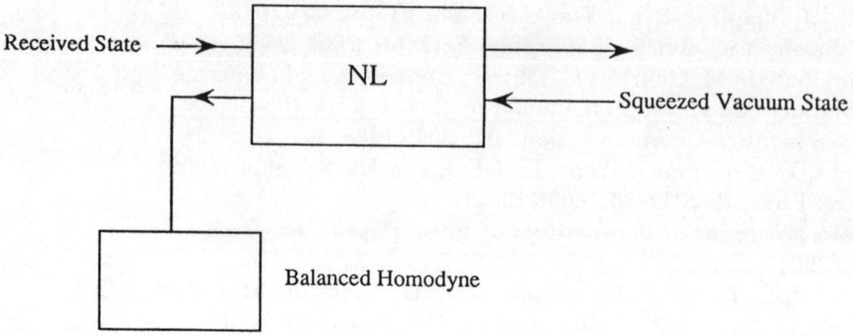

Fig3. Realization of RQSC by nonlinear beam splitter

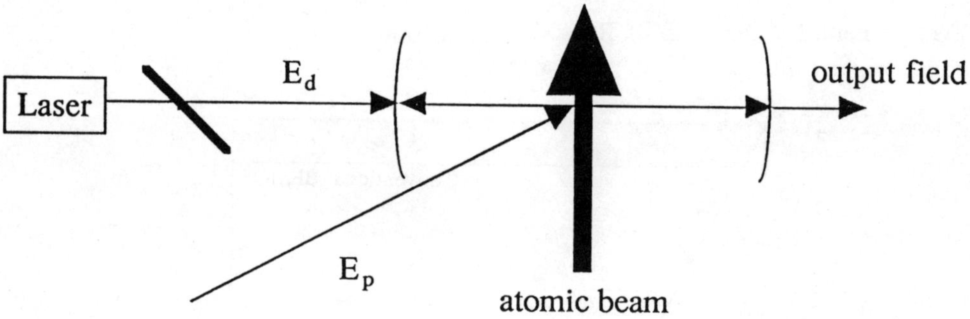

Fig4. Configuration of Quantum State Switch System

A NEW METHOD FOR IMPROVING THE BANDWIDTH UTILIZATION EFFICIENCY IN OPTICAL PPM

G. Cancellieri, F. Chiaraluce, E. Gambi
Dipartimento di Elettronica ed Automatica, Università di Ancona
Via Brecce Bianche 60131 ANCONA (Italy)

ABSTRACT

The performances of different sets of suitably encoded symbols over an ideal photon counting channel are investigated, on the basis of Shannon theorem for discrete channels. The possibility of reaching a transmission efficiency higher than that of a PPM transmission, which represents the most efficient practical transmission scheme suggested till now, is shown. In particular, this efficiency is comparable with that of a binary transmission with different a priori probabilities, with the advantage of a practical feasibility.

1. INTRODUCTION

M-ary orthogonal PPM has been demonstrated to be a very efficient transmission technique over an ideal photon counting channel [1,2]. It is characterized by an average information per symbol which tends to lg_2M and, correspondingly, by an efficiency, expressed in bit/photon, which increases indefinitely with increasing M. As a drawback, we have an enormous spread of bandwidth according to the factor M/lg_2M. The band utilization efficiency of optical PPM can be improved by modifying the modulation method as it is done in multipulse PPM (M-PPM) [3]. This method allows a relevant reduction of the bandwidth in comparison with that of conventional optical PPM at the same transmission efficiency.

The aim of the present paper is to show that further improvements, in this sense, can be obtained by changing the transmission alphabet, so as to include a very large set of suitably encoded symbols. The analysis is developed considering the transmission efficiency, expressed in bit/photon, whose maximum limit is estimated by means of Shannon theorem for discrete channels [4].

Lecture Notes in Physics, Vol. 378
C. Bendjaballah, O. Hirota, S. Reynaud (Eds.),
Quantum Aspects of Optical Communications

A comparison with the performance obtainable by a binary transmission with different a priori probabilities [5] is carried out and an outline of the possibility of using the proposed alphabet in conjunction with efficient coding schemes, like those applied to the PPM transmission [6], is also given.

2 . THEORY

Subdividing the symbol time duration into n time slots, we can obtain a set of up to 2^n different sequences of low and high transmission levels. Excluding the all-zero sequence, we are left with (2^n-1) sequences. For any integer M, such as $n \leq M \leq 2^n-1$, there is an optimum choice of the M sequences forming the transmission alphabet in order to minimize the bandwidth expansion factor obtaining, at the same time, a transmission efficiency sufficiently high. Although several choices are possible, in principle, the alphabet we have considered in the present paper consists of all the sequences having at most k high transmission levels per symbol time duration, with $1 \leq k \leq n$. In this way, the number of different sequences forming the transmission alphabet is

$$M = \sum_{i=1}^{k} \binom{n}{i} \tag{1}$$

It is important to stress that this alphabet, which will be called in the following complete multipulse PPM (CM-PPM), is different from the M-PPM one where we have, exactly and only, k high transmission levels per symbol time duration and, consequently,

$$M_{M\text{-}PPM} = \binom{n}{k} \tag{2}$$

different sequences. On the other hand the results valid for the M-PPM case can be easily obtained by means of simple modifications of the present analysis, while the PPM transmission is a particularization of the alphabet we propose to the case k=1.

The amount of information per symbol carried on average in the channel according to Shannon [4] can be written as

$I = H(Y) - H(Y|X)$, with

$$H(Y) = \sum_{j=1}^{M} P_j' \lg_2\left(\frac{1}{P_j'}\right) \quad \text{and} \quad H(Y|X) = \sum_{i=1}^{M} P_i \sum_{j=1}^{M} p_{ij} \lg_2\left(\frac{1}{p_{ij}}\right) , \qquad (3)$$

where H(Y) is the entropy and H(Y|X) the conditional entropy. In the above expressions, P_i and P_j' represent, respectively, the input and the output probabilities of the M symbols while with p_{ij} we have noted their transition probabilities, according to the scheme of Fig. 1. In the same figure, p_{ic} represents the erasure probability, i.e. the transition probability to a symbol which does not belong to the transmission alphabet. It must be noted that in the most general case the erasure symbol is not unique and, as in the M-PPM case, different erasure probabilities must be provided.

Both expressions (3) can be easily computed leading to the following compact expressions in the case of CM-PPM transmission:

$$H(Y) = \frac{1}{M} \sum_{j=1}^{k} \binom{n}{j} \sum_{i=0}^{k-j} \binom{n-j}{n-j-i} [1-\exp(-N)]^j [\exp(-N)]^i$$

$$\lg_2 \frac{M}{\sum_{i=0}^{k-j} \binom{n-j}{n-j-i} [1-\exp(-N)]^j [\exp(-N)]^i} , \qquad (4)$$

$$H(Y|X) = \frac{1}{M} \sum_{j=1}^{k} \binom{n}{j} \sum_{i=0}^{j-1} \binom{j}{j-i} [1-\exp(-N)]^{j-i} [\exp(-N)]^i$$

$$\lg_2 \frac{1}{[1-\exp(-N)]^{j-i} [\exp(-N)]^i} , \qquad (5)$$

and, in the case of M-PPM transmission,

$$H(Y) = [1-\exp(-N)]^k \lg_2 \frac{M}{[1-\exp(-N)]^k} , \qquad (6)$$

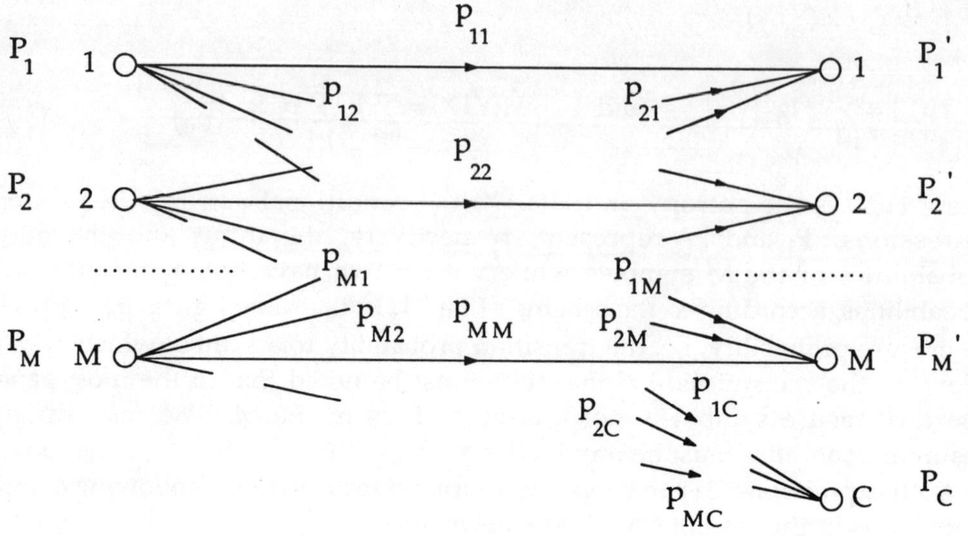

Fig. 1 - Scheme of a general M-ary channel

$$H(Y|X) = [1\text{-}\exp(\text{-}N)]^k \ \lg_2 \frac{1}{[1\text{-}\exp(\text{-}N)]^k} \tag{7}$$

In (4)÷(7) N represents the average number of received photons when the high level is transmitted. In order to maintain a fixed information rate, we have to consider a bandwidth expansion factor, whose minimum value is

$$\beta = \frac{n}{I} \tag{8}$$

while the transmission efficiency, expressed in bit/photon, turns out to be

$$\rho = \frac{I}{nP_{high}N} \tag{9}$$

where

$$P_{high} = \frac{\sum_{i=1}^{k} i \binom{n}{i}}{n \sum_{i=1}^{k} \binom{n}{i}} \tag{10}$$

is the input probability of the high transmission level. The quantities defined till now can be computed for different values of the parameters. In particular, Figs. 2 and 3 show the transmission efficiency ρ as a function of the minimum bandwidth expansion factor β, in the case of PPM transmission, of M-PPM transmission, with k=2, and of CM-PPM transmission, again with k=2. Along each curve n is taken variable. We have assumed, in Fig. 2, N=1 and, in Fig. 3, N=10.

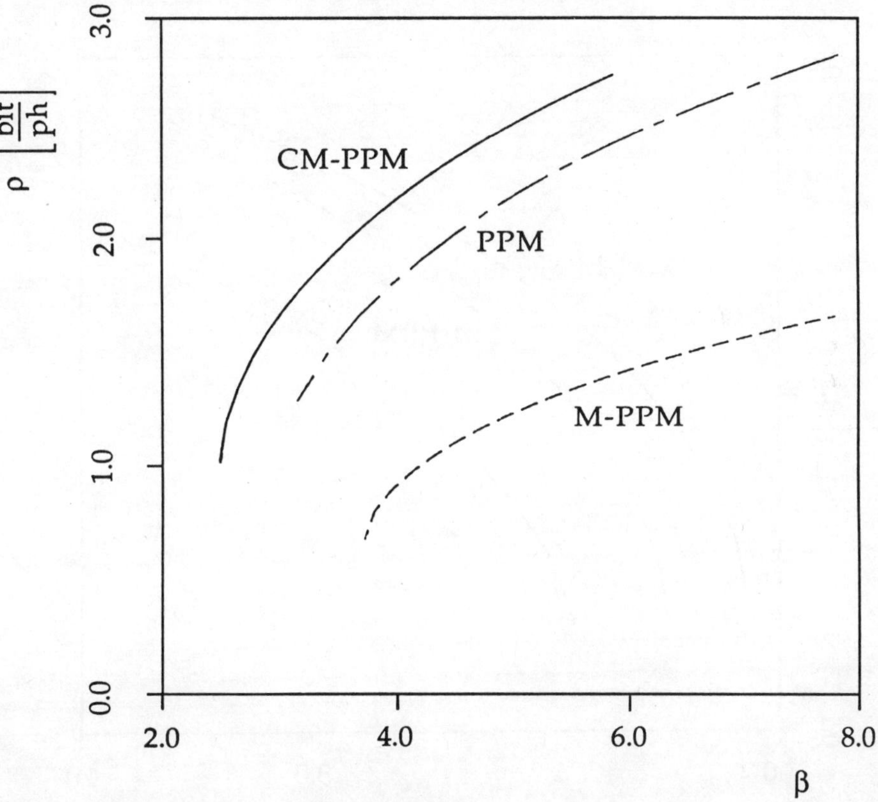

Fig. 2 - Transmission efficiency ρ as a function of the minimum bandwidth expansion factor β, with N=1

In the latter case the values of ρ are lower because of the presence of N at the denominator of expression (9). As it is clearly shown in the Figures, M-PPM transmission is efficient for reducing the bandwidth expansion factor of a PPM transmission only in the case of N sufficiently high.

On the contrary, the curves resulting from the use of CM-PPM are always over that of the PPM one, so confirming how it is effective in order to reduce the bandwidth occupancy for a fixed transmission efficiency.

A qualitative explanation of this behaviour can be found in the observation that the transmission here proposed must satisfy to weaker constraints with respect to the PPM and M-PPM transmissions. A similar result was recently observed [5] in the case of the binary transmission with different a priori probabilities for the two transmitted symbols. In practice, the objective of realizing a binary transmission characterized by the probability of the high transmission level different from that of the low transmission level can be achieved just by means of a set of encoded symbols like those here proposed.

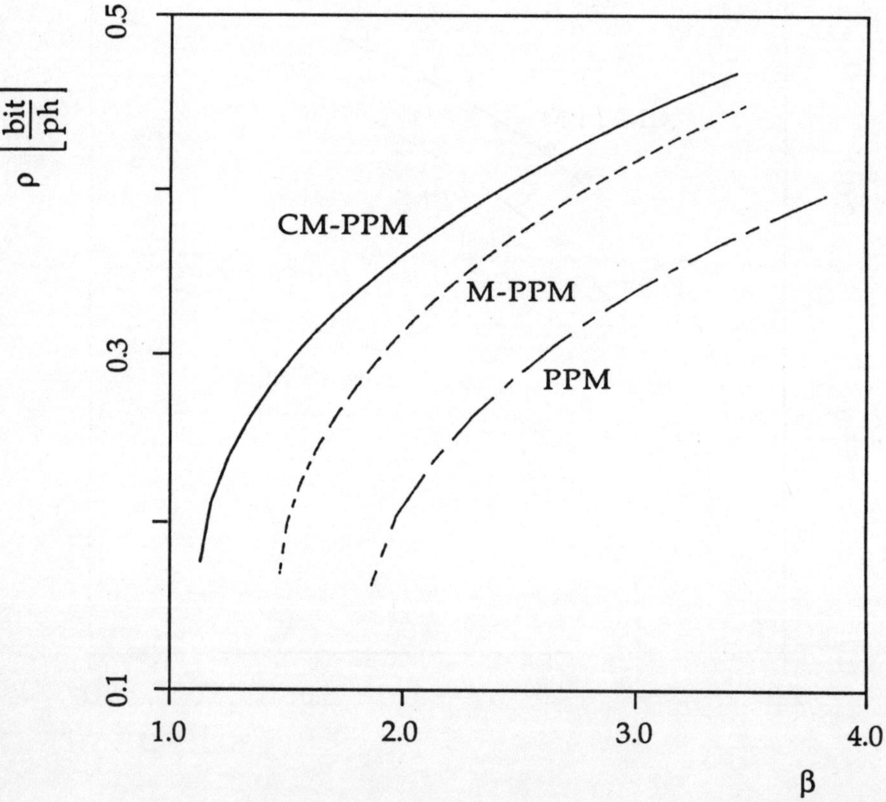

Fig. 3 - Transmission efficiency ρ as a function of the minimum bandwidth expansion factor β, with N=10

In particular, it is easy to understand that the performances of these two transmission systems become closer and closer when n tends to infinity .

For confirming this statement, we have reported in Fig. 4 the transmission efficiency ρ as a function of β for four different values of k, assuming N=1, in the case of an M-ary PPM transmission (dashed line), of a binary transmission with different a priori probabilities (continuous line) and of a CM-PPM transmission (dots). Though the CM-PPM dots are always lower the line of the binary transmission with different a priori probabilities, which seems to be a sort of limit curve, they tend to it when k is increased, that is when, for a fixed value of β, a larger alphabet is considered.

As regards the bit error rate characteristics of the alphabet here proposed, we can say that a slight degradation, with respect to PPM or M-PPM transmission is expected. Nevertheless, the possibility of applying codes which exploit the asymmetry peculiar to the photon counting channel, whose effects are combined with those of CM-PPM in a multiplicative way, seems to be very promising.

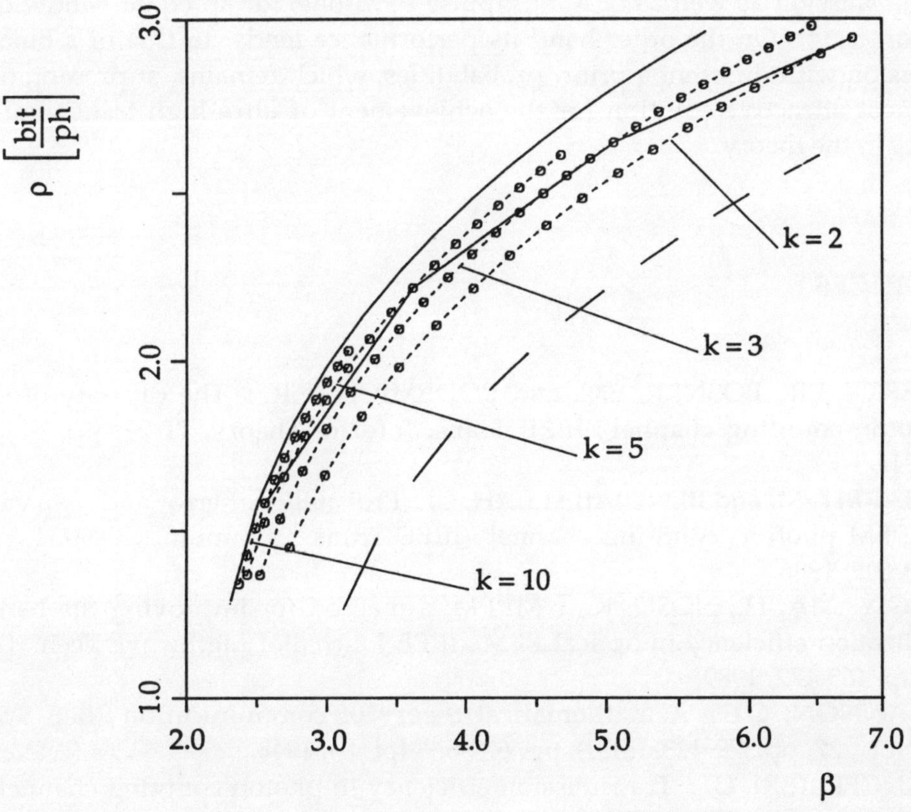

Fig. 4 - Transmission efficiency ρ as a function of the minimum bandwidth expansion factor β, with N=1

The objective is to synthesize new coding schemes that, minimizing the number of control bits, as it is made possible by the characteristics of the asymmetric channel, allow to reach bit error rate values at least comparable with those resulting, for example, by the application of a Reed-Solomon code to a PPM transmission [6].

3 . CONCLUSIONS

The possibility of improving the bandwidth utilization efficiency over an ideal photon counting channel has been analyzed in the ideal situation represented by Shannon theorem for discrete channels. A new structure of the transmission alphabet, including a very large set of suitably encoded symbols has been proposed. It is characterized by a higher efficiency than that of an orthogonal PPM transmission as well as of a multipulse PPM one, for an equal bandwidth expansion factor. On the other hand its performance tends to that of a binary transmission with different a priori probabilities, which remains, at present, one of the most attractive solution for the achievement of ultra-high transmission efficiency in the theory.

REFERENCES

[1] PIERCE, J.R., POSNER, E.C. and RODEMICH, E.R. : 'The capacity of the photon counting channel', IEEE Trans. Inform. Theory, IT-27, pp. 61-75, 1981.

[2] CHARBIT, M. and BENDJABALLAH, C. : 'Probability of error and capacity of PPM photon counting channel', IEEE Trans. Commun., COM-34, pp. 600-605, 1986.

[3] SUGIYAMA, H., NOSU, K., : 'MPPM: a method for improving the band-utilization efficiency in optical PPM', IEEE Journal of Ligthwave Tech., LT-7, pp. 465-472, 1989.

[4] SHANNON, C.E.: 'A mathematical theory of communication', Bell Syst. Tech.J. , 27, pp.379-423, (pt. I); pp. 623-656 (pt. II), 1948.

[5] CANCELLIERI, G. : 'Transmission efficiency in photon counting channels', IEEE Trans. Commun., COM-37, pp. 183-187, 1989.

[6] McELIECE, R.J.: 'Practical codes for photon communication', IEEE Trans. Inform. Theory, IT-27, pp. 393-398, 1981.

Properties of System using Spatial Multi-Mode Squeezer

Seitetsu BUN[†] Tsuyoshi SASAKI[††]

[†]Faculty of Science and Technology Keio University, Yokohama, Kanagawa, 223 Japan
[††]Faculty of Engineering, Tamagawa University, Machida, Tokyo, 194 Japan

1. Introduction

Recently, optical communications based on the quantum theory have received much attention.[1],[2] Especially, the quantum control system is one of important topics in the field of quantum communication theory. By quantum control system, we mean an application of non-standard quantum state (squeezed state, photon number state and so on) to optical communications. So far theoretical and experimental analysis have been reported for basic properties and generations of such non-standard quantum states.[3] At present time, problems of realization of advanced communication system are most important.

We are concerned with the realization of the quantum communication utilizing non-standard quantum states, especially, squeezed state.[4] However, a serious problem arises in such a system for practical use, namely, effect of the energy loss. To cope with this serious problem, a mathematical model of received quantum state control (RQSC) has been proposed by our group.[5] According to the theory, the advantage of squeezed state does not suffer the effect of the energy loss. As a result, the realization problem is close up.

In this paper, we give a proposal of quantum state switch system using a spatial multi-mode squeezer for realization of RQSC system. Particularly, it is verified that a property of error rate of the proposed system can overcome the standard quantum limit. Then we compare the properties of proposed RQSC system using three types of optical receiver as quantum measurement process, namely, homodyne, photon counting receiver and quantum optimum receiver.

2. Proposed Quantum State Switch System in RQSC System

In RQSC system, a transmitter is a conventional laser. The light wave of the transmitter is modulated by On-Off Keying (OOK), namely, vacuum state $|0>$ as signal "0" and coherent state $|\alpha>$ as signal "1" are sent, respectively. It is well known that the coherent state is only the quantum state which maintains the pure state passing through the transmission line with energy loss. So the received signals are the coherent state signals.

On the other hand, the proposed RQSC receiver is basically constructed by three parts: spatial multi-mode squeezer, spatial mode filter and quantum measurement. Fig.1 shows a block diagram of a proposed received quantum state control (RQSC) receiver part. The spatial multi-mode squeezer and spatial mode filter operates as

Lecture Notes in Physics, Vol. 378
C. Bendjaballah, O. Hirota, S. Reynaud (Eds.),
Quantum Aspects of Optical Communications

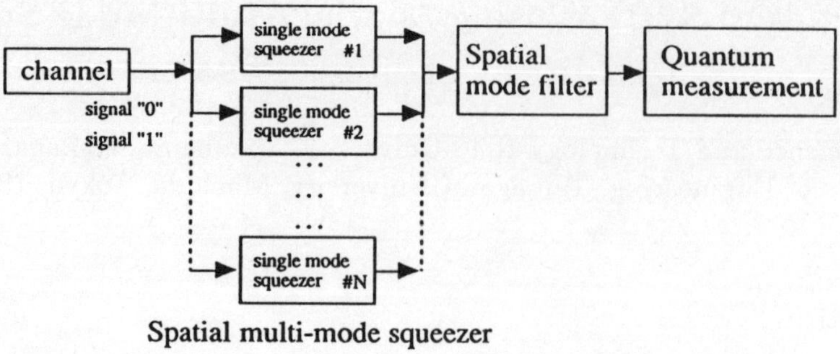

Figure 1: Block diagram of RQSC system using spatial multi-mode squeezer

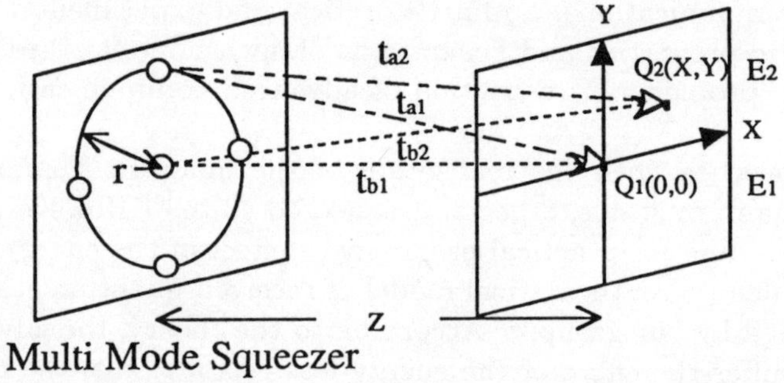

Figure 2: Construction of spatial multi-mode squeezer

quantum state switch system at the receiver. In this section, we explain feature of each part of quantum state switch system.

(a) Spatial Multi-Mode Squeezer

The spatial multi-mode squeezer carries out the quantum state control of the received coherent state signals. The spatial multi-mode squeezer is constructed by bundle of the single mode squeezer. Fig.2 shows a bundle construction of the spatial multi-mode squeezer.

The binary received signals are devided and injected to each single mode squeezer. If binary received signals are vacuum state (signal "0") and coherent state (signal "1"), output signals of each single mode squeezer produce a squeezed vacuum state and squeezed state, respectively.

In our squeezer system, each squeezer is spatial independent. So output squeezed states signals of the squeezer are given by a composition light of each mode squeezer. When the input is signal "0" or "1", the spatial multi-mode squeezer emits the amplified coherent signal or zero coherent signal, respectively. That is,

$$\tau_j = \frac{r_{2j} - r_{1j}}{c} \tag{8}$$

where c is the speed of light and r_{ij} is a distance from Q_i on screen (i=1 or 2) to single mode squeezer. By the same methods, the average powers of E_1 and E_2 in case of signal "0" are given as follows:

$$\begin{aligned}\left\langle |E_1|^2 \right\rangle = \left\langle |E_2|^2 \right\rangle &= \left\langle |E_a|^2 \right\rangle + \left\langle |E_b|^2 \right\rangle + \ldots.. \\ &= \sum_{j=a} \Gamma_j(0)\end{aligned} \tag{9}$$

From Eqs.(4), (7) and (9), the spatial coherence of the spatial multi-mode squeezer results in

$$|\gamma_{12}| = \left| \frac{\sum_{j=a} \Gamma_j(\tau_j)}{\sum_{j=a} \Gamma_j(0)} \right| \tag{10}$$

Eq.(10) means that the spatial coherence of the spatial multi-mode squeezer depends on the auto-correlation of each single mode squeezer.

It is known that the single mode squeezer in case of signal "0" can realize degenerate parametric amplifier (DPA)[6]. The auto-correlation function of DPA was given as follows:[7]

$$\Gamma(\tau) = \left\{ \frac{1}{4}\varepsilon\gamma\exp(-\frac{1}{2}\gamma|\tau|) \left[\frac{\exp(\varepsilon|\tau|)}{\frac{1}{2}\gamma - \varepsilon} - \frac{\exp(-\varepsilon|\tau|)}{\frac{1}{2}\gamma + \varepsilon} \right] + |\alpha_{\text{out}}|^2 \right\} \times \exp(i\omega_s\tau) \tag{11}$$

where ϵ, γ are the driving rate and damping factor of DPA, respectively.

Fig.3 shows numerical examples of the coherence functions when the signal "0" is injected into our receiver system. In Fig.3, the coherence function for signal "0" rapidly varies on the screen. This property means that the spatial multi-mode squeezer emits spatial in-coherent light in the case of signal "0".

If an arrangement of each mode squeezer is irregular, moreover, spatial differences between output squeezed states signals becomes large. So we can consider that spatial coherence of signal "0" from spatial multi-mode squeezer are perfect in-coherent perfectly.

On the other hand, a phase of output squeezed state emitted by each mode squeezer in case of signal "1" is locked by an injected signal. So there exists a perfect correlation between each output squeezed state signal. It is evident that output squeezed state in case of signal "1" is spatially coherent light.

[Signal"0"]

$$< a_1 >_{out} = 0 \tag{1}$$

[Signal"1"]

$$< a_1 >_{out} = \sum_{n=1}^{N} (\mu_n + \nu_n) < a_1 >_{in} \tag{2}$$

where N and (μ_n,ν_n) are number of squeezer and squeezing parameters, respectively. On the other hand, output squeezed states of both signals have the following quantum noise for cosine component:

$$< (\Delta a_1)^2 >_{out} = \sum_{n=1}^{N} \frac{1}{4}(\mu_n + \nu_n)^2 \tag{3}$$

(b) Spatial Coherence of Output Squeezed States

Moreover, these output squeezed states signals have spatial differences because of spatial independence and correlation between output squeezed state signal of each mode squeezer. In general, it is well known that a degree of spatial coherence $|\gamma_{12}|$ is defined as follows:

$$|\gamma_{12}| = \left| \frac{\langle E_1 E_2^* \rangle}{\left\{ \langle |E_1|^2 \rangle \langle |E_2|^2 \rangle \right\}^{1/2}} \right| \tag{4}$$

If there exists a screen beyond the spatial multi-mode squeezer (Fig.2), composition light waves E_1 and E_2 at certain places Q_1 and Q_2 on the screen are given as follows:

$$\mathrm{E}_1 = E_a(t_{1a}) + E_b(t_{1b}) + ... \tag{5}$$

$$\mathrm{E}_2 = E_a(t_{2a}) + E_b(t_{2b}) + ... \tag{6}$$

where t_{1j} and t_{2j}j are the time to reach at Q_1 and Q_2, respectively. (j:mode number of each squeezer, j = a,b,...)

By nature of the independence of signal "0" emitted by each single mode squeezer, a mutual coherence function $< E_1 E_2^* >$ is given as follows:

$$\begin{aligned} \langle \mathrm{E}_1 E_2^* \rangle &= \langle \{E_a(t_{1a}) + E_b(t_{1b}) + ...\} \{E_a(t_{2a}^*) + E_b^*(t_{2b}) + ...\} \rangle \\ &= \langle E_a(t_{1a}) E_a(t_{2a}^*) \rangle + \langle E_b(t_{1b}) E_b(t_{2b}^*) \rangle + ... \\ &= \sum_{j=a} \Gamma_j(\tau_j) \end{aligned} \tag{7}$$

where $\Gamma_j(\tau)$ is auto-correlation function ($\Gamma_j(\tau) =< E_j(t) E_j^*(t+\tau) >$) and τ_j is as follows:

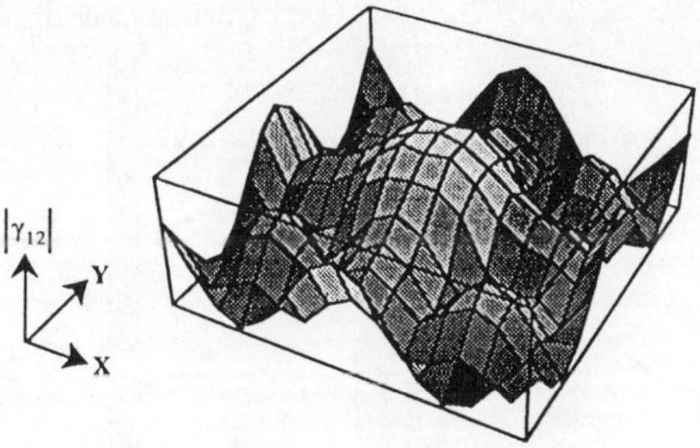

Figure 3: Spatial coherence of signal"0"

(c) Spatial Mode Filter

When binary signals emitted by the spatial multi-mode squeezer is forcused by lens, each forcused signal produces a different beam area, because of the difference of spatial coherence between signals. By utilizing this difference of beam area, a spatial mode filter which is pin-spot type filter can exclude quantum noise of others modes except for signal mode: spatial in-coherent light.

Let A_c and A_i be this produced beam areas of the spatial coherent light part and the in-coherent light part after passing through a lens, respectively. When there exists an optical lens between the spatial multi-mode squeezer and spatial mode filter as shown in Fig.4, each beam area produced by optical lens is defined as follows:

$$\mathrm{A}_c = \pi r_c^2, A_i = \pi r_i^2 \tag{12}$$

where r_c and r_i are radius of beam area, respectively. Moreover, it is considered that spatial coherence of signal "0" and signal "1" from spatial multi-mode squeezer are perfect in-coherent and coherent. So r_c and r_i is given as follows:

$$\mathrm{r}_c = f_2 \frac{\lambda}{D}, r_i = f_2 \frac{2r}{f_1} \tag{13}$$

where f_1 and f_2 are distance from lens to spatial multi-mode squeezer and spatial mode filter, respectively. λ and D are a wavelength of output squeezed state and a diameter of lens. As a result, the efficiency of the spatial mode filter which has the same spot size as A_c is given as follows:

$$\eta = \frac{\mathrm{A_c}}{\mathrm{A_i}} = (\frac{\lambda \mathrm{f_1}}{\mathrm{2rD}})^2 \tag{14}$$

Fig.5 shows numerical example of efficiency of spatial mode filter and beam area of spatial coherent light (signal"1"). In Fig.5, each signal has a different beam area by lens ($A_c << A_i$), so that the efficiency of the spatial mode filter becomes very small. This property means that the spatial mode filter can eliminate the spatial in-coherent light, namely, quantum noise of signal "0". So the squeezed state of each signal passing

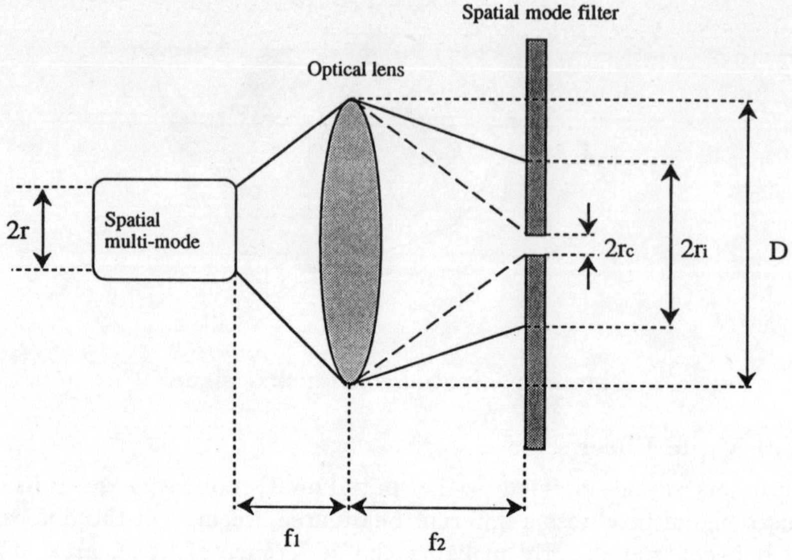

Figure 4: Schematic diagram of spatial mode filter

through this spatial-mode filter has the following signal power and quantum noise of cosine component:

[Signal"0"]

$$< a_1 >_{out} = 0 \tag{15}$$

$$< (\Delta a_1)^2 >_{out} = \frac{\eta}{4} \sum_{n=1}^{N} (\mu_n + \nu_n)^2 + \frac{1-\eta}{4} \tag{16}$$

[Signal"1"]

$$< a_1 >_{out} = \sum_{n=1}^{N} (\mu_n + \nu_n) < a_1 >_{in} \tag{17}$$

$$< (\Delta a_1)^2 >_{out} = \frac{1}{4} \sum_{n=1}^{N} (\mu_n + \nu_n)^2 \tag{18}$$

The third term of the right hand side in Eq.(16), is the additive vacuum noise by spatial-mode filter. These equations mean that the output beams provide the different signal and noise powers for signal "0" and signal "1".

Furthermore, Fig.5 shows that the efficiency of the spatial mode filter is close to zero. By this effect of the spatial mode filter, output quantum state of signal "0"

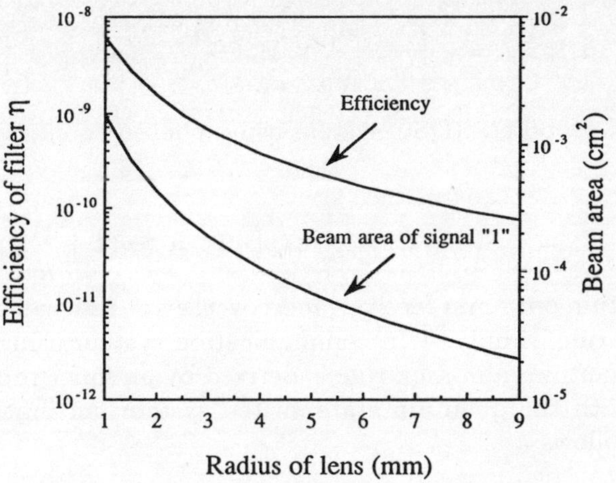

Figure 5: Efficiency of spatial mode filter and beam area of signal "1"

and signal "1" become close to the vacuum state $|0>$ and squeezed state $|\alpha_{sq};\mu,\nu>$, respectively. Namely, received quantum states of signal "0" and "1" are transformed into vacuum state and squeezed state by the spatial multi-mode squeezer and spatial mode filter, respectively. This transformation of signal means the spatial multi-mode squeezer and spatial mode filter operating as quantum state switch system. However, these operations in the proposed RQSC receiver are based on the difference of the spatial coherence of binary signals. So a modulation of the transmitted signal is to be On-Off Keying in the proposed RQSC system.

3. Effect of Various Measurements

In proposed RQSC system, the binary signals with unbalanced quantum noises produced by quantum switch system are detected by quantum measurements. In case of the balanced homodyne receiver as one of quantum measurements, a new signal decision rule, so called double thresholds rule is used. This signal decision rule is obtained by applying the Bayes decision rule to such unbalanced noises.[8]

Here, we analyze the error property of the proposed RQSC system using other types of receivers instead of the balanced homodyne receiver, namely, photon counting and quantum optimum receiver as quantum measurement. As described in Section 2, we can consider that received quantum states of signal "0" and "1" are transformed into the vacuum state $|0>$ and squeezed state $|\alpha_{sq};\mu,\nu>$ by quantum state switch system, respectively.

In case of photon counting receiver, error rate is given by photon distribution of signal "1", namely, squeezed state, because channel model is Z channel by the quantum state switch system. It is known that the photon distribution P(n) of squeezed state is Hermite distribution as follows:

$$P(n) = |< n|\alpha_{out}; \mu, \nu >|^2 = \frac{1}{|\mu| n!} (\frac{|\nu|}{2|\mu|})^n \left| H_n(\frac{\alpha_{in}}{\sqrt{2\mu\nu}}) \right|^2 \exp[-\frac{|\alpha_{in}|^2 + |\alpha_{out}|^2}{2|\mu|^2}] \quad (19)$$

So error rate of proposed RQSC system using photon counting receiver is given as follows:

$$P_e = \frac{1}{2} P(n = 0) = \frac{1}{2\mu} exp[-2 < n >_R (1 + \frac{\nu}{\mu})] \quad (20)$$

In case of quantum optimum receiver, moreover, error rate can be given by applying quantum mini-max rule, simply.[9] In communication system using binary signals, the error rate by the quantum mini-max rule is derived by an inner-product of each signal. In RQSC system with the quantum state switch system, an inner-product of binary signals is given as follows:

$$< 0|\alpha_{sq} > = \frac{1}{\sqrt{\mu}} \exp[-\frac{1}{2} |\alpha_{in}|^2 (1 + \frac{\nu}{\mu})] \quad (21)$$

So error rate of the proposed RQSC system using the quantum optimum receiver is given as follows:

$$P_e = \frac{1}{2} [1 - \{1 - \frac{1}{\mu} exp[-2 < n >_R (1 + \frac{\nu}{\mu})]\}^{1/2}] \quad (22)$$

These equations mean that the properties of the proposed RQSC system depends on received average photon number and squeezing parameter, while that of the general quantum receiver only depends on received average photon number.

Fig.6 and Fig.7 shows the error property of the proposed RQSC system using three types of receivers with respect to squeezing parameter and received average photon number. Fig.6 shows error rate of RQSC system is improved by the effect of squeezing. Moreover, Fig.7 shows that the proposed RQSC system can overcome the standard quantum limit achieved by the conventional binary PSK homodyne system using coherent state signals. This means that the proposed system can provide an advantage of quantum state controls regardless of the energy loss. Moreover, the error rate property of the proposed RQSC system with the quantum state switch system hardly depends on receiver types in Fig.7. In other words, the proposed RQSC system using the quantum optimum receiver can provide again superior property, but the difference among the other measurement processes is reduced.

4. Conclusions

In this paper, we have proposed a realization of the quantum state switch system by using spatial coherence property. More the properties of the received quantum state control system using three types of quantum measurements have been clarified. As a result, it was shown that the quantum optimum receiver provides again superior property, but the difference among the other measurement processes is reduced.

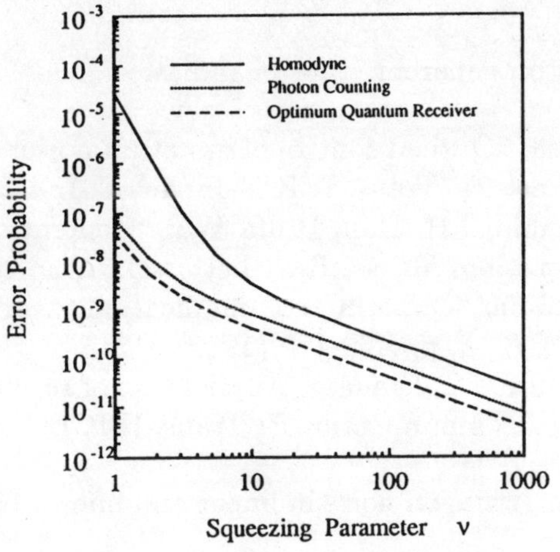

Figure 6: Error rate properties of RQSC system (1)

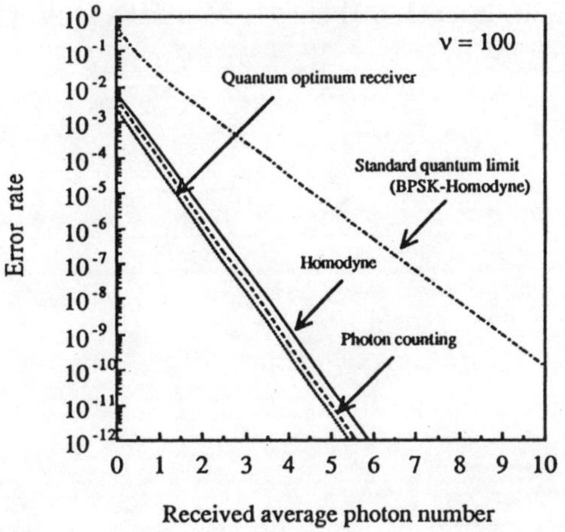

Figure 7: Error rate properties of RQSC system (2)

Acknowledgment

The authors would like to acknowledge O.Hirota (Tamagawa University), M.Nakagawa (Keio University), O.Kagami (NTT) and K.Yamazaki (Tamagawa University) for their helpful advice. The authors also wish to thank all members studying in our research group for their guidance and helpful advice.

References

[1] H.P Yuen: "Two-photon coherent state of radiation fields", Phys. Rev. A-13, pp.-2226-2243 (1976)

[2] O.Hirota and S.Ikehara: "Optical control of quantum noise; Optical communication by two photon laser:Part 2", Trans. IEICE Japan, E61, 4, pp173-179 (1978)

[3] Ling-An Wu, H.J.Kimble, J.H. Hall, Huifa Wu: "Generation of squeezed state by parametric down conversion", Phys. Rev. Lett. vol.57, no.20, pp.2520-2523 (1986)

[4] O.Hirota and H.Tsushima: "Quantum communication theory and its applications", Trans. IEICE Japan, E72, 5, pp.460-470 (1989)

[5] O.Hirota, O.Kagami and M.Takahara: "Application of received quantum state control for coherent optical communications", Trans. IEICE Japan, E70, 9, pp.801-803 (1987)

[6] C.M.Caves: "Quantum limits on noise in linear amplifier", Phys. Rev. D-8, pp.1817-1839 (1982)

[7] M.J.Collet and R.Loudon: "Output property of parametric amplifiers in cavities", J. Opt. Soc.Am.B, 4, 10, pp.1525-1534 (1987)

[8] S.Bun, T.Sasaki. H.Tsushima, O.Hirota and M.Nakagawa: "A realization of optical communication using received quantum state control", Trans.IEICE Japan, E73,10 (1990)

[9] O.Hirota: "Optical communication theory", Morikita Pub. (1985)

Achievement of New Quantum Limit by Switching of Pump Frequency of Squeezer

Hidenori Koyano† Norihiro Yoshida†† Tomohiko Uyematsu†

† Tokyo Institute of Technology †† Tamagawa University

1 Introduction

Recently, the performance of optical fiber communication systems is approaching to the standard quantum limit (SQL). Thus, as a method to overcome the SQL, the quantum communications[1] have received much attention in the field of practical engineering. So far, several systems have been proposed to overcome the SQL. However, almost all systems do not provide the ultimate performance, since these systems transmit the squeezed states in order to gain high performance, but this advantage of the squeezed state is diminished by the transmission loss of the channel[2]. To cope with such a problem, the theory of received quantum state control (RQSC) has been proposed by Hirota's group[3]. Now, the realization of such a RQSC system has been studied. And, we find that the key to realize RQSC systems is how to construct the quantum state switch mechanism by using practical squeezers. In general, the squeezer cannot improve the signal to quantum noise ratio, but it is possible to realize the quantum state switch by changing the operation of the squeezer depending on the incident light signals, namely the incoming states of light. Thus, the quantum state switch system is realized by the squeezer and can improve the error performance of the communication system. Furthermore, the SQL may be overcome.

In this paper, we first propose practical method to realize the quantum state switch system. Then, we show that the RQSC system utilizing the quantum state switch system can overcome the SQL. Furthermore, we discuss the realization conditions on the proposed RQSC system, and show an method of system design.

2 Realization of RQSC system

2.1 Principle of system

The squeezer is represented by the unitary operator in the mathematical physics. This means that the squeezer corresponds to the noiseless amplifier or attenuator. Hence, the signal to quantum noise ratio (SNR) is not improved by the squeezer. However, the error performance of the binary digital transmission system can be improved by the

Lecture Notes in Physics, Vol. 378
C. Bendjaballah, O. Hirota, S. Reynaud (Eds.),
Quantum Aspects of Optical Communications

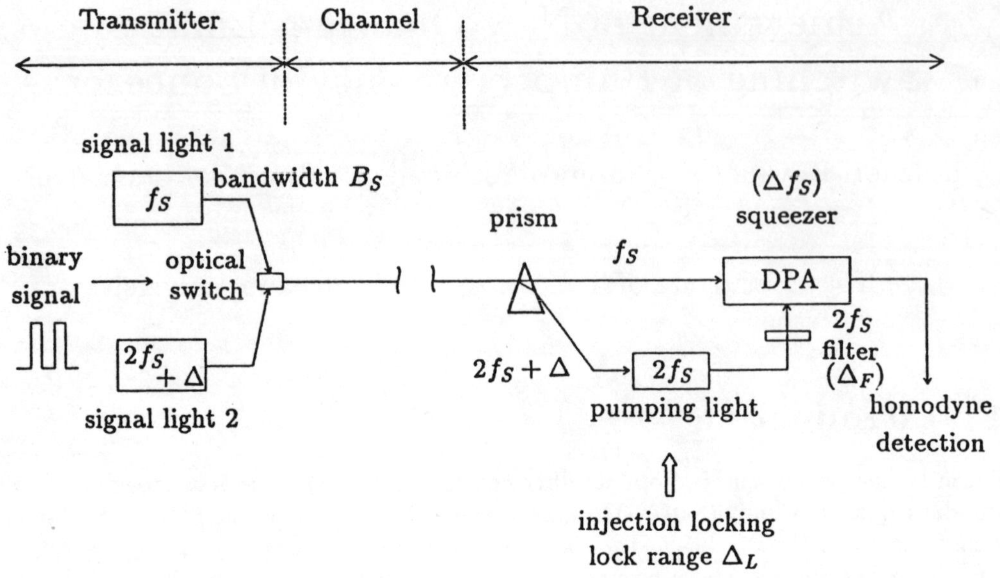

Figure 1: The quantum coherent communication system

application of quantum state switch using the squeezer and double thresholds decision process at receiver systems. As is well known, the best error performance, namely the SQL can theoretically be achieved by the optical communication system consists of the binary phase shift keying (BPSK) and homodyne detection. Hence, we propose a system which improves the error performance of the above communication system.

Fig.1 shows our proposed system. Assume that the symbols of the binary signals are "1" and "0". Transmitter consists of two optical sources with their light frequencies f_S and $2f_S+\Delta$. Moreover, we assume that these lights are in coherent states (represented as $|\alpha>$ in quantum mechanics). Here, we employ frequency shift keying (FSK) as modulation scheme. Thus, input binary signals switch two optical sources. That is, the signal "1" corresponds to the transmitted light with its frequency f_S and "0" corresponds to that with $2f_S + \Delta$. The proposed receiver consists of a prism, a squeezer as a type of a degenerate parametric amplifier (DPA) with pumping light source with the frequency $2f_S$, and a balanced homodyne detector. The homodyne detector is adjusted so as to detect the light of frequency f_S. This implies that the frequency of the local laser is f_S.

When the signal "1" is transmitted, the signal light 1, which is in a coherent state $|\alpha>$ with its frequency f_S, is injected into the squeezer directly. Then, it is amplified by the nonlinear optical phenomena of DPA, if both frequency matching and phase matching conditions are satisfied. Thus, output of the squeezer becomes a squeezed state $|\alpha : \mu, \nu>$ as shown in fig.2.

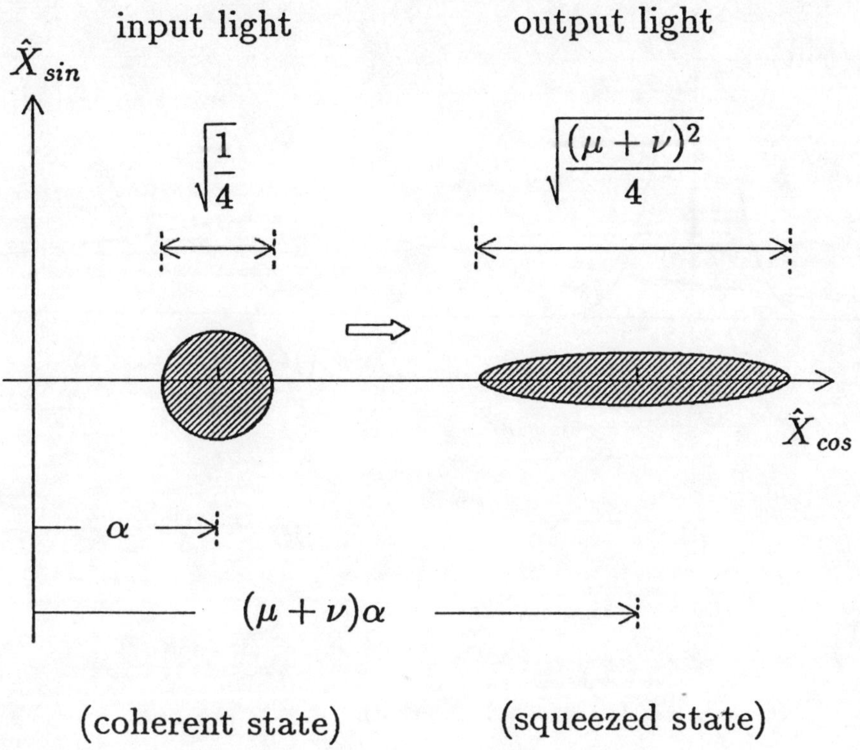

Figure 2: The function of squeezer

On the other hand, when the signal "0" is transmitted, the signal light 2 is branched by the prism and is injected into the pumping laser. In this case, the input of the squeezer is the vacuum state $|0>$. Since the pumping laser is locked by the injected signal, its frequency is shifted to $2f_S + \Delta$ from $2f_S$ by the injection locking phenomena. If the frequency shift goes beyond the bandwidth Δ_F of the optical filter before DPA, the pumping light is rejected by the filter and is not injected into DPA. Hence, DPA does not act as a squeezer and emits the vacuum state $|0>$ of frequency f_S, since no pumping light is injected into DPA. Consequently, we can conclude that the proposed system can realize the quantum state switch by using injection locking phenomena, namely, the frequency shift of the pumping light of DPA.

2.2 Error performance

In the proposed system, the output of balanced homodyne detector gives different noise properties for the symbols "1" and "0". That is, when signal "1" is transmitted, the squeezed state $|\alpha : \mu, \nu>$ with amplified quadrature amplitude (cosine component $\hat{X}_{COS}$) is detected. On the other hand, when "0" is transmitted,the vacuum state $|0>$ is detected. The probability densities of each case are Gaussian distributions as shown in Fig.3, and

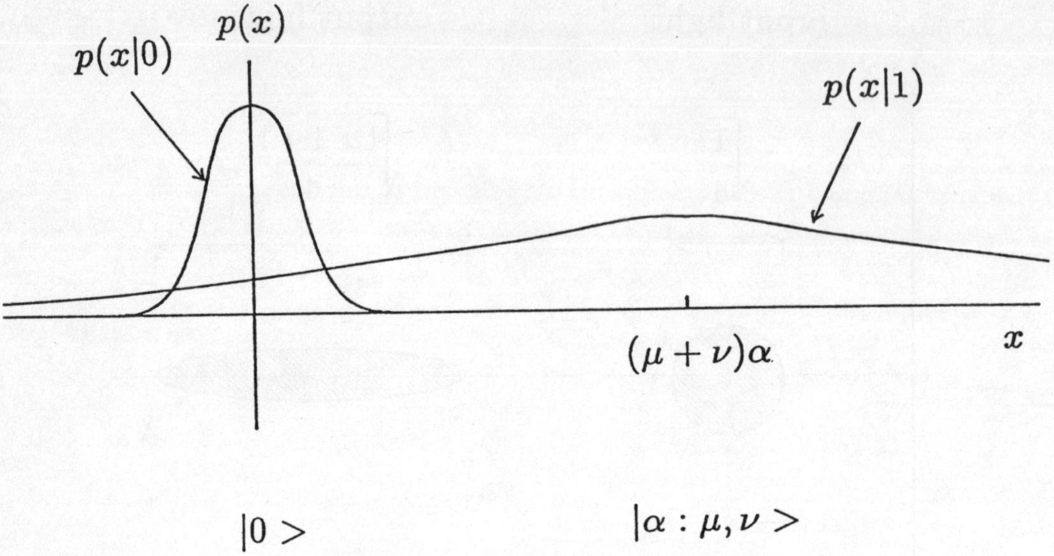

Figure 3: The probability densities

written as

(i) "1" is transmitted ($|\alpha : \mu, \nu >$ is received)

$$p(x|1) = \sqrt{\frac{2}{\pi(\mu+\nu)}} \exp\left[-\frac{2\{x-(\mu+\nu)\alpha\}^2}{(\mu+\nu)^2}\right] \tag{1}$$

$$< \hat{X}_{cos} > = (\mu+\nu)\alpha \tag{2}$$

$$< \Delta\hat{X}^2_{cos} > = \frac{1}{4}(\mu+\nu)^2 \tag{3}$$

(ii) "0" is transmitted ($|0 >$ is received)

$$p(x|0) = \sqrt{\frac{2}{\pi}} \exp(-2x^2) \tag{4}$$

$$< \hat{X}_{cos} > = 0 \tag{5}$$

$$< \Delta\hat{X}^2_{cos} > = \frac{1}{4} \tag{6}$$

where α denotes the average quadrature amplitude of input light of the squeezer, and μ, ν are called *squeezing parameters*, which represent the degree of squeezing. These parameters satisfy the relation

$$\mu^2 - \nu^2 = 1. \tag{7}$$

Moreover, $< \hat{X}_{cos} >$ and $< \Delta\hat{X}^2_{cos} >$ represent average and variance of the quadrature amplitude.

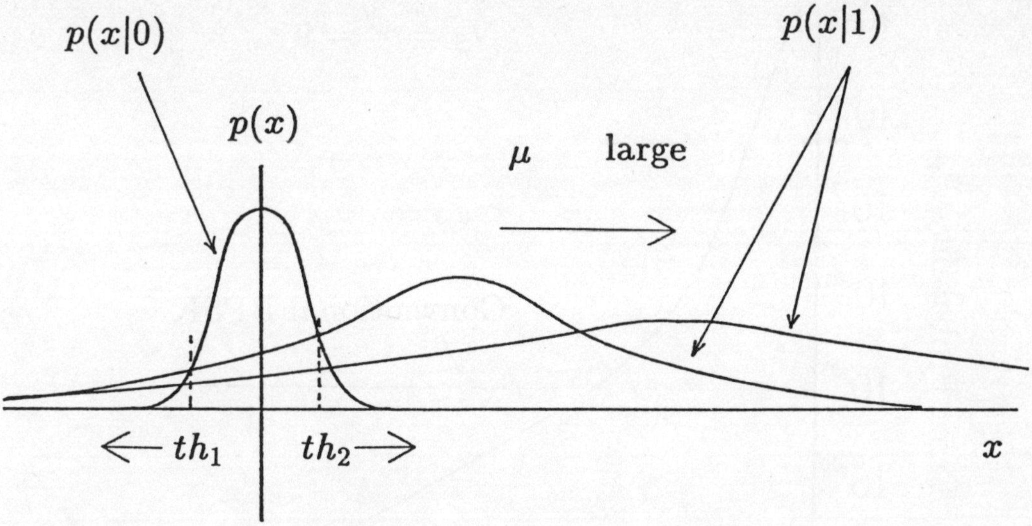

Figure 4: Double thresholds decision

The best decision rule for these noise distributions is known as the double thresholds decision rule as shown in Fig.4. The crossing points of $p(x|0)$ and $p(x|1)$ are threshold levels th_1 and th_2. Assume that the symbols "1" and "0" are transmitted with equal probability, the error probability is given by

$$Pe = \frac{1}{2}\int_{th1}^{th2} p(x|0)dx \quad (8)$$

$$+ \frac{1}{2}\left[\int_{-\infty}^{th1} p(x|1)dx + \int_{th2}^{\infty} p(x|1)dx\right]. \quad (9)$$

Fig.5 shows the behavior of the error probability P_e with respect to the squeezing parameter μ, when the transmitted photon number equals to 9. For comparison, this figure includes the SQL (the error rate of the conventional BPSK homodyne detection system equal to 10^{-9}) by the dotted line. From this error rate characteristics, we can concluded that our system can overcome the SQL if the squeezing parameter μ satisfies the following condition:

$$\mu > 22. \quad (10)$$

2.3 Design of proposed system

In this section, we discuss the various conditions required for the realization of proposed system, and show an design method for the physical parameters in the system satisfying the conditions.

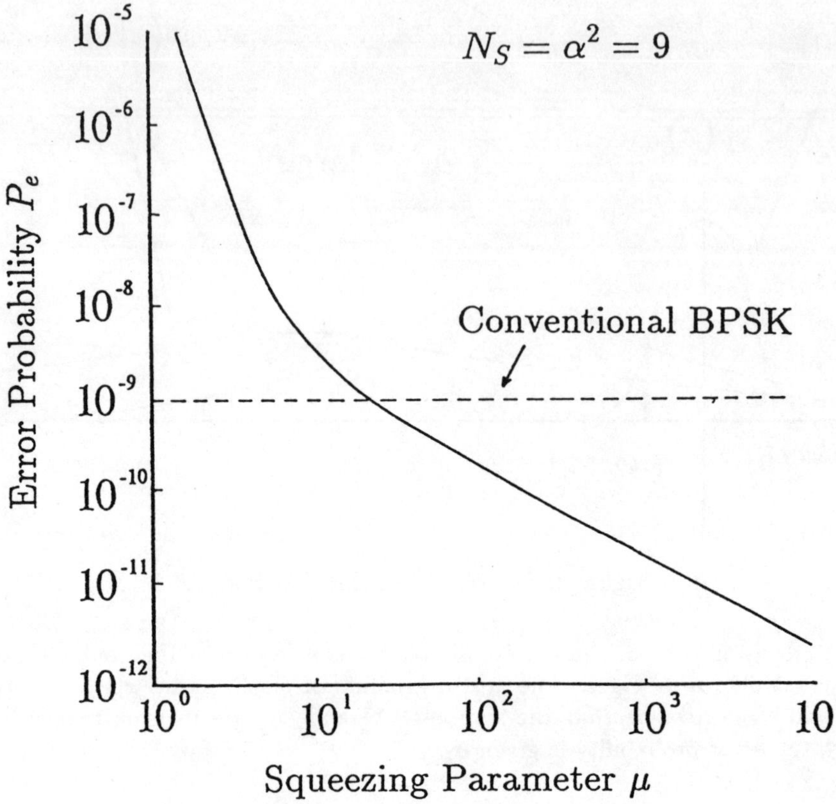

Figure 5: The behavior of error rate

(a) Parametric gain

As previously shown, the proposed system can overcome the SQL, as long as the squeezing parameter μ is larger than 22. The squeezing parameter μ is given by

$$\mu = \cosh(g_0 z_L), \tag{11}$$

where g_0 denotes the parametric gain, and z_L is the interaction length of DPA, which determines the gain of amplitude and the degree of squeezing. The parametric gain g_0 is given by

$$g_0 = \kappa\sqrt{P_P}, \tag{12}$$

and depends on nonlinear optical constant (or coupling constant) κ and power of pumping light P_P. The nonlinear optical constant κ is defined by

$$\kappa = \frac{4\pi^2\chi^{(2)} f_S}{n_0^{3/2}}\sqrt{\frac{8\pi}{c^3\sigma}}, \tag{13}$$

where $\chi^{(2)}$ represents the second order nonlinear susceptibility of the material, n_0 is the refractive index on index matching between signal light (ordinary ray) and pumping light (extraordinary ray), c is the velocity of light and σ is the cross area of interaction. From eq.(10),(11) and (12), in order that the proposed system overcomes the SQL, it must satisfy the following condition:

$$\boxed{\text{Condition 1}} \qquad \kappa\sqrt{P_P z_L} > 3.8. \tag{14}$$

(b) Squeezing bandwidth

If both frequency matching and phase matching conditions are satisfied in DPA, we can obtain the maximum gain of DPA by perfect squeezing. However, if the frequency of signal light with frequency f_S shifts to $f_S + \epsilon$, the refractive index matching Δn between the index of ordinary ray (n_o) and extraordinary ray (n_e) causes the phase mismatching Δk (see Fig.6). Since the efficiency of coupling by the interaction is degraded by the phase mismatching Δk, the frequency shift ϵ degrades the gain of DPA. This implies that DPA has its bandwidth for the squeezing. If the frequency $f_S + \epsilon$ of the signal light lies in the phase matching bandwidth, the signal light is regarded as a perfect squeezed state. From this fact, in what follows, the phase matching bandwidth is called by "*squeezing bandwidth*". Similarly, the frequency shift of pumping light also causes the degradation of squeezing.

Concerning the squeezing bandwidth, Caves and Crouch have derived the squeezing bandwidth Δf_S by the following equations[4][5]:

$$\Delta f_S = \sqrt{\frac{4\pi f_S c g_0}{\pi^2 |4\pi f_S n' + 4\pi^2 f_S^2 n''|}}, \tag{15}$$

$$\Delta f_P = \frac{c g_0}{2\pi f_S |n'|}, \tag{16}$$

where n' and n'' denote the second and third order differential coefficients of the curve describing the dependence of the refractive index on frequency, which are approximated by Sellmeir's equation. Fig.7 shows the characteristics of the squeezing bandwidth for the parametric gain g_0 when $LiNbO_3$ is employed as a nonlinear medium in DPA, and the wave length of optical source is 1.5μm (f_S is 200THz).

In optical communication systems, the signal light must have its bandwidth B_S around carrier frequency f_S due to the modulation. In our proposed system, signal light with bandwidth B_S must be squeezed by DPA. Thus, required squeezing bandwidth Δf_S is larger than the signal bandwidth B_S:

$$\boxed{\text{Condition 2}} \qquad \Delta f_S > B_S. \tag{17}$$

Thus, eq.(17) is the second realization condition for our proposed system.

(c) Lock range

In the proposed system, injection locking phenomena plays very important role in quantum state switching. When the symbol "0" is transmitted, signal light 2 with the frequency $2f_S + \Delta$ is injected into the pumping source with its frequency $2f_S$. In this case, by optical

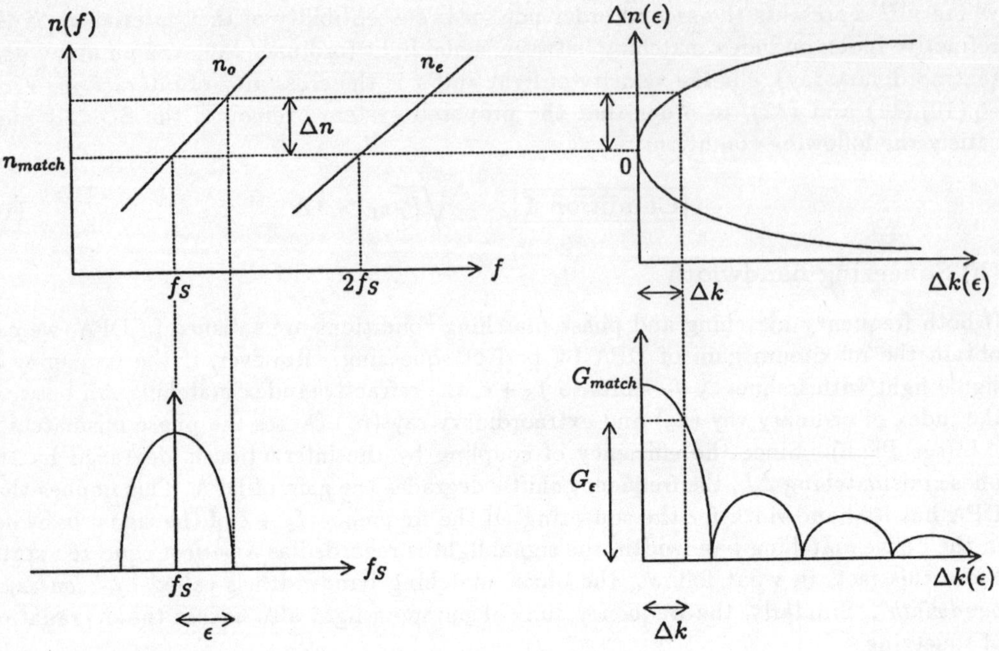

Figure 6: Degradation of gain due to index mismatching

injection locking phenomena, the signal laser acts as an injection laser and the pumping laser acts as a slave laser. Therefore, the frequency of the pumping light is locked to that of the signal light $2f_S + \Delta$. The lock range Δ_L of the pumping source is given by

$$\Delta_L = \frac{1}{2\pi\tau_P}\sqrt{\frac{N_S}{N_P}}, \tag{18}$$

where N_S is the photon number of the injection laser corresponding to the signal light 2, while N_P is that of the pumping light at free running. Moreover, τ_P denotes the life time of photon in the pumping cavity.

Next, the output of pumping source with the frequency locked to $2f_S + \Delta$ must be rejected by the optical filter with the bandwidth Δ_F. Therefore, the injection locking system must satisfy the following condition:

$$\boxed{\textbf{Condition 3}} \qquad \Delta_L > \Delta > \Delta_F. \tag{19}$$

So far, we have obtained three conditions in order to realize the proposed RQSC system achieving new quantum limit.

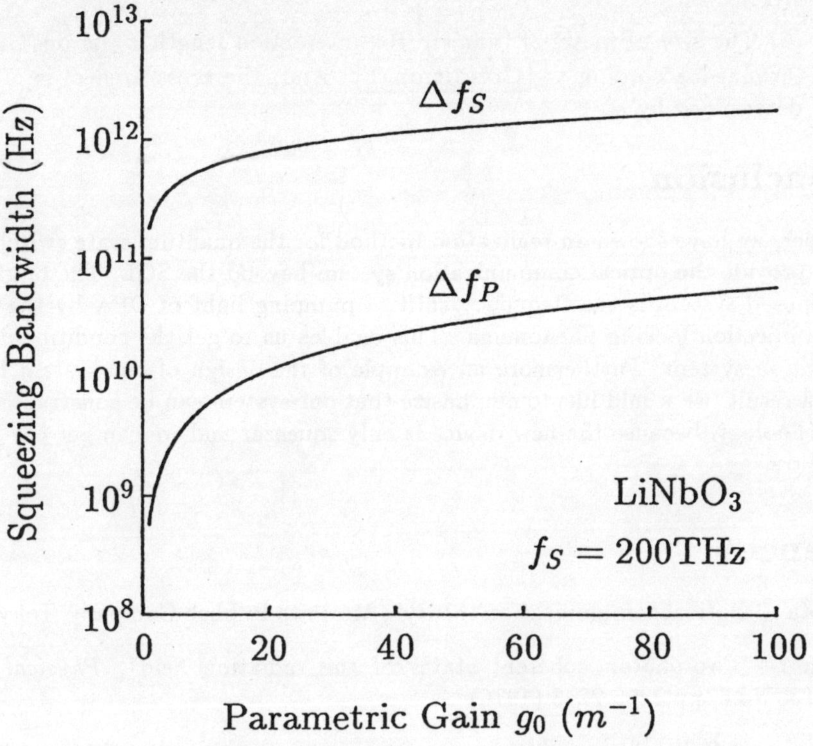

Figure 7: The characteristics of squeezing bandwidth

(d) Determination of physical parameters

Now, we show a method to determine the physical parameters in the proposed system satisfying these realization conditions.

An example of the procedure to determine the physical parameters is shown as follows:

(1) From **Condition 2**, determine the squeezing bandwidth Δf_S by signal bandwidth B_S.

↓

(2) For the material to be used in DPA, $\chi^{(2)}$, n' and n'' can be found. Then, parametric gain g_0 is determined.

↓

(3) According to **Condition 3**, the lock range Δ_L is determined by the bandwidth Δ_F.

↓

(4) τ_P and P_P are determined by Δ_L. Then a type of laser for the pumping light is determined.

↓

(5) Coupling constant κ is determined.

↓
(6) The size of material, namely the interaction length z_L is determined according to **Condition 1**. And, the cross area σ is determined by κ.

3 Conclusion

In this paper, we have shown an realization method for the quantum state switch system which can provide the optical communication system beyond the SQL. The original idea of our proposed system is the frequency shift of pumping light of DPA by the application of the injection locking phenomena. This enables us to get the conditional unitary communication system. Furthermore an example of the design of our system has been given. As a result, we would like to emphasize that our system can be constructed by the current technology, because the new device is only squeezer and we can get the squeezer in near future.

References

[1] O.Hirota : "Optical communication theory", Morikita Publish Company, Tokyo (1985)

[2] H.Yuen : "Two-photon coherent states of the radiation field", *Physical Review*, **vol.A13**, No.6, pp.2226-2243 (1976)

[3] O.Hirota and H.Tsushima : "Quantum communication theory and its application", *Journal of IEICE of Japan*, **vol.E72**, No.5, pp.460-470 (1989)

[4] C.Caves and D.Crouch : "Quantum wideband traveling-wave analysis of a degenerate parametric amplifier", *Optical Society of America* , **vol.B4**, No.10, pp.1535-1545 (1987)

[5] D.Crouch : "Broadband squeezing via degenerate parametric amplification", *Physical Review* , **vol.A38**, No.1, pp.508-511 (1988)

[6] A.Yariv : "Quantum electronics", John Wiley and sons (1989)

Session V

– Squeezed and Nonclassical States –

Chairman : S. Reynaud

INFORMATION TRANSMISSION WITH SQUEEZED LIGHT

Malvin C. Teich
Columbia Radiation Laboratory and Department of Electrical Engineering
Columbia University, New York, New York 10027, USA

and

Bahaa E. A. Saleh
Department of Electrical and Computer Engineering
University of Wisconsin, Madison, Wisconsin 53706, USA

INTRODUCTION

The nature of photon-number-squeezed light (also called sub-Poisson light) has been elucidated in recent years.[1-13] This type of light is expected to find use in the study of optical interactions in various disciplines, ranging from the behavior of the human visual system at the threshold of seeing[14] to optical precision measurement.[15] In this paper we consider the potential use of photon-number-squeezed light in direct-detection lightwave communication systems and other information-carrying applications.[16]

All lightwave communication systems that have been developed to date make use of Poisson (or super-Poisson) light.[17] For Poisson light, the variance of the photon number is identically equal to its mean for all values of the counting time T. Photon-number-squeezed light, on the other hand, has a photon-number variance that is less than its mean for all or some values of T.[1-3] Such light is intrinsically nonclassical in nature. The earliest sources of photon-number-squeezed light exhibited only a slight reduction of the variance.[1,2] Far stronger photon-number squeezing has been produced in recent years,[18] and continuing advances promise further improvement. It is therefore of interest to examine the advantages to be gained in using photon-number-squeezed light in a direct-detection lightwave communication system.

There are essentially two classes of mechanisms by means of which unconditionally photon-number-squeezed light may be generated. In the first class, squeezed photons are produced from a beam of initially Poisson (or super-Poisson) photons. This can be achieved in a number of ways, e.g., by mixing coherent light with quadrature-squeezed vacuum photons or by making use of correlated photon beams.[19] An experiment of

Lecture Notes in Physics, Vol. 378
C. Bendjaballah, O. Hirota, S. Reynaud (Eds.),
Quantum Aspects of Optical Communications

this kind was carried out by Tapster, Rarity, and Satchell.[20] Squeezed photons were generated from the pair of correlated photon beams produced in parametric downconversion; one of the twin beams was then fed back to control the pump.

The second class of mechanisms relies on the direct generation of squeezed photons from a beam of **initially sub-Poisson excitations** (e.g., electrons).[2,21] This technique was first used by Teich and Saleh in a space-charge-limited version of the Franck-Hertz experiment.[3] Perhaps the simplest implementation of this principle is achieved by driving a light-emitting diode (LED) with a sub-Poisson electron current,[22] but it is most effectively achieved by the use of a semiconductor injection laser.[7,12,18]

We discuss calculations of the channel capacity of a lightwave communication system based on the observation of the **photoevent point process**, demonstrating that it **cannot** in principle be increased by the use of photon-number-squeezed light.[16] We also discuss calculations that show that the channel capacity of a **photon-counting** system **can** be increased by the use of photon-number-squeezed light.[16] The channel capacity is the maximum rate of information that can be transmitted through a channel without error. The capacity of the photon channel has been the subject of a number of studies over the years.[23,5] We also discuss an example in which the use of photon-number-squeezed light produced from Poisson light either degrades or enhances the **error performance** of a simple binary ON-OFF keying photon-counting system, depending on where the average power constraint is placed.

COMMUNICATING WITH MODIFIED POISSON PHOTONS

Consider the transformation of a Poisson beam of photons (represented by a Poisson point process N_t of rate μ_t) into a sub-Poisson beam of photons represented by a point process M_t of rate λ_t, as illustrated in Fig. 1.

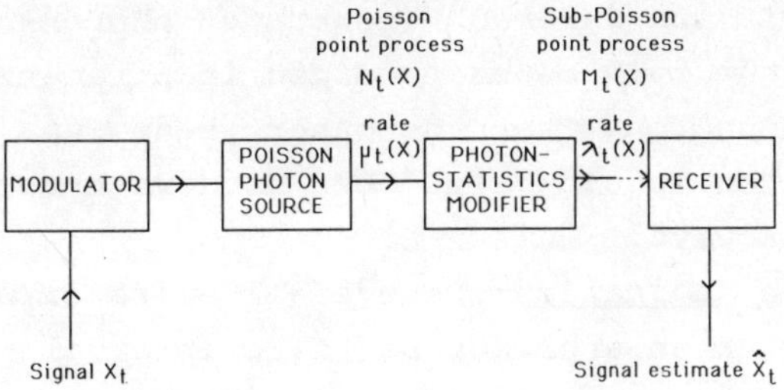

Fig. 1. Idealized lightwave communication system employing a Poisson photon source and a photon-statistics modifier. [After M. C. Teich and B. E. A. Saleh, in Progress in Optics, vol. 26, E. Wolf, ed., North-Holland, Amsterdam (1988)].

The events of the initial process N_t are assumed to be observable [e.g., by the use of correlated photon beams or a quantum-nondemolition (QND) measurement] and their registrations used to operate a mechanism which, in accordance with a specified rule, leads to the events of the transformed photon process M_t. The rate λ_t of the process M_t is thereby rendered a function of the realizations of the initial point process N_t at prior times, i.e., $\lambda_t = \lambda_t(N_{t'}; t' \le t)$.

Several examples of transformations of this kind that have been suggested for use in quantum optics are illustrated in Fig. 2 and discussed below. It is assumed for simplicity (but without loss of generality) that the various conversions can be achieved in an ideal manner.

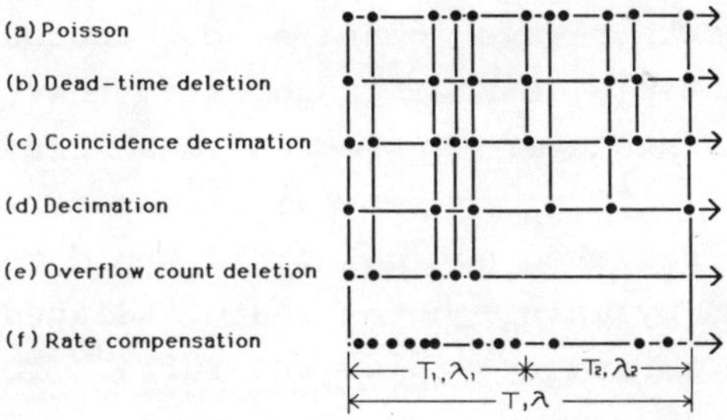

Fig. 2. Several transformations of Poisson photons into sub-Poisson photons that have been suggested for use in quantum optics. [After M. C. Teich and B. E. A. Saleh, in Progress in Optics, vol. 26, E. Wolf, ed., North-Holland, Amsterdam (1988)].

(1) Dead-time deletion: Delete all photons within a prescribed fixed (nonparalyzable) dead time τ_d following the registration of a photon.[24]

Rarity, Tapster, and Jakeman[25] generated photon-number-squeezed light by using one of the twin beams produced in parametric downconversion to selectively gate photons from the other beam via dead-time control. Dead-time deletion could also be used with correlated photon beams produced in other ways.

(2) Coincidence decimation: Remove all pairs of photons separated by a time shorter than a prescribed time interval τ'. This is achieved, for example, in second-harmonic generation (SHG); two photons closer than the intermediate-state lifetime of the SHG process are exchanged for a third photon (which is at twice the frequency and therefore easily eliminated).[26]

(3) Decimation: Select every rth photon (r = 2,3,...) of an initially Poisson photon process, deleting all intermediate photons. Saleh and Teich[27] suggested using correlated photon beams to implement this technique. In cascaded atomic emissions from ^{40}Ca, for example, sequences of correlated photon pairs (green and violet) are emitted. The green photons can be detected and used to operate a gate that passes every rth violet photon. Decimation control could also be used in conjunction with parametric-downconversion photon twins.

(4) Overflow count deletion: The number of photons occurring in preselected time intervals $[0,T_0]$, $[T_0,2T_0]$,..., is counted, retaining the first n_0 photons in each time interval (without changing their ocurrence times) and deleting the remainder. If the average number of photons in $[0,T_0]$ of the initial process is $\gg n_0$, then the transformed process will almost always contain n_0 photons within this time interval. As an example, Mandel[28] suggested that if a collection of n_0 atoms in the ground state are subjected to a brief, intense, incoherent excitation pulse, all n_0 atoms will become excited with high probability; the radiated optical field would then be describable, to good approximation, by an n_0-photon state. Related schemes have been proposed by Yuen[29] and by Stoler and Yurke[30] for use with parametric processes.

We proceed to illustrate that none of these modifications can increase the channel capacity of a communication system based on photoevent point-process observations.

If a constraint is placed on the rate of the initial Poisson process $\mu_t \le \mu_{max}$, then it is obvious that C cannot be increased by the modification $N_t \longrightarrow M_t$. This is simply a consequence of the definition of channel capacity: it is the rate of information carried by the system without error, maximized over all coding, modulation, and **modification** schemes.

However, can the modification $N_t \longrightarrow M_t$ increase the channel capacity if the constraint is instead placed on the rate of the modified process λ_t (i.e., $\lambda_t \le \lambda_{max}$)? We address this question for an arbitrary self-exciting point process in the next section.

COMMUNICATING WITH PHOTONS DESCRIBED BY A SELF-EXCITING POINT PROCESS

Consider a self-exciting point process M_t of rate $\lambda_t(M_{t'};\ t' \le t)$. This is a process that contains an inherent feedback mechanism in which present event occurrences are affected by the previous event occurrences of the same point process. Of course, the modified Poisson processes $N_t \longrightarrow M_t$ introduced above are special cases of self-exciting point processes.

An example of a system that generates a self-exciting point process is that of rate compensation (by linear feedback) of a source which, without feedback, would produce a Poisson process. Let each photon registration at time t_i cause the rate of the process to be modulated by a factor $h(t-t_i)$ (which vanishes for $t < t_i$). In linear negative feedback the rate is $\lambda_t = \lambda_0 - \Sigma_i h(t-t_i)$, where λ_0 is a constant. If the instantaneous photon registration rate happens to be above the average then it is reduced, and vice versa. This process is schematically illustrated in Fig. 2(f) for two adjacent sub-intervals T_1 and T_2. Yamamoto, Imoto, and Machida[31] suggested the use of rate compensation in conjunction with a QND measurement (using the optical Kerr effect) but rate compensation could be used just as well, for example, with correlated photon pairs. Dead-time deletion can be viewed as a special case of rate compensation in which the occurrence of an event zeros the rate of the process for a specified time period τ_d after the registration.[19]

Now consider a communication system that uses a point process $M_t(X)$ whose rate $\lambda_t(X)$ is modulated by a signal X_t. The process $M_t(X)$ can be an arbitrary self-exciting point process (e.g., it can be photon-number-squeezed) which includes processes obtained by the feedforward- or feedback-modification of an otherwise Poisson process.[19]

Neither feedforward nor feedback transformations can increase the capacity of this channel, as provided by Kabanov's theorem[32] and its extensions[33]:

Kabanov's Theorem — The capacity of the point-process channel cannot be increased by feedback. Under the constraint $\lambda_0 \le \lambda_t \le \lambda_{max}$, the channel capacity C is

$$C = \lambda_0 \left[\frac{1}{e} \left(1 + \frac{s}{\lambda_0}\right)^{1+\lambda_0/s} - \left(1 + \frac{\lambda_0}{s}\right) \log\left(1 + \frac{s}{\lambda_0}\right) \right], \tag{1}$$

where $s = \lambda_{max} - \lambda_0$. When $\lambda_0 = 0$ (no dark counts), this expressions reduces to

$$C = \lambda_{max}/e. \tag{2}$$

When the capacity is achieved, the output of the zero-dark-count point-process channel is a Poisson process with rate $\lambda_t = \lambda_{max}/e$ (the base e has been used for simplicity). The channel capacity has also been determined under added constraints on the mean rate. A coding theorem has also been proved. Kabanov's theorem is analogous to the well-known result that the capacity of the white Gaussian channel cannot be increased by feedback.[34]

In summary, no increase in the channel capacity of a point-process lightwave communication system may be achieved by using photons that are first generated with Poisson statistics and subsequently converted into sub-Poisson statistics, regardless of whether the power constraint is placed at the Poisson photon source or at the output of the conversion process. Nor may an increase in channel capacity be achieved by using feedback to generate a self-exciting point process.

COMMUNICATING WITH SQUEEZED PHOTON COUNTS

These conclusions are valid only when there are no restrictions on the receiver structure. The conlusion is different if the receiver is operated in the photon-counting regime, in which information is carried by the random variable n representing the number of photoevents registered in time intervals of prescribed duration T (rather than by the photon occurrence times).

The capacity of the photon-counting channel is given by[23]

$$C = B[\bar{n}\ln(1 + 1/\bar{n}) + \ln(1 + \bar{n})], \tag{3}$$

where $\bar{n}$ is the mean number of counts in T and B = 1/T is the bandwidth. Two limiting expressions emerge:

$$\begin{aligned} C &= B\bar{n}\ln(1/\bar{n}), && \bar{n} \ll 1 \\ C &= B\ln(\bar{n}), && \bar{n} \gg 1. \end{aligned} \tag{4}$$

If an added constraint is applied to the photon counts, such that they must obey the Poisson counting distribution, the capacity is further reduced. In that case, the limiting results analogous to Eq. (4) are

$$\begin{aligned} C &= B\bar{n}\ln(1/\bar{n}), && \bar{n} \ll 1 \\ C &= \tfrac{1}{2}B\ln(\bar{n}), && \bar{n} \gg 1. \end{aligned} \tag{5}$$

The capacity in the region $\bar{n} \gg 1$ is a factor of 2 smaller in Eq. (5) than in Eq. (4). The capacity-to-bandwidth ratio C/B is plotted versus $\bar{n}$, for both the unrestricted and Poisson photon-counting channels, in Fig. 3.

In the case of photon counting, therefore, an increase in the channel capacity <u>can</u> in principle be realized by using photon-number-squeezed light. However, in the small mean-count limit $\bar{n} \ll 1$ (when the counting time T is very short), the capacity of the Poisson counting channel approaches that of the unrestricted counting channel, and the advantage of photon-number squeezing disappears. This is not unexpected in view of the result obtained from Kabanov's theorem for the point-process channel.

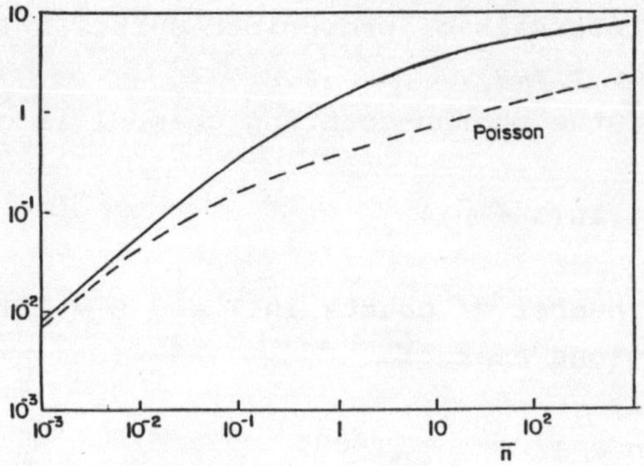

Fig. 3. Comparison of the capacity of the unrestricted photon-counting channel (solid curve) with the Poisson photon-counting channel (dashed curve).

PERFORMANCE OF A BINARY OOK PHOTON-COUNTING RECEIVER

The channel capacity provides a limit on the maximum rate of error-free information transmission for all codes, modulation formats, and receiver structures.[23] As such, it does not specify the performance (error probability) achievable by a communication system with prescribed coding, modulation, and receiver structure.

It is therefore of interest to discuss the performance of a system with specified structure. We consider a binary on-off keying (OOK) photon-counting system.[17] The information is transmitted by selecting one of two values for the photon rate λ_t, in time slots of (bit) duration T. The receiver operates by counting the number of photons received during the time interval T and then deciding which rate was transmitted in accordance with a likelihood-ratio decision rule (threshold test). For simplicity, it is assumed that background light, dark noise, and thermal noise are absent so that photon registrations are not permitted when the keying is OFF (i.e., false-alarms are not possible). Furthermore, the detector quantum efficiency is initially taken to be unity so that system performance is limited only by the quantum fluctuations of the light.

A measure of performance for a digital system such as this is the error probability P_e. In the simplified system described above, errors are possible only when the keying is ON and 0 photons are received (a miss). For a Poisson transmitter, P_e is[17]

$$P_e(\text{Poisson}) = \tfrac{1}{2}\exp(-\bar{n}), \quad (6)$$

where $\bar{n}$ denotes the mean number of emitted photons. To minimize P_e, $\bar{n}$ is made equal to its maximum allowed value $\bar{n}_{max}$. This result is now compared with those obtained for photon-number-squeezed light derived from an initially Poisson source. The outcome will depend on where the mean photon-number constraint is placed. Two transformations are explicitly considered: dead-time deletion and decimation.

It will become evident from these examples that system performance can be enhanced by the use of photon-number-squeezed light, provided that the constraint is applied to the squeezed light. No enhancement of system performance emerges in converting Poisson photons into squeezed photons when the constraint is at the Poisson source.

Dead-Time-Modified-Poisson Photon Counts — For a nonparalyzable dead-time modifier that is always **blocked** for a dead time period τ_d at the beginning of the counting interval T, the passage of 0 photons arises from the emission of 0 photons in the time $T - \tau_d$, independent of the number of emissions during τ_d. The error probability for this system is therefore

$$P_e(\text{dead-time}) = \tfrac{1}{2}\exp[-\bar{n}(1 - \tau_d/T)]. \quad (7)$$

To minimize error under the mean photon-number constraint $\bar{n} \le \bar{n}_{max}$, we take $\bar{n} = \bar{n}_{max}$. The error is larger than that for the Poisson channel [Eq. (6)], as illustrated in Fig. 4, so no performance enhancement can be achieved by use of this modifier.

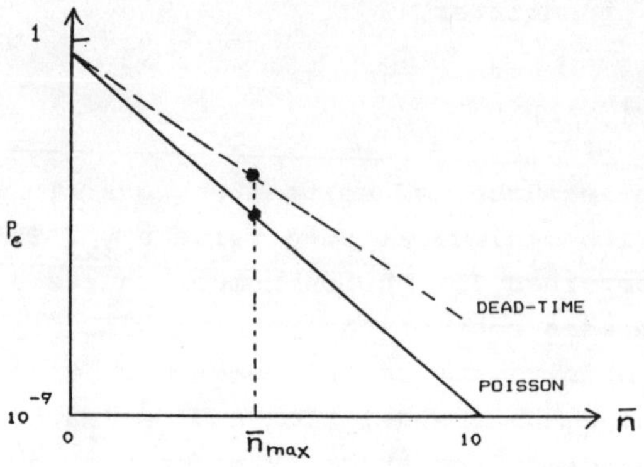

Fig. 4. Comparison of the error probabilities for the Poisson (solid curve) and blocked dead-time (dashed curve) channels with the constraint at the input. No performance improvement is possible.

If, instead, the dead-time modifier is always **unblocked** at the beginning of each bit then the passage of 0 photons arises from the emission of 0 photons in the time T, and the dead-time has no effect on the error rate in this simple system. Calculations for the unblocked counter in the presence of false alarms, however, demonstrate that the presence of dead time always does, in fact, degrade system performance with such a constraint.[35] Although the detailed calculations were carried out for electrical dead time, the results are also applicable for optical dead-time when the photon detection efficiency $\eta = 1$.

On the other hand, if the constraint is placed on the mean photon count $\bar{m}$ after dead-time modification ($\bar{m} \leq \bar{m}_{max}$), it can be shown that there exists a value of $\bar{m}_{max}$ below which performance is degraded, and above which performance is improved, relative to the Poisson channel. This is illustrated in Fig. 5.

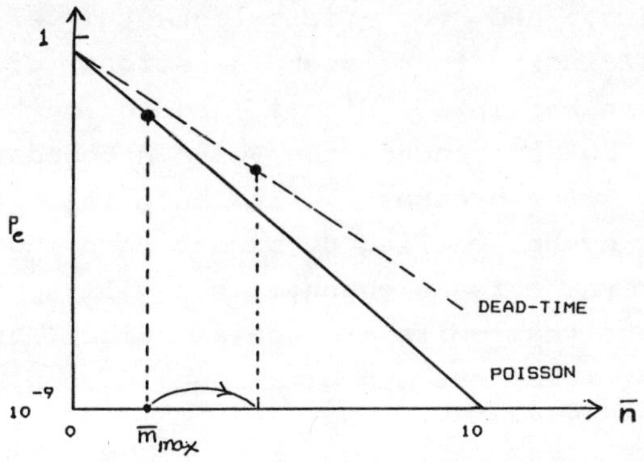

Fig. 5. Comparison of the error probabilities for the Poisson (solid curve) and blocked dead-time (dashed curve) channels with the constraint at the output. Performance improvement is possible.

Decimated-Poisson Photon Counts — We assume that the decimation parameter is $r = 2$ (i.e., every other photon of a Poisson sequence of events is selected) and that the decimation process is reset at the beginning of each bit (i.e., the first photon in each bit is not selected). The error probability is then

$$P_e(\text{decimation}) = \tfrac{1}{2}(1 + \bar{n})\exp(-\bar{n}), \tag{8}$$

which again represents a degradation of performance in comparison with the Poisson channel (under a constraint $\bar{n} \le \bar{n}_{max}$). In this case, the error rate is increased because there are two ways for the passage of 0 photons to arise in the time T: from the emission of 0 photons or from the emission of 1 photon. However, if the constraint is placed on the modified process then, once again, there exists a value of $\bar{m}_{max}$ below which performance is degraded and above which it is improved, relative to the Poisson channel.

PERFORMANCE DEGRADATION ARISING FROM PHOTON LOSS

We conclude by discussing the effects of photon loss (random deletion).[8] We do this in the context of an ideal source that generates a deterministic photon number. This is an important consideration because random photon deletion is inevitable; it results from absorption, scattering, and the finite quantum efficiency of the detector.[8] It is well-known that such deletions will transform a deterministic photon number into a binomial photon number, which always remains sub-Poisson but approaches the Poisson boundary as the random deletion probability η decreases.[36] It has been shown that the information rate per symbol carried by such a counting channel will be greater than that for the Poisson channel, but will approach the latter as $\eta \longrightarrow 0$.[28] A source that emits a binomial number at the outset[37] retains its binomial form, but exhibits reduced mean, in the presence of random deletion.[36]

The performance of such a binary OOK photon-counting receiver, in the absence of background, is limited by the binomial fluctuations of the detected photons. In this case, it is easily shown from the binomial distribution that[9,38]

$$P_e(\text{binomial}) = \tfrac{1}{2} F_n^{[2\langle n'\rangle/(1-F_n)]} \tag{9}$$

where $F_n = 1 - \eta$ is the Fano factor of the photon-counting distribution and where $\langle n'\rangle$ represents the mean number of photons/bit (note that $2\langle n'\rangle = \langle n\rangle$ since there are two bits per pulse in OOK). The Poisson result in Eq. (6) is recovered as $F_n \longrightarrow 1$. The probability of error represented by Eq. (9) is plotted as a function of the mean number of photons per bit $\langle n'\rangle$, with the Fano factor F_n as a parameter, in Fig. 6. System performance improves as F_n decreases.

Solving Eq. (9) for the mean number of photons per bit $\langle n'\rangle$ provides

$$\langle n'\rangle = \tfrac{1}{2}[(1-F_n)/\ln(1/F_n)]\ln(1/2P_e), \tag{10}$$

which leads to a direct-detection quantum limit that is < 10 photons/bit (< 20 photons per pulse) for OOK, if $F_n < 1$ and $P_e = 10^{-9}$. The mean number of photons per bit $\langle n'\rangle$ is plotted as a function of F_n in Fig. 7. The usual quantum limit ($\langle n'\rangle = 10$ photons/bit) emerges in the limit $F_n = 1$ where the binomial distribution goes over to the Poisson.

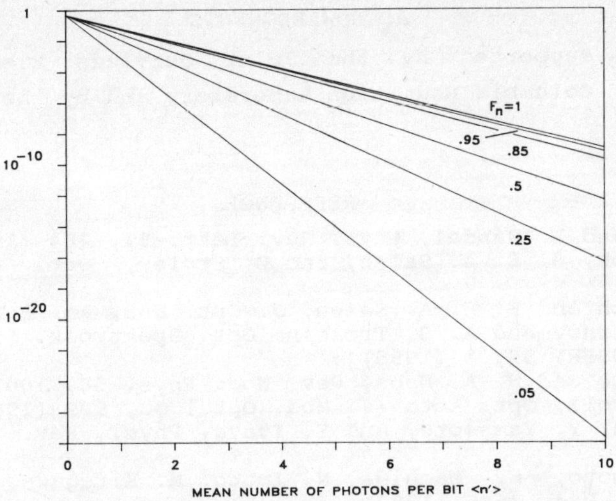

Fig. 6. Error probability (P_e) vs mean number of photons per bit <n'> for the binomial channel, with the Fano factor F_n as a parameter. System performance clearly improves as F_n decreases below unity.

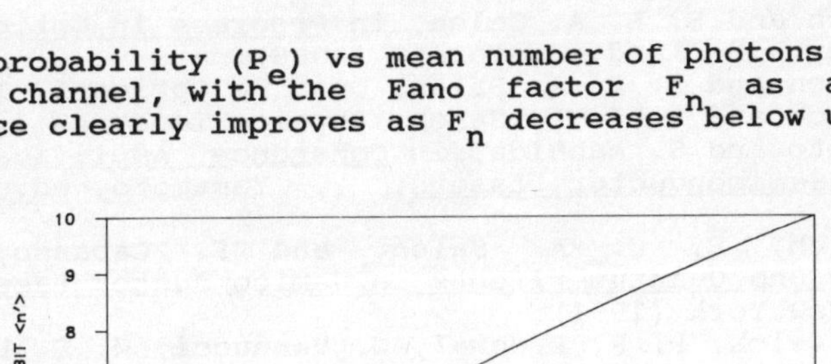

Fig. 7. Mean number of photons per bit <n'> as a function of the Fano factor F_n for the binomial channel. The well-known "quantum limit" (10 photons/bit) emerges as the binomial distribution goes over to the Poisson distribution (F ——> 1).

ACKNOWLEDGMENTS

This work was supported by the Joint Services Electronics Program through the Columbia Radiation Laboratory and by the National Science Foundation.

REFERENCES

1. R. Short and L. Mandel, Phys. Rev. Lett. **51**, 384 (1983).
2. M. C. Teich, B. E. A. Saleh, and D. Stoler, Opt. Commun. **46**, 244 (1983).
3. M. C. Teich and B. E. A. Saleh, J. Opt. Soc. Am. B **2**, 275 (1985).
4. D. F. Smirnov and A. S. Troshin, Opt. Spektrosk. **59**, 3 (1985) [Opt. Spectrosc. (USSR) **59**, 1 (1985)].
5. Y. Yamamoto and H. A. Haus, Rev. Mod. Phys. **58**, 1001 (1986).
6. J. E. Carroll, Opt. Acta (J. Mod. Opt.) **33**, 909 (1986).
7. S. Machida, Y. Yamamoto, and Y. Itaya, Phys. Rev. Lett. **58**, 1000 (1987).
8. Y. Yamamoto, S. Machida, N. Imoto, M. Kitagawa, and G. Björk, J. Opt. Soc. Am. B **4**, 1645 (1987).
9. M. C. Teich and B. E. A. Saleh, in Progress in Optics, vol. 26, E. Wolf, ed., North-Holland, Amsterdam (1988).
10. M. C. Teich and B. E. A. Saleh, Quantum Optics **1**, 153 (1989).
11. M. C. Teich and B. E. A. Saleh, Physics Today **43** (#6), 26 (1990).
12. Y. Yamamoto and S. Machida, in Coherence, Amplification and Quantum Effects in Semiconductor Lasers, Y. Yamamoto, ed., Wiley, New York (1991).
13. M. C. Teich, B. E. A. Saleh, and F. Capasso, in Coherence, Amplification and Quantum Effects in Semiconductor Lasers, Y. Yamamoto, ed., Wiley, New York (1991).
14. M. C. Teich, P. R. Prucnal, G. Vannucci, M. E. Breton, and W. J. McGill, Biol. Cybern. **44**, 157 (1982); B. E. A. Saleh and M. C. Teich, Biol. Cybern. **52**, 101 (1985).
15. E. Jakeman and J. G. Rarity, Opt. Commun. **59**, 219 (1986).
16. B. E. A. Saleh and M. C. Teich, Phys. Rev. Lett. **58**, 2656 (1987).
17. P. S. Henry, IEEE J. Quantum Electron. **QE-21**, 1862 (1985).
18. W. H. Richardson, S. Machida, and Y. Yamamoto, "Observation of 10 dB Squeezing in the Amplitude Fluctuations of Light from a Diode Laser," postdeadline paper presented at the International Quantum Electronics Conference, 1990.
19. J. H. Shapiro, G. Saplakoglu, S.-T. Ho, P. Kumar, B. E. A. Saleh, and M. C. Teich, J. Opt. Soc. Am. B **4**, 1604 (1987).
20. P. R. Tapster, J. G. Rarity, and J. S. Satchell, Phys. Rev. A **37**, 2963 (1988).
21. M. C. Teich, B. E. A. Saleh, and J. Peřina, J. Opt. Soc. Am. B **1**, 366 (1984).
22. P. R. Tapster, J. G. Rarity, and J. S. Satchell, Europhys. Lett. **4**, 293 (1987).
23. T. E. Stern, IRE Trans. Inf. Theory **IT-6**, 435 (1960); J. P. Gordon, Proc. IRE **50**, 1898 (1962); J. R. Pierce, E. C. Posner, and E. R. Rodemich, IEEE Trans. Inf. Theory **IT-27**, 61 (1981).
24. M. C. Teich and G. Vannucci, J. Opt. Soc. Am. **68**, 1338 (1978).
25. J. G. Rarity, P. R. Tapster, and E. Jakeman, Opt. Commun. **62**, 201 (1987).
26. R. Loudon, The Quantum Theory of Light, 2nd ed., Clarendon, Oxford, (1983).

27. B. E. A. Saleh and M. C. Teich, Opt. Commun. 52, 429 (1985).
28. L. Mandel, J. Opt. Soc. Am. **66**, 968 (1976).
29. H. P. Yuen, Phys. Rev. Lett. **56**, 2176 (1986).
30. D. Stoler and B. Yurke, Phys. Rev. A **34**, 3143 (1986).
31. Y. Yamamoto, N. Imoto, and S. Machida, Phys. Rev. A **33**, 3243 (1986).
32. Yu. M. Kabanov, Theory Prob. Appl. **23**, 143 (1978).
33. M. H. A. Davis, IEEE Trans. Inf. Theory **IT-26**, 710 (1980); A. A. Lazar, in Proc. 14th Annual Conf. on Information Sciences and Systems, March 26-28, 1980, Princeton, NJ; P. Bremaud, "A coding theorem for continuous-time point-process channels," unpublished, Sept. 1983.
34. T. T. Kadota, M. Zakai, and I. Ziv, IEEE Trans. Inform. Theory IT-17, 368 (1971).
35. M. C. Teich and B. I. Cantor, IEEE J. Quantum Electron. **QE-14**, 993 (1978).
36. M. C. Teich and B. E. A. Saleh, Opt. Lett. **7**, 365 (1982).
37. D. Stoler, B. E. A. Saleh, and M. C. Teich, Opt. Acta (J. Mod. Opt.) 32, 345 (1985).
38. K. Yamazaki, O. Hirota, and M. Nakagawa, Trans IEICE (Japan) 71, 775 (1988).

TWIN PHOTONS AND SQUEEZED LIGHT

A. Heidmann
Laboratoire de Spectroscopie Hertzienne de l'Ecole Normale Supérieure
Centre National de la Recherche Scientifique
Université Pierre et Marie Curie F 75252 Paris France

ABSTRACT: Squeezing in various nonlinear interactions can be theoretically evaluated using a new semi-classical method which is a linear input-output transformation for quantum fluctuations. We present experimental results concerning the generation of twin photon beams using a non degenerate Optical Parametric Oscillator and the generation of a high intensity sub-Poissonian light using an optoelectronic feedforward correction of one of these twin beams.

1. INTRODUCTION

Quantum fluctuations are present in every measurement and it has long been thought that they were an insuperable limit to the accuracy of experiments. In the last decade, theoretical studies have shown that this limit could be circumvented and several nonlinear optical systems have experimentally generated squeezed states of light. The amount of noise reduction in several experiments now ranges in the 70% to 80%. In most cases the involved phenomena are parametric processes, coming from a second or third order nonlinearity. The most relevant experiments (noise reduction larger than 60%) are:

– Generation of squeezed vacuum using a degenerate Optical Parametric Amplifier[1]. A nonlinear crystal inserted in an optical cavity is pumped with a single mode doubled YAG laser. The observation of squeezing in the subharmonic field emitted by the medium is done by heterodyne detection: the light exiting the parametric amplifier is combined on a beamsplitter with the light at the fundamental frequency from the YAG laser which acts as a strong local oscillator. The noise in the observed signal is then related to the fluctuations of one quadrature component of the subharmonic field. The maximum noise reduction observed is equal to 63%.

– Twin photon beams generation[2-5]. In this case, the parametric medium emits two quantum correlated beams which differ by either their frequencies or their polarizations and which can be separated. Several experiments have been done where the correlations have been observed either between intense laser-like twin beams using an Optical Parametric Oscillator[2-3], in the fluorescence light of the parametric

Lecture Notes in Physics, Vol. 378
C. Bendjaballah, O. Hirota, S. Reynaud (Eds.),
Quantum Aspects of Optical Communications

medium[4], or between pulsed twin beams[5]. The noise reduction in these experiments is of the same order, from 67% up to 80%.

– Sub-Poissonian intense beam generation using a laser diode[6]. The laser converts the electrons into photons with a good efficiency so that the noise is mainly limited by the electron noise which can be lower than the standard photon noise. The maximum noise reduction recently observed is of the order of 80%.

In next sections, we present a semi-classical method which is useful to describe the noise properties of such systems (§2) and experimental results concerning the twin photon beams generation (§3) and the sub-Poissonian light beam generation using a feedforward correction mechanism (§4).

2. SEMI-CLASSICAL TRANSFORMATION FOR THE QUANTUM FLUCTUATIONS

The theoretical methods of quantum optics[7] are based upon a Glauber representation of the field. The evolution of this distribution is given by a Fokker Planck equation with non positive diffusion coefficients. The field correlation functions are then calculated using a linearization procedure.

In the semi-classical approach[8], the field fluctuations are described by the Wigner distribution. The system is considered as a "noise processor" with input fields in the vacuum or in a coherent state, and which generates at its output a squeezed field. For most systems, the evolution of the Wigner distribution can be considered as classical. In particular, when the mean fields are much larger than the field fluctuations, a linearization procedure can be done and the quantum terms in the Wigner evolution become negligible. The method is thus similar to the linear stability analysis[9] where the evolution of the field fluctuations is described by classical equations linearized in the vicinity of a working point. More precisely, the fluctuations inside the system are driven by classical random fluctuations associated with the incident fields which are chosen to fit the Wigner representation of the input quantum fields. The input fields are thus described by classical stochastic variables:

$$E^{in} = \overline{E^{in}} + \delta E^{in} \tag{1}$$

where $\overline{E^{in}}$ is the mean field and δE^{in} a classical random field associated with the quantum field fluctuations. The output fields are obtained from the input ones using the evolution equations of classical electrodynamics inside the system:

$$E^{out} = \mathcal{F}(E^{in}) \tag{2}$$

where $\mathcal{F}$ is the transfer function of the system. A linearization of this equation around the mean field values finally gives a linear input-output transformation for the fluctuations:

$$\delta E^{out} = \mathcal{G}(\delta E^{in}) \tag{3}$$

which relates the output noise characteristics to the input fields statistics and to the system dynamics. The advantage of this approach is that it corresponds to the intuitive idea that the outgoing field fluctuations are obtained from the incident ones through a filtering by the noise processor. In particular, the noise sources are associated with the incident field fluctuations rather than with non positive diffusion coefficients.

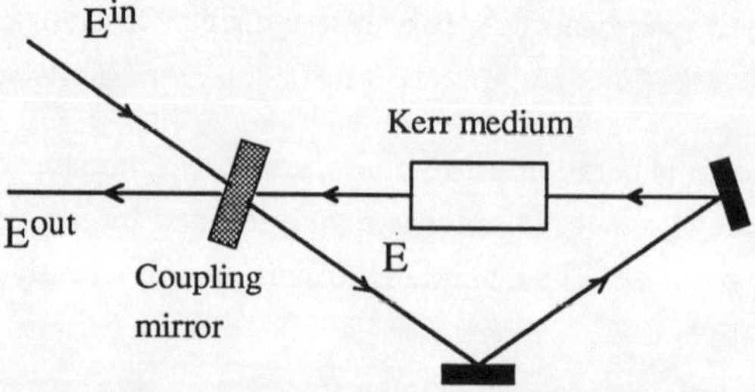

Figure 1: Scheme of a bistable device made of a Kerr medium inside a single port optical cavity. E^{in}, E and E^{out} are respectively the input, intracavity and output fields.

As an example, we can compute the photon noise reduction which is the squeezing of the field amplitude component, in optical bistability[10]. The semi-classical method is particularly interesting since it leads to simple geometrical pictures. The bistable device is composed of a Kerr medium inserted in a cavity (see figure 1). The input field E^{in}, the intracavity field E and the output field E^{out} are related through the roundtrip phase shift Φ which includes the Kerr effect:

$$t E^{in} = (1 - r e^{-i\Phi}) E \tag{4a}$$

$$t E^{out} = (e^{-i\Phi} - r) E \tag{4b}$$

$$\Phi = \Phi_0 + f |E|^2 \tag{4c}$$

where r and t are the reflection and transmission coefficients of the coupling mirror. One finds that the input and output intensities are equal and that the relation between the intracavity and the input intensities exhibits a bistable behavior.

In a phase space representation based on the Wigner distribution, the input field, assumed to be in a coherent state, is described by a disk shape (see figure 2). From eqs. (4), the output field distribution is

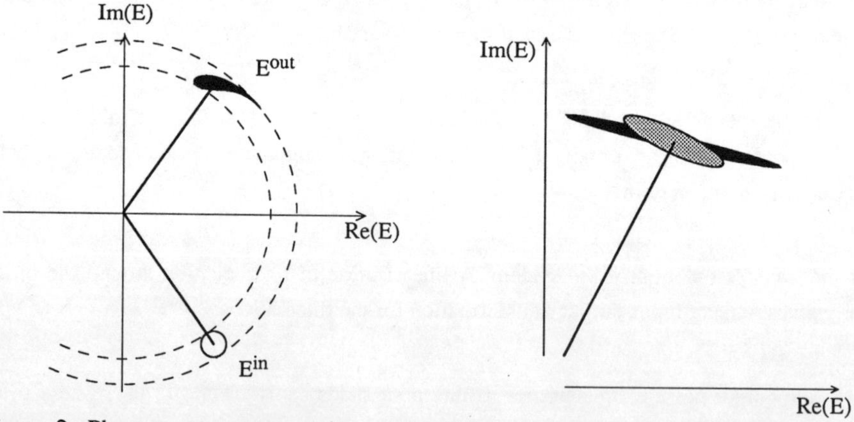

Figure 2: Phase space representation of the fields in the bistable device. The left part corresponds to a static analysis where the input field (white disk) is transformed into a bean shape (black distribution). The right part represents the linearized dynamical analysis. Starting from the static distribution (black ellipse), the ellipse rotates when frequency is increased so that the amplitude quadrature is squeezed (gray ellipse).

obtained from the input one by a rotation around the origin with an intensity dependant angle. Near a bistability turning point, a critical divergence of the fluctuations in one direction occurs. Since the geometrical transformation preserves the area, another dimension of the distribution has to be decreased. In other words, squeezing is expected in one particular quadrature component of the output field.

This picture is only valid in the static limit and describes the field fluctuations on long time scale. In particular, the fact that the intensity distribution is not changed in the rotation corresponds to the conservation of the number of photons on long times.

The semi-classical approach can be considered as a dynamical extension of this picture. At zero frequency, it gives an elliptical shape which is the limit of the previous distribution when the geometrical transformation is approximated as a linear transformation for the fluctuations. When the analysis frequency is increased, the elliptical shape goes back to a disk shape (see figure 2). However, it also rotates so that the minor axis of the ellipse may become aligned with the mean field. In this case, the amplitude component is squeezed and the photon noise is reduced below the shot noise although the input beam is assumed to be at the shot noise. This photon noise reduction is associated with a temporal redistribution of the incident photons by the cavity.

3. TWIN PHOTON BEAMS GENERATION

The basic phenomenon in parametric downconversion is the annihilation of one pump photon and the creation of two signal photons at the same time. In the nondegenerate operation, the two signal fields are different and the pair emission is responsible for photon correlation between the two beams. When the nonlinear crystal is inserted in an optical cavity resonant for the signal fields and is pumped with a high enough intensity, the system oscillates like a laser. Such an Optical Parametric Oscillator (OPO) emits twin laser-like beams. The statistics of the beams going out of the cavity retain part of the properties of the photon pair emission, provided that the measurement is done on times long compared to the cavity storage time[11]. As a consequence, the noise in the difference between the intensities of the two output beams is reduced below the standard shot noise level for frequencies within the cavity bandwidth.

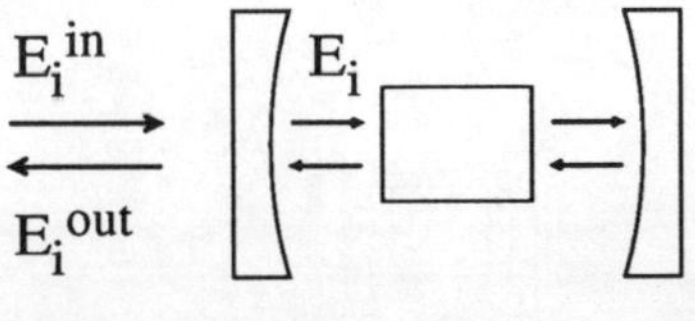

Figure 3: Scheme of the Optical Parametric Oscillator. The nonlinear medium is inserted in a high finesse cavity. E_i^{in}, E_i and E_i^{out} are respectively the input, intracavity and output fields (i=0 for the pump, i=1,2 for the signals).

The photon noise spectrum can be theoretically determined using the semi-classical formalism. Noting E_1, E_2 the two signal intracavity fields and E_0 the pump field (see figure 3), the classical equations of motion for E_1 and E_2 are[12]:

$$\tau\,\dot{E}_1 = -\gamma\,E_1 - \chi\,E_2^*\,E_0 + \sqrt{2\gamma}\,E_1^{in} \tag{5a}$$

$$\tau\,\dot{E}_2 = -\gamma\,E_2 - \chi\,E_1^*\,E_0 + \sqrt{2\gamma}\,E_2^{in} \tag{5b}$$

where τ is the cavity roundtrip time, γ the damping coefficient of the signal fields in the cavity (γ is related to the amplitude transmission t of the coupling mirror by $t = \sqrt{2\gamma}$); χ is the parametric coupling coefficient and E_i^{in} are the vacuum fluctuations entering the cavity through the coupling mirror. The outgoing fields E_i^{out} are related to the intracavity fields and to the incoming vacuum fields by the reflection-transmission relation on the coupling mirror:

$$E_i^{out} = \sqrt{2\gamma}\,E_i - E_i^{in} \tag{6}$$

where the reflection coefficient $r = 1-\gamma$ has been approximated by one. By linearizing these equations around a particular steady state, one can write the evolution of the fluctuations for the various components of the field. Since the mean fields can be assumed to be real, we define the amplitude components as $q_i = (E_i + E_i^*)/\sqrt{2}$ and the difference between the two amplitude components as $q = (q_1 - q_2)/\sqrt{2}$. Taking the Fourier transform of the equations of motion, one gets the input-output relation for the fluctuations $\delta\tilde{q}(\Omega)$ at frequency Ω:

$$\delta\tilde{q}^{out}(\Omega) = \frac{\gamma + \chi\,\overline{E_0} - i\,\Omega\,\tau}{\gamma - \chi\,\overline{E_0} + i\,\Omega\,\tau}\;\delta\tilde{q}^{in}(\Omega) \tag{7}$$

From eqs. (5), one gets $\gamma = -\chi\,\overline{E_0}$ when the OPO operates above threshold, so that the fluctuations in the difference between the amplitudes of the signal fields go to zero at zero noise frequency. The noise spectrum $S_{I_1-I_2}(\Omega)$ in the intensity difference, normalized to the shot noise level, is then given by:

$$S_{I_1-I_2}(\Omega) = \frac{\Omega^2\,\tau^2}{\Omega^2\,\tau^2 + 4\,\gamma^2} \tag{8}$$

One gets a perfect noise reduction for frequencies well inside the cavity bandwidth. This effect does not depend on the noise of the pump beam whose contribution cancels out in the difference. The only processes which may degrade the squeezing are extra losses. The correlation between the twin beams will be degraded each time a photon is lost inside the cavity. As a result, the predicted squeezing at zero frequency is not perfect any more. The semi-classical method can be used for a complete calculation of the OPO characteristics, including various imperfections (extra losses, imbalance,...)[13].

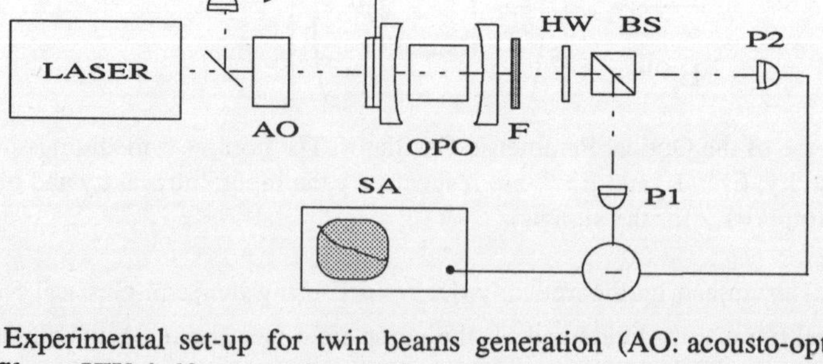

Figure 4: Experimental set-up for twin beams generation (AO: acousto-optic modulator, F: green filter, HW: half-wave plate, BS: polarizing beamsplitter, P1,P2: photodiodes, SA: spectrum analyzer).

The experimental set-up[2-3] comprises a nonlinear KTP crystal placed in an optical cavity and pumped by an Ar^+ laser optically isolated by an acousto-optic modulator (see figure 4). The cavity has a high finesse for the signal fields and a low finesse for the pump. When the pump power is large enough, the OPO emits two laser-like beams with orthogonal polarizations. The weak signal light that leaks through the input mirror is used in a feedback loop to control the output intensities. At the output of the OPO, a filter blocks the green light and a polarizing beamsplitter separates the two signal beams.

The intensities of the two beams are detected and subtracted. The resulting spectrum reflects the fluctuations in the intensity difference. To determine the noise reduction factor, it is necessary to measure the shot noise level which is obtained when the two beams are not correlated at all. To do so, a half-wave plate (HW in figure 4) is inserted before the beamsplitter with its axes in such a position that the polarization of each beam is rotated by 45°. Then the polarizing beamsplitter acts like a semi-reflecting plate for each beam. Since the beams have different wavelengths, the interference effects are unobservable and the noise detected in the difference is the standard shot noise level. The resulting spectrum, normalized to the shot noise level, is shown in figure 5. The maximum noise reduction is equal to 80% around 4 MHz.

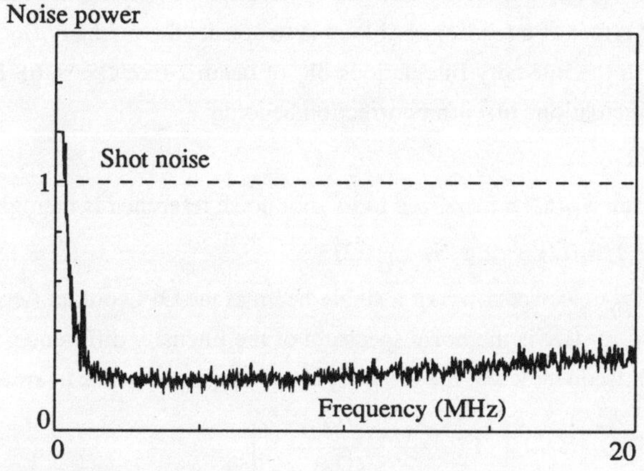

Figure 5: Observed noise spectrum as a function of the frequency. The maximum noise reduction factor is equal to 80%.

4. SUB-POISSONIAN LIGHT GENERATION

The quantum correlation between twin beams can be used to generate amplitude squeezed beams by electronic control. A common technique to reduce the intensity fluctuations of a laser beam consists in using an optoelectronic servo loop. A beamsplitter reflects off a part of the beam, providing a feedback signal to stabilize the laser intensity. However, this scheme cannot be used to reduce the quantum fluctuations: photons are randomly distributed by the beamsplitter so that the quantum fluctuations of the reflected and transmitted beams are not correlated.

To reduce the quantum fluctuations, one can replace the beamsplitter by a twin beam generator. One beam provides a feedback signal to stabilize the other beam below the shot noise level. A quantum noise reduction of 22% has been observed in low intensity incoherent light using feedback from spontaneous parametric down-conversion[14]. The twin beams at the output of an OPO are particularly convenient for the generation of high intensity sub-Poissonian light using configurations where the control signal from one beam is either fed back to the pump or fed forward directly to the second beam (see figure 6). The latter system is preferable to avoid self-oscillation problems.

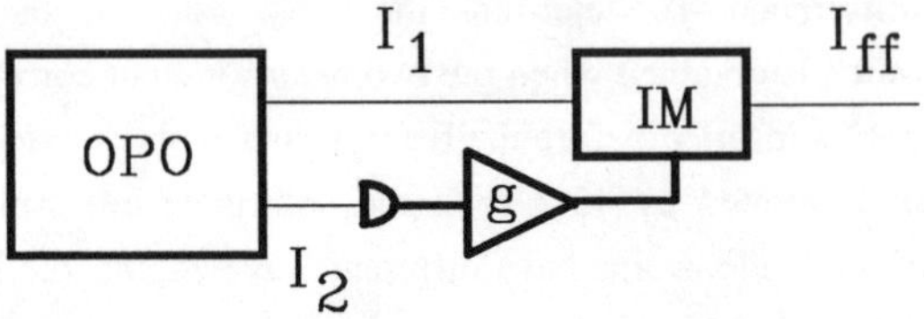

Figure 6: Feedforward control mechanism (IM: intensity modulator, g: variable gain amplifier).

The theoretical noise reduction attainable with such a system can be determined using the semi-classical approach. The significant effect of a feedforward loop is to correct the intensity fluctuations δI_1 of beam 1 by a term proportional to the intensity fluctuations δI_2 of beam 2 (see figure 6). Neglecting the losses of beam 1, the intensity fluctuations δI_{ff} after correction become:

$$\delta I_{ff} = \delta I_1 - g\ \delta I_2 \tag{9}$$

The intensity noise spectrum $S_{ff}(\Omega)$ normalized to its shot noise reference is then given by:

$$S_{ff}(\Omega) = (1-g^2)\ S_I(\Omega) + 2\,g\ S_{I_1-I_2}(\Omega) \tag{10}$$

where $S_I(\Omega)$ is the intensity noise spectrum of a single beam at the OPO output (assumed to be the same for the two beams) and $S_{I_1-I_2}(\Omega)$ is the noise spectrum of the intensity difference, as it is defined in the previous section. For each frequency Ω_0, the intensity noise $S_{ff}(\Omega_0)$ reaches its minimum value:

$$S_{ff}^{opt}(\Omega_0) = 2\ S_{I_1-I_2}(\Omega_0)\left(1 - \frac{S_{I_1-I_2}(\Omega_0)}{2\ S_I(\Omega_0)}\right) \tag{11}$$

when the gain g is set to the value:

$$g^{opt} = 1 - \frac{S_{I_1-I_2}(\Omega_0)}{S_I(\Omega_0)} \tag{12}$$

When the noise $S_I(\Omega)$ is large compared to one, the optimum gain g^{opt} is equal to one and the optimally corrected intensity noise is simply $S_{ff}(\Omega) = 2\ S_{I_1-I_2}(\Omega)$. In particular, when the two beams at the OPO output are not correlated at the quantum level ($S_{I_1-I_2}(\Omega) \geq 1$), the feedforward loop cannot reduce the intensity noise below the shot noise: only classical excess noise of beam 1 can be corrected. When excess noise becomes smaller ($S_I(\Omega) \approx 1$), noise levels less than $2\ S_{I_1-I_2}(\Omega)$ can be reached for gains smaller than one.

In our experiment[15], the intensity modulator is an electro-optic modulator followed by a polarizer and acts as an analog shutter with a transmission proportional to the applied voltage from the electronic amplifier. A double-balanced detector is used to measure the noise spectra, made of a semi-reflecting

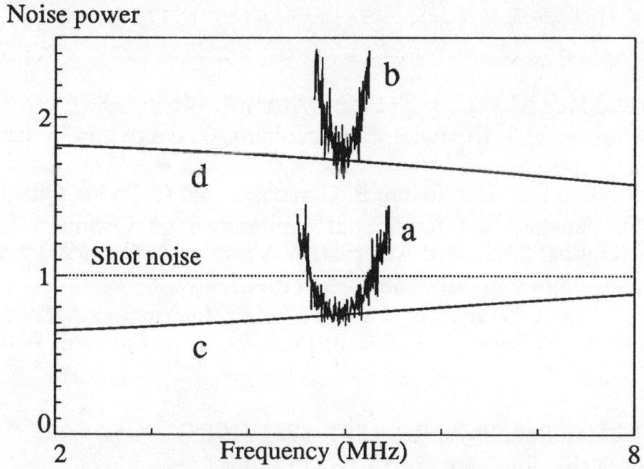

Figure 7: Intensity noise reduction factor, when the two beams at the OPO output are quantum correlated (a) and decorrelated (b). Theoretical curves (c) and (d) are the optimum noise reduction factor for these cases.

plate followed by two photodiodes. The photocurrents can be either added or subtracted so that both the total noise, obtained by summing the signals, and the associated shot noise level, obtained by subtracting the signals, are measured almost simultaneously.

Figure 7 shows the photon noise spectrum, normalized to the shot noise level (curve a). The noise reduction is sensitive both to the gain and phase lag of the correction signal, which are properly matched in a frequency range centered here around 5 MHz. The reduction in intensity noise is up to 24% below the shot noise level. Curve (c) represent the theoretical optimum intensity noise $S_{ff}^{opt}(\Omega_0)$ as a function of the frequency Ω_0, when total losses of beam 1 are taken into account.

Curve (b) is the experimental noise spectrum when the two beams at the OPO output are not quantum correlated and curve (d) is the corresponding theoretical optimum intensity noise. This situation is attainable experimentally by inserting a half-wave plate before the polarizing beamsplitter (HW in figure 4), as explained in the previous section. As expected, the feedforward loop does not reduce the noise even to the shot noise.

In conclusion, very large noise reductions have been experimentally obtained these last years, in particular with Optical Parametric Oscillators. The twin photon beams generated by an OPO can be used to control the intensity fluctuations of a single laser-like beam. Such a technique seems attractive for potential applications, particularly with the advent of monolithic OPO technology[16].

Applications of quantum noise reduction are now foreseen. They concern ultrasensitive measurements (detection of gravitational waves, parity violation in atomic physics, new spectroscopic techniques) and new physical phenomena (quantum nondemolition measurements, two-photon interferences, inhibited relaxation). They may also be useful in communication systems and in gyrolasers.

This work has been supported in part by the EEC contract ESPRIT BRA 3186.

REFERENCES

1 L.A. Wu, Min Xiao and H.J. Kimble, J. Opt. Soc. Am. **B4**, 1465 (1988)

2 A. Heidmann, R.J. Horowicz, S. Reynaud, E. Giacobino, C. Fabre and G. Camy, Phys. Rev. Lett. **59**, 2555 (1987)

3 T. Debuisschert, S. Reynaud, A. Heidmann, E. Giacobino and C. Fabre, Quantum Opt. **1**, 3 (1989)

4 J.G. Rarity and P.R. Tapster, in International Conference on Quantum Electronics Technical Digest Series Vol. 8 (Optical Society of America, Washington D.C., 1990) p. 8

5 O. Aytur and P. Kumar, Phys. Rev. Lett. **65**, 1551 (1990)

6 W.H. Richardson, S. Machida and Y. Yamamoto, in International Conference on Quantum Electronics Technical Digest Series Vol. 8 (Optical Society of America, Washington D.C., 1990) p. 395

7 See for example:
C.W. Gardiner and M.J. Collett, Phys. Rev. **A31**, 3761 (1985)
M.J. Collett and D.F. Walls, Phys. Rev. **A32**, 2887 (1985)

8 S. Reynaud and A. Heidmann, Opt. Commun. **71**, 209 (1989)

9 H.M. Gibbs, "Optical bistability: controlling light with light", Academic Press, New York (1965)

10 S. Reynaud, C. Fabre, E. Giacobino and A. Heidmann, Phys. Rev. **A40**, 1440 (1989)

11 S. Reynaud, Europhys. Lett. **4**, 427 (1987)

12 S. Reynaud, C. Fabre and E. Giacobino, J. Opt. Soc. Am. **B4**, 1520 (1987)

13 C. Fabre, E. Giacobino, A. Heidmann and S. Reynaud, J. de Physique **50**, 1209 (1989)

14 P.R. Tapster, J.G. Rarity and G.S. Satchell, Phys. Rev. **A37**, 2963 (1988)

15 J. Mertz, A. Heidmann, C. Fabre, E. Giacobino and S. Reynaud, Phys. Rev. Lett. **64**, 2897 (1990)

16 C.D. Nabors and R.M. Shelby, Phys. Rev. **A42**, 556 (1990)

A NEW DEFINITION AND GENERALIZATION OF SQUEEZED STATES

Roman S. Ingarden
Institute of Physics, Nicholas Copernicus University
Grudziądzka 5, PL 87-100 Toruń, Poland

SUMMARY: A new, generalized definition of squeezed states is given. The author abstracts from the quantum oscillator or the vacuum state, where this concept is usually defined, and defines it with respect to any probability measure, classical or quantum. A classical case occurs, e. g., in generalized information thermodynamics of the author, with two thermal temperatures: first-order T_1 and second-order T_2. The reference variance in this case is that of the Boltzmann-Gibbs state with one temperature $T_1 = T$.

1. INTRODUCTION

The squeezed states are usually defined as special cases of coherent light, cf. /1/,/2/,/3/. As R. Loudon and P. L. Knight write in /3/ p. 710: "Squeezed states of light were first studied by theorists interested in their properties as generalized minimum-uncertainty states. These properties were discovered indpendently by several workers using different terminologies and have been described variously as 'pulsating wave packets', 'new coherent states', 'two-photon coherent states' and 'ideal squeezed states'." / We omit the historical references which may be found in /3/./ The point is, however, that not only different words and descriptions have been used, but also that the content of this concept was not always the same. Namely, "ideal" and "nonideal" squeezed states differ by requiring or not the condition that the states are of minimum uncertainty; as was said above, they are, in general, only "generalized minimum-uncertainty states".

M. M. Nieto in /2/ tried to explain "what are squeezed states really like?" in the spirit of the "Bethe philosophy", i. e., "if you really understand it, then you've got to be able to explain it simple." This explanation trying to give the real essence of the concept of squeezed states goes as follows, p. 288 :

"Here we go, folks. A coherent state, very simply, is a state in a harmonic oscillator potential which, at t = 0, i/ is a Gaussian that

Lecture Notes in Physics, Vol. 378
C. Bendjaballah, O. Hirota, S. Reynaud (Eds.),
Quantum Aspects of Optical Communications

is displaced from the origin by x_0, ii/ has a phase which is proportional to the position x, and iii/ the width of the Gaussian is that of the ground state, σ_0. In particular, it is

$$\psi_{cs} = (2\pi\sigma_0^2)^{-1/4} \exp\left[-\left(\frac{x - x_0}{2\sigma_0}\right)^2 + \frac{i}{\hbar} p_0 x\right] , \qquad (1.1)$$

$$\sigma_0^{\,2} = \frac{\hbar}{2m\omega} = \frac{1}{2a_0^2} . \qquad (1.2)$$

A squeezed state is the same Gaussian but the width is different than that of the ground state,

$$\sigma_0 \rightarrow \sigma = s\sigma_0 . \qquad (1.3)$$

That's all it is, honest to God. You can go out to eat lunch in Santa Fe." And then at p. 282: "σ_0 is the width of the ground state or the coherent state. "s" is what I call the "squeeze factor".

This shows you why I like the name "squeeze". What you're doing by putting that "s" in there is that you're either squeezing in the Gaussian /$s < 1$/ or you're squeezing it out /$s > 1$/, the momentum-space wave function being squeezed oppositely. You're taking the Gaussians and you're squeezing them. When "s" is equal to 1 you have a coherent state. When "s" is not equal to 1 you still have s squeezed state."

In the present paper we propose a generalization of the above definition which seems to be even simpler, but which can be applied in principle both to classical and quantum states, and not necessarily is connected with the vacuum fluctuations. Actually, because of the technical reasons of simplicity, we start from a somewhat special case of generalized information thermodynamics /4/ formulated for a classical ideal model in 3 dimensions, but the idea can be in principle easily applied to any case of such thermodynamics or information dynamics, classical or quantum. Therefore, also the above case considered by Nieto can be included as a special case.

2. INFORMATION SECOND-ORDER THERMODYNAMICS

In the recent paper of the author /4/ the ideas have been developed which for the first time were published in /5/ and /6/ and then discussed step by step in many other papers, cf., e. g., /7/ and /8/, and the references in /4/. In /4/ we discussed 1D and 3D models with 1 and 2 higher-order thermal temperatures. The general philosophy is to resign from the assumption by Nieto /2/ in the sentence "A squeezed state is the same Gaussian but ...". In /4/ and in the present paper we resign from this assumption and consider <u>generalized</u> Gaussian states of the form

$$f(p)=\left[Z(\beta_1,\beta_2)\right]^{-1}\exp\left[-\beta_1 H(p)-\beta_2 H^2(p)\right],\quad H(p)=\frac{1}{2m}\sum_{j=1}^{N}\underline{p}_j^2, \tag{2.1}$$

where $\underline{p}_j=(p_{xj},p_{yj},p_{zj})$ and

$$\beta_1=\frac{1}{k_BT_1},\quad \beta_2=\frac{1}{(k_BT_2)^2}\geqslant 0,\quad T_2>0, \tag{2.2}$$

$$Z(\beta_1,\beta_2)=\frac{V^N}{N!}\frac{(2m\pi)^{3N/2}}{(2\beta_2)^{3N/4}}\exp\left(\frac{\beta_1^2}{8\beta_2}\right)U\left(\frac{3N}{2}-\frac{1}{2},\frac{\beta_1}{\sqrt{2\beta_2}}\right),\quad U(a,x)=D_{-a-\frac{1}{2}}(x), \tag{2.3}$$

while T_1 and T_2 are first-order and second-order temperatures, respectively, both expressed in the energetic units, k_B is the Boltzmann constant and $U(a,x)$ or $D_p(x)$ are different notations of Weber's parabolic cylinder function, cf. /9/. The other constants correspond to parameters of an ideal gas of N identical particles of mass m and momenta $\underline{p}_j$ in 3D volume V. Formula (2.1) is obtained by maximizing entropy (information) under the constraints

$$U=U_1=\langle H\rangle=-\frac{\partial\log Z}{\partial\beta_1},\quad U_2^2=\langle H^2\rangle=-\frac{\partial\log Z}{\partial\beta_2},U_1,U_2>0, \tag{2.4}$$

where U_1 and U_2 are also expressed in the energetic units and are considered as given by a macroscopic or mesoscopic measurement, assuming that both are observable with accuracy of the respective instruments. So in our generalized thermodynamics we have thermodynamical parameters U_1, U_2, T_1, T_2, connected by conditions (2.4). It is not easy to solve these equations with respect to β_1 and β_2, although we know that (in spite of transcendental character of function (2.3)) the solution is unique because of the convexity of entropy function.

To simplify the solution we assume that N is large, but not infinite. In this case we can apply C. G. Darwin's (1949) assymptotic expansion of $U(a,x)$ for $a>0$ and large $X^2=x^2+4a\gg 1$. For large N we have in our case

$$a=\frac{3N}{2}>0,\quad X^2=\frac{\beta_1^2}{2\beta_2}+6N\gg 1, \tag{2.5}$$

so these conditions are satisfied. Darwin's approximation is then

$$U(a,x)=(2\pi)^{1/4}\,\Gamma(a+\tfrac{1}{2})^{-1/2}\exp\left[-\frac{1}{4}xX-a\log\frac{x+X}{2a^{1/2}}\right], \tag{2.6}$$

where we neglected the higher order terms. Using this simplification and the notation D for the fluctuation of energy

$$D^2 = U_2^2 - U_1^2 = -\frac{\partial U}{\partial \beta_1} = \frac{\partial^2 \log Z}{\partial \beta_1^2} \geqslant 0, \quad D \geqslant 0, \tag{2.7}$$

we obtain finally the solution in the form

$$\beta_1 = -\frac{1}{U}\left(\frac{U^2}{D^2} - 3N\right), \quad \beta_2 = \frac{1}{U^2}\left(\frac{U^2}{D^2} - \frac{3N}{2}\right) \geqslant 0, \tag{2.8}$$

while U and D^2 are now expressed by β_1 and β_2 in the form

$$U = (4\beta_2)^{-1}\left[(2\beta_2)^{1/2}\left(\frac{\beta_1^2}{2\beta_2} + 6N\right)^{1/2} - \beta_1\right] \geqslant 0, \tag{2.9}$$

$$D^2 = (4\beta_2)^{-1/2}\left[1 - \beta_1 (2\beta_2)^{-1/2}\left(\frac{\beta_1^2}{2\beta_2} + 6N\right)^{-1/2}\right] \geqslant 0. \tag{2.10}$$

We have to add that we took only the leading terms of expression (2.6). Another important comment is that solution (2.8) exists if and only if

$$\frac{D}{U} \leqslant \sqrt{\frac{2}{3N}} \,. \tag{2.11}$$

Condition (2.11) is very important physically. Namely, for $N \to \infty$, in the thermodynamic limit, relative fluctuation D/U and β_2 vanish (the exact cancelation of two infinities in (2.8) occurs for the equality sign in (2.11), but we have to remember that our calculation is approximative and corresponds to near vicinity of this point). We see therefore that in the usual thermodynamic limit, for homogeneous media and $V \to \infty$ under the condition $V/N = \text{const.} < \infty$, the higher-order temperature coefficient β_2 disappears ($T_2 = \infty$) and we have the usual thermodynamics with one temperature $T = T_1 > 0$. So the correspondence with the conventional physics is satisfactory.

But we have to point out that the conventional philosophy of physics disregards cases which, although rather rare, are nevertheless important. These are all the cases when fluctuations of physical quantities are macroscopically or mesoscopically measurable, as in the vicinity of critical points of phase change, in turbulence, and in small systems as macromolecules, microclusters, colloids, quantum optics, etc. More detail discussion of these questions can be found in /4/. Roughly we can estimate that the order of magnitude of N for which assumption (2.1) can be reasonable is in the range 10^5-10^6, very far from 10^{23} of the Avogadro number, cf./4/.

3. SQUEEZED STATES

It is clear from the above discussion that the case of the exactly vanishing second-order temperature coeffcient $\beta_2 = 0$, i. e., that of the Gaussian or Boltzmann-Gibbs distribution, corresponds to the equality sign in (2.11) and gives for fluctuation (variance) D of energy the value

$$D_o = U \left(\frac{2}{3N} \right)^{1/2} . \qquad (3.1)$$

Hence we can write (2.11) in the form

$$D = s\, D_o \,, \qquad 0 \leqslant s \leqslant 1 \,, \qquad (3.2)$$

where s is a squeeze factor as in (1.3). In such a way we obtain a far reaching analogy to the usual theory of squeezed states in quantum optics. Actually, also in optics we can apply the methods of higher-order information thermodynamics, cf. /9/ /10/ /11/ /12/, also in the quantum domain. Therefore, the above generalization seems to be applicable to a large field of cases of practical importance.

Interesting difference between cases (3.2) and (1.3) is that our squeeze factor s should be less or equal 1 while in (1.3) is no such condition. What is the reason ? In our case the inequality is connected with the existence of universal laws of high numbers, called also the limiting laws of probability theory. No such laws exist in classical or quantum mechanics. But, on the other hand, our theory is much more complete than the previous theory of squeezed states. Namely, we can calculate from our theory statistical moments of any order and compare them with experiment, if it is possible. In particular, it is important that for a continuous energy spectrum, as is the case of the classical gas discussed above and also for the quantum gas, we obtain a possibility of negative T_1 by positive T_2, i. e., an "inverted population" or "optical pumping", cf. /4/. So it seems that a further development of our ideas may be interesting in the domain of quantum optics. The point is to resign to limit ourselves only to discussion of pure states which are more familiar to physicists of modern age than generalized thermodynamical methods since thermodynamics at all looks sometimes as outdated and, may be, even clumsy. But such a pessimistic point of view, as it seems to the author, has no reasonable foundation and should better be avoided.

It is clear that the above method can be generalized to yet higher order terms, and also to other observables, cf./4/, but it is connected with many mathematical complications. That is the reason why we avoided it here. Nevertheless, it is a strong analogy of such generalizations with the higher-order squeezed states, cf. /13/ /14/, and we hope that in future also they will be investigated.

REFERENCES

/1/ D. F. Walls, Squeezed states of light, Nature 306:141-146 (1983).

/2/ M. M. Nieto, What are squeezed states really like? In: Frontiers of Nonequilibrium Statistical Physics (G. T. Moore and M. O. Scully eds.), Plenum Press, New York 1986, 287-307.

/3/ R. Loudon and P. L. Knight, Squeezed light, J. Mod. Optics 34:709-759 (1987).

/4/ R. S. Ingarden, Higher-order information dynamics in comesoscopic compact systems, J. Stat. Phys., to be published.

/5/ R. S. Ingarden, Information theory and variational principles in statistical physics, Bull. Acad. Polon. Sci., Ser. math. astr.phys. 11:541-547 (1963).

/6/ R. S. Ingarden, Information theory and thermodynamics of light, II. Principles of information thermodynamics, Fortschr. Phys. 13:755-805 (1965).

/7/ R. S. Ingarden and A. Kossakowski, On the connection of non-equilibrium information thermodynamics with non-hamiltonian quantum mechanics of open systems, Ann. Phys. NY 89:451-485 (1975).

/8/ H. Haken, Information and Self-Organization, A Macroscopic Approach to Complex Systems, Springer, Berlin 1988.

/9/ R. S. Ingarden, Higher order temperatures and coherence of light, Bull. Acad. Polon. Sci., Ser. math. astr. phys. 16:57-59 (1968).

/10/ R. S. Ingarden, Generalized thermodynamics of the electromagnetic radiation in a cavity, I. A 5-temperatures thermodynamics of one mode, Acta Phys. Polon. 36:855-885 (1969).

/11/ R. S. Ingarden, An information-theoretical approach to the theory of lasers, Bull. Acad. Polon. Sci., Ser. math. astr. phys.19:77-82 (1971).

/12/ R. S. Ingarden, Generalized irreversible thermodynamics and its application to lasers I,II, Acta Phys. Polon. A43:3-35 (1973).

/13/ C. K. Hong and L. Mandel, Generation of higher-order squeezing of quantum electromagnetic fields, Phys. Rev. A32:974-982 (1985).

/14/ C. Gerry and S. Rodrigues, Higher-order squeezing from an anharmonic oscillator, Phys. Rev. A35:4440-4442 (1987).

Squeezing of a filtered thermal field in a degenerated parametric optical amplifier

G. Oliver and O. Perrot

Laboratoire de Photophysique Moleculaire du CNRS
Bat. 213 – Université de Paris–Orsay
91405–Orsay Cedex (France)

Abstract: We present the results of a computer simulation of the statistical properties of light at the output of a lossless degenerated parametric optical amplifier *DPOA*. If the fundamental mode input is a coherent Glauber field and the harmonic mode input is a monochromatic filtered thermal field, then the harmonic mode output can be a squeezed state. The possibilities of intensive computer simulation in quantum optics are pointed out

Introduction

The degenerated parametric optical amplifier is one of the most promising device for production of squeezed states. The non-linear crystal mixes one optical mode and its first harmonic. The process which can lead to the generation of squeezed states is a quantum process and much care must be used in a theoretical work. The processes which use semiclassical approximations or decorrelation must be carefully considered. Other hand-made calculations are valid only for a short time.So we have attempted a numerical simulation of the behaviour of a *DPOA* with the two criterions of treating the exact quantum operator and making a very accurate calculations using precision control methods.

1 Theory

We consider the case of a crystal without losses and without external field pump. Then the evolution of the system of the two field modes of frequencies ω_1 and $\omega_2 = 2\omega_1$ is governed by the Hamiltonian

$$\mathbf{H} = \hbar\omega_1 \mathrm{a}_1^\dagger \mathrm{a}_1 + \hbar\omega_2 \mathrm{a}_2^\dagger \mathrm{a}_2 + i\hbar\chi \quad (\mathrm{a}_1^{\dagger 2}\mathrm{a}_2 - \mathrm{a}_1^2 \mathrm{a}_2^\dagger) \tag{1}$$

Lecture Notes in Physics, Vol. 378
C. Bendjaballah, O. Hirota, S. Reynaud (Eds.),
Quantum Aspects of Optical Communications

where χ is a coupling coefficient whose value is depending on the physical properties of the crystal.

Our aim is to study the statistical properties of the electrical fields of the two modes at the output of the $DPOA$ for any input initial conditions. The best method is to compute the elements of the density matrix of the two-modes field. In fact this density matrix let know all the properties of the field for the one–time measurements.

Using the density matrix approach allows us to introduce any initial conditions and particularly states of the system which cannot be described by a wave function. The system of differential equations governing the time evolution of this density matrix $\rho(t)$ under the Fock states basis is

$$\begin{aligned}\frac{\mathbf{d}\rho_{n_1,n_2,n_3,n_4}}{\mathbf{d}t} = \chi \quad (&\sqrt{n_1(n_1-1)(n_2+1)} \quad \rho_{n_1-2,n_2+1,n_3,n_4} \\ &-\sqrt{(n_3+1)(n_3+2)n_4} \quad \rho_{n_1,n_2,n_3+2,n_4-1} \\ &-\sqrt{(n_1+1)(n_1+2)n_2} \quad \rho_{n_1+2,n_2-1,n_3,n_44} \\ &+\sqrt{n_3(n_3-1)(n_4+1)} \quad \rho_{n_1,n_2,n_3-2,n_4+1}) \end{aligned} \tag{2}$$

Among all the physical quantities which can be computed when knowing the density matrix, the following are of particular interest :

(i) the energy of each mode, defined as $< \mathbf{n} >$

(ii) the variances of each component of the electric field for each mode, defined by

$$< \Delta\mathbf{P}^2 > = < \mathbf{E}_+^2 > - < \mathbf{E}_+ >^2 \tag{3}$$

$$< \Delta\mathbf{Q}^2 > = < \mathbf{E}_-^2 > - < \mathbf{E}_- >^2 \tag{4}$$

where

$$\mathbf{E}_+ = \mathbf{a} + \mathbf{a}^\dagger \tag{5}$$

$$\mathbf{E}_- = \frac{(\mathbf{a} - \mathbf{a}^\dagger)}{i} \tag{6}$$

are the two components of the electric field operator.

(iii) the bunching coefficient is defined by

$$\mathbf{Q} = < \mathbf{n}^2 > - < \mathbf{n} >^2 - < \mathbf{n} > \tag{7}$$

With these definitions the variances of the electric field are 1 for a coherent Glauber field and the variance of one component is smaller than 1 for a squeezed state; the bunching coefficient is negative for an antibunched field. Particularly $\mathbf{Q} = 0$ for a coherent field and $\mathbf{Q} = < \mathbf{n} >$ for a filtered thermal field.

2 Computational method

According to eq. 2, we must integrate a system of N linear differential equations with constant coefficients.The particular problem which occurs in our case comes from the very great number of equations. Indeed for an accurate result the Fock-states basis must range from $|0,0>$ to $|m_1, m_2>$ with m_1 and m_2 defined by the relation :

$$1 - \epsilon < \sum_{i=0}^{i=m_1} \sum_{j=0}^{j=m_2} \rho_{i,j,i,j} \leq 1 \tag{8}$$

So even a system of 10^5 equations allows only for $m_1 = 30, m_2 = 15$. Moreover this system is stiff, as the real parts of its eigenvalues are large negative numbers. We have integrated the system using a variable step Runge-Kutta method of order 7. At each step the precision is controlled by comparing the results obtained by the Runge-Kutta method of order 7 to the results obtained with a method of order 8. The relative precision required is 10^{-8}. The probabilities of photon-counting for each mode are computed to ensure the values of the diagonal elements of the density matrix are really probability distributions. Particularly the conservation of trace is monitored using $\epsilon \leq 10^{-7}$. The whole precision control process let us claim that the final precision of the physical quantities computed from the density matrix element is at least 10^{-3}. All the computations have been performed on the $CRAY-1$ Computer of the Centre de Calcul Vectoriel pour la Recherche $C.C.V.R.$

3 Results

We have studied by this simulation method, the squeezing of the two components of the output field for various input conditions. We give here the results obtained in the most interesting case : The input field of the fundamental mode is a coherent Glauber field and the input field of the harmonic mode is a monochromatic filtered thermal field. The thermal field is weaker than the coherent field. Then at the output the thermal field is turned into an amplified squeezed field. This is shown in Figs. 1 and 2. The energy of the coherent field is 5 photons in the two cases whereas the energies of the thermal field is 0.05 photon in Fig. 1 and 0.5 photon in Fig. 2. The two figures show also the gain and the bunching coefficient of the thermal field. In the two cases when sufficient energy has been transfered from the fundamental mode the harmonic mode is squeezed and antibunched. The squeezing is stronger (better than 0.8) when the ratio of the input field energies is greater.

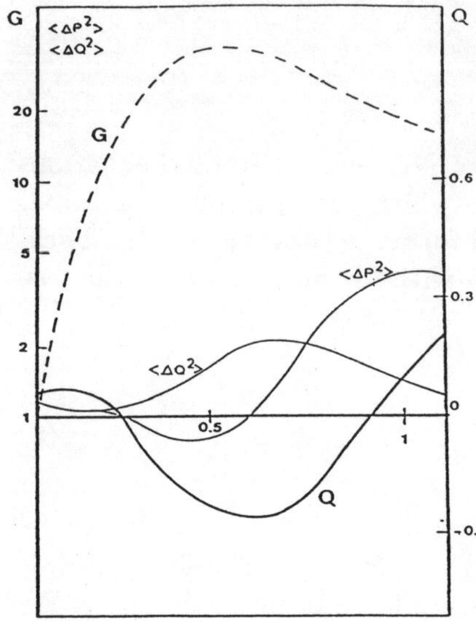

Fig.1– $< n_1 > = 5, < n_2 > = 0.05$

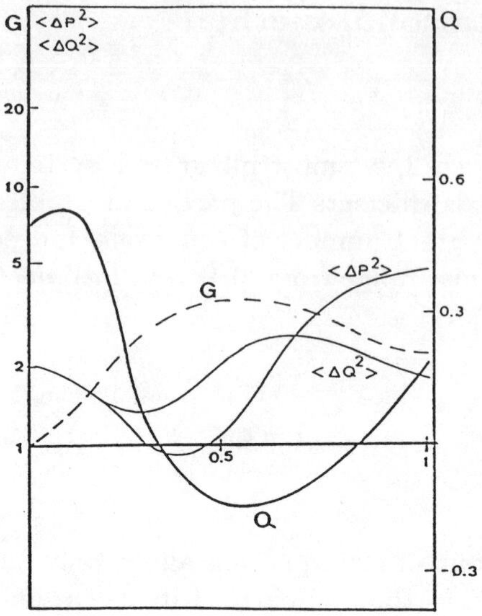

Fig.2– $< n_1 > = 5, < n_2 > = 0.5$

Conclusion

The two advantages of our simulation method are its precision and its capacity to accept any initial conditions.Its major drawback, beside the cost of computing,is the limitation at very weak fields which do not fit experimental conditions. This is particularly worrying when stuying quantum phenomenoms as they can disappears at classical energy levels. Fortunately, we can be certain that the squeezing of a field by a *DPOA* is not a quantum phenomenon. When the initial state condition can be described by a wave function $|\Psi(0)>$ the expectation value of any operator $\mathbf{O}(t)$ at time t is

$$< \mathbf{O}(t) > = < \Psi(0)| \quad \mathbf{O}(t) \quad |\Psi(0) > \tag{9}$$

The Taylor series of a time dependent operator can be easily computed. Using the **Reduce** symbolic computation langage, the Taylor series of the electric field operators, has been computed at order 28. Then the Taylor series of the expectation value of the electric field is computed and resummed, using quadratic Padé approximants technique. This method has been used with a coherent Glauber input field and an input energy of 1000 photons. It has been found that the squeezing already obtained by our numerical method for a few photons is growing with the photon number. Although 1000 is yet a very small number of photons it is a sufficiently large number to ensure that the squeezing of a field by a *DPOA* is not a quantum phenomenon. This fact allows us to claim that the results we have obtained for a very weak thermal field remain valid for a real thermal field and the squeezing of such a field can be important.

Finally we wish to point out the capabilities of the computer simulation and especially of symbolic calculation in the domain of Quantum Optics, where these methods are not yet widely used.

Displaced Number States

V.Bužek*, W.K.Lai, and P.L.Knight
Optics Section, The Blackett Laboratory
Imperial College, London SW7 2BZ, England

1 Introduction

Recent developments in quantum optics have led to suggestions of how number states of the electromagnetic field may be prepared [1]. At microwave frequencies the Rydberg-atom micromaser [2] is highly sensitive to the quantized nature of the radiation field in a cavity and has "trap" states where the field approaches a number state with a large degree of sub-Poisson photon statistics [3]. These trap states are those in which the quantum field has the required photon number to generate multiples of full Rabi cycles in subsequent atoms entering the cavity, leaving the field, in consequence, unchanged. A distribution of interaction times, analogous to normal laser pump fluctuations, will wash out such effects [4]. Experimental observation of sub-Poissonian trap states, with the cavity field prepared to a good approximation in a number state, has been reported when care was taken to ensure uniform atom-field interaction times in the micromaser [5].

Localized one-photon states have been constructed by optical shutter techniques using photons generated in pairs in parametric down conversion [6], where the signal photon opens a photoelectric detection gate to the idler photon. Related aspects of such

*Permanent address: Institute of Physics, Slovak Academy of Sciences, Dúbravská cesta 9, 842 28 Bratislava, Czechoslovakia.

Lecture Notes in Physics, Vol. 378
C. Bendjaballah, O. Hirota, S. Reynaud (Eds.),
Quantum Aspects of Optical Communications

conditioned measurements have been proposed for photon number-state preparation in nondegenerate parametric amplification [7].

Given that photon number states (at least those with modest occupation numbers) can be generated, it is natural to ask whether they can be amplified. Rather than address this problem directly, we turn our attention to the simpler problem of *displacing* a number state by a translation operator $\hat{D}(\alpha)$ [8], the generator of coherent states from the vacuum. We will call the state derived by acting on the number state with a displacement operator the *displaced number state.*

In our lecture we will address two problems. First we show that displaced number states have interesting and unusual physical properties. Secondly, we will analyze the possibility of generating these states in a linear directional coupler.

2 Properties of displaced number states

In one sense, the displaced number state is obtained from a number state by adding a nonzero value to the field amplitude. The number state is determined by its photon number while the phase is completely random. The amplitude of the field has a zero expectation value. By displacing in phase space, a field amplitude is added to this state, and the photon number has now a contribution from the coherent component of the field. The state becomes phase dependent because of the phase of the displacement, is centred around a new origin located at the coherent amplitude position, and is invariant by a rotation around this point, with phase-independent fluctuations. Despite this apparent simplicity, several interesting consequences emerge from this study.

The displaced number state is defined by

$$|\alpha, n\rangle = \hat{D}(\alpha, \hat{a})|n\rangle, \tag{1}$$

where $\hat{D}(\alpha)$ is the displacement operator [8], given by

$$\hat{D}(\alpha, \hat{a}) = \exp\left(\alpha\hat{a}^\dagger - \alpha^*\hat{a}\right). \tag{2}$$

For $n = 0$, the displaced number state, Equation (1), reduces to the well-known coherent state. Some properties of the displaced number state can be derived using the transformation of the annihilation and creation operators under a displacement

$$\hat{D}(\alpha, \hat{a})\hat{a}\hat{D}^\dagger(\alpha, \hat{a}) = \hat{a} - \alpha,$$

$$\hat{D}(\alpha, \hat{a})\hat{a}^\dagger\hat{D}^\dagger(\alpha, \hat{a}) = \hat{a}^\dagger - \alpha^*. \tag{3}$$

We define the quadrature operators by

$$\hat{X}_1 = \hat{a} + \hat{a}^\dagger, \tag{4}$$

$$\hat{X}_2 = -i(\hat{a} - \hat{a}^\dagger). \tag{5}$$

With the use of Equations (1)-(5), the variances $\langle(\Delta\hat{X}_i(t))^2\rangle = \langle\hat{X}_i^2\rangle - \langle\hat{X}_i\rangle^2$ of the quadrature operators in the state $|\alpha, n\rangle$ are

$$\langle(\Delta\hat{X}_i(t))^2\rangle = \langle\alpha, n|(\Delta\hat{X}_i)^2|\alpha, n\rangle = (2n+1) \quad \text{for} \quad i = 1, 2. \tag{6}$$

These variances are independent of the coherent amplitude of the state, but linearly dependent of the initial photon number n. The photon number variance $\langle(\Delta\hat{n}(t))^2\rangle$ is

$$\langle(\Delta\hat{n}(t))^2\rangle = (2n+1)|\alpha|^2, \tag{7}$$

and is always greater than that for a number state and a coherent state. With the use of Equations (3) and (7), we find the Mandel Q parameter [9], measuring the deviation from a Poission statistics for the state $|\alpha, n\rangle$

$$Q = \frac{\langle(\Delta\hat{n}(t))^2\rangle - \langle\hat{n}\rangle}{\langle\hat{n}\rangle} = n\left(\frac{2|\alpha|^2 - 1}{n + |\alpha|^2}\right), \tag{8}$$

where the average photon number

$$\langle\hat{n}\rangle = n + |\alpha|^2, \tag{9}$$

in which the contributions from the numberlike and coherentlike characters of the field are explicitly displayed. For Poissonian statistics, $Q = 0$. If $Q < 0$, the light is said to be sub-Poissonian, otherwise, it is super-Poissonian. From Equation (8), we find that the state $|\alpha, n\rangle$ has sub-Poissonian photon statistics if $|\alpha|^2 < 1/2$. In other words, if the coherent contribution adds more than half a photon to the average photon number, that state is super-Poissonian irespective of the initial photon number n.

The coordinate space wave function of the number state is that of the harmonic oscillator

$$\Psi_n(x) = \frac{1}{\pi^{1/4}} \frac{1}{(2^n n!)^{1/2}} H_n(x) \exp(-\frac{1}{2}x^2), \tag{10}$$

where H_n is a Hermite polynomial. The displacement operator for a *real* displacement translates $x \to x - \sqrt{2}\alpha$ [10], thus the wave function $\Psi_{dn}(x)$ for the displaced number state is

$$\Psi_{dn}(x) = \frac{1}{\pi^{1/4}} \frac{1}{(2^n n!)^{1/2}} H_n(x - \sqrt{2}\alpha) \exp[-\frac{1}{2}(x - \sqrt{2}\alpha)^2], \tag{11}$$

The number distribution $P_{dn}(m)$ of the displaced number state is one of the quantities of most interest. For the displaced number state, the number distribution is defined by

$$P_{dn}(l) = \left| \int_{-\infty}^{\infty} dx \Psi_{dn}(x) \Psi_n(x) \right|^2, \tag{12}$$

which is directly analogous to the Franck-Condon overlap in molecular transitions responsible in that context for the oscillations which are such a noticeable feature of vibrational substructure in molecular electronic transitions. Assuming $l \geq n$, it is found that

$$P_{dn}(l) = \frac{n!}{l!} \alpha^{2(l-n)} e^{-\alpha^2} \left[\mathcal{L}_n^{(l-n)}(\alpha^2)\right]^2, \tag{13}$$

where $\mathcal{L}_n^{(l-n)}(x)$ is an associated Laguerre polynomial. From Equation (10), it is readily seen that this number distribution is symmetrical in n and l, i.e., the probability of l photons in state $|\alpha, n\rangle$ is equal to the probability of n photons in state $|\alpha, l\rangle$. The photon number distribution (13) can be rewritten in the following way:

$$P_{dn}(l) = \frac{e^{-|\alpha|^2} |\alpha|^{2(l-n)}}{n! l!} \left| \sum_{k=0}^{n} \frac{n! l! (-1)^k |\alpha|^{2(n-k)}}{k!(n-k)!(l-k)!} \right|^2. \tag{14}$$

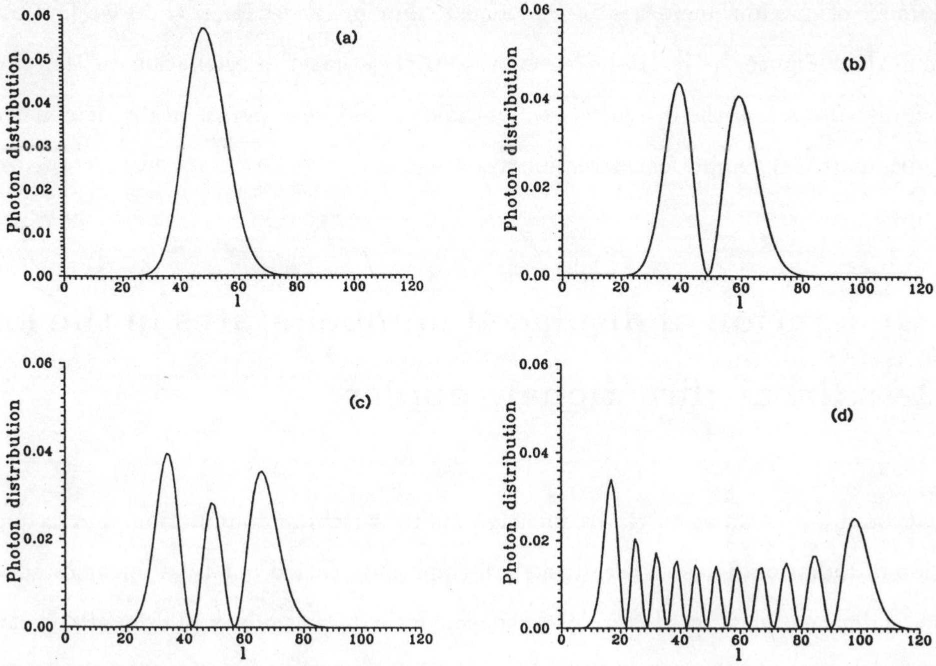

Figure 1: Photon distribution for the displaced number state $|7, n\rangle$: (a) $n = 0$; (b) $n = 1$; (c) $n = 2$; (d) $n = 10$.

For $\alpha = 0$, $P_{dn}(l) = \delta_{ln}$, as expected. In the squared modulus, we have a polynomial in l of degree n, so extending this function for real values of l, one would have eventually n zeros. Therefore it can be expected that $P_{dn}(l)$ have up to n minima between $(n+1)$ maxima. In Figures 1(a)-1(d), we plot $P_{dn}(l)$ for $\alpha = 7$, and $n = 0, 1, 2, 10$. We see that there is a strong similarity between these functions and the Hermite polynomials of order n. Let us discuss the case with $n = 1$. In this case, it is readily seen that Equation (14) can be rewritten as

$$P_{dn}(l) = (l!)^{-1} e^{-|\alpha|^2} |\alpha|^{2(l-n)} \left(l - |\alpha|^2\right)^2 . \tag{15}$$

This distribution has a zero at $l = |\alpha|^2$ if $|\alpha|^2$ is an integer. It is interesting to note that adding a coherent amplitude to a one-photon state, the probability of having zero photons becomes nonzero. We see in Figure 1(b) that there are two peaks, around the minimum, as opposed to a coherent state [see Figure 1(a)], where there is only one peak value for the number of photons. For $n = 2$, we have a second minimum appearing [Figure 1(c)], and the number of maxima increases as n increases. For instance, for $n = 10$ we find eleven maxima. [see Figure 1(d)]. It has been shown [11] that the oscillation in the photon number distribution of the displaced number state could be experimentally demonstrated even with a strongly imperfect detection system.

3 Generation of displaced number states in the lossless linear directional coupler

Optical communication networks frequently require switching, modulation, and frequency selection of radiation fields. These tasks are commonly caried out by a linear directional coupler, a device consisting of two closely spaced optical wave guides. The overlap between the mode functions in the respective guides allows a controlled transfer of power from one guide to the other. Classical treatments of the coupler based on coupled mode theory have been given by many authors (see, for instance, [12] and references quoted therein). While the classical treatment is sufficient for many purposes, a quantum treatment becomes essential when the statistical properties of the radiation field are important [13].

The quantum lossless linear directional coupler can be modelled as a system of two coupled quantum oscillators obeying the following equation of motion:

$$i\frac{d\hat{a}_1}{dz} = \beta_1\hat{a}_1 + \chi\hat{a}_2;$$

$$i\frac{d\hat{a}_2}{dz} = \beta_2\hat{a}_2 + \chi\hat{a}_1, \tag{16}$$

where $\hat{a}_i$ and $\hat{a}_i^\dagger$ $(i = 1, 2)$ are the photon annihilation and creation operators of mode i respectively; β_i are the propagation constants of the relevant modes. We assume, for simplicity, equal coupling constants χ. Including mode losses and mode coupling through scattering is straighhtforward; we discuss the consequences of these effects elsewhere.

The evolution operator $\hat{T}(L)$ corresponding to the equations (16) can be writen as:

$$\hat{T}(L) = \exp\left\{-i[\beta_1 \hat{a}_1^\dagger \hat{a}_1 + \beta_2 \hat{a}_2^\dagger \hat{a}_2 + \chi(\hat{a}_1^\dagger \hat{a}_2 + \hat{a}_2^\dagger \hat{a}_1) - (\beta_1 + \beta_2)/2]\right\}, \tag{17}$$

where L is the length of the coupler. The photon operators at the output of the coupler can be found using the relation

$$\hat{a}_1(L) = \hat{T}^\dagger(L)\hat{a}_1(0)\hat{T}(L) = u(L)\hat{a}_1(0) + v(L)\hat{a}_2;$$

$$\hat{a}_2(L) = \hat{T}^\dagger(L)\hat{a}_2(0)\hat{T}(L) = v(L)\hat{a}_1(0) + u^*(L)\hat{a}_2, \tag{18}$$

where

$$u(L) = \cos\bar{\chi}L - i\frac{\Delta\beta}{\bar{\chi}}\sin\bar{\chi}L;$$

$$v(L) = -i\frac{\chi}{\bar{\chi}}\sin\bar{\chi}L;$$

$$\Delta\beta = \frac{\beta_1 - \beta_2}{2}; \tag{19}$$

$$\bar{\chi} = \sqrt{\chi^2 + (\Delta\beta)^2}.$$

Let now suppose that the n-photon state $|n\rangle_1$ enters port 1 and the coherent state $|\alpha\rangle_2$ enters port 2 of the coupler, respectively. In this case we can write the input state of the coupler as

$$|in\rangle = |n\rangle_1|\alpha\rangle_2 = \frac{(\hat{a}_1^\dagger(0))^n}{\sqrt{n!}}\hat{D}(\alpha, \hat{a}_2(0))|0\rangle_1|0\rangle_2. \tag{20}$$

Using the evolution operator $\hat{T}(L)$ we obtain for the output state of the coupler the following expression:

$$|out\rangle = \hat{T}(L)|in\rangle, \tag{21}$$

or

$$|out\rangle = \exp[\alpha(v\hat{a}_1^\dagger + u^*\hat{a}_2^\dagger) - \alpha^*(v^*\hat{a}_1 + u\hat{a}_2)]$$
$$\times \frac{1}{\sqrt{n!}}(u\hat{a}_1^\dagger + v\hat{a}_2^\dagger)^n|0\rangle_1|0\rangle_2, \tag{22}$$

where we have used the notation: $\hat{a}_i(0) = \hat{a}_i$; $v(L) = v$ and $u(L) = u$. Finally we can rewrite the expression for the output state in the form:

$$|out\rangle = \sum_{k=0}^{n} \begin{pmatrix} n \\ k \end{pmatrix}^{1/2} (v)^k(u)^{n-k}|\alpha v, n-k\rangle_1|\alpha u^*, k\rangle_2; \tag{23}$$

from which we see that the linear directional coupler with the input state (20) can serve as a generator of mixtures of displaced number states in each of the output ports. It is easy to show, that for high enough intensity of the coherent field and supposing that

$$|v| \ll |u| \tag{24}$$

(i.e. $\bar{\chi}L \simeq \pi$) we obtain from (23) the relation for the output state of the coupler

$$|out\rangle \simeq |\alpha v, n\rangle_1 |\alpha u^*\rangle_2. \tag{25}$$

In other words we can say that under particular conditions (see above) the linear directional coupler can serve as a generator of displaced number states.

It should be mentioned here that relations (17-19) can be identified as the beam-splitter transformations [14], so, we can say that the lossless coupler behaves in essentially the same way as a beam-splitter. Therefore, if we suppose that the strong coherent field enters the second (unused) port, the coupler behaves in the same way as the homodyne detector. Moreover, the condition (24) corresponds to ordinary homodyning, which means that if the input field in the ordinary homodyne detection is in the n-photon state then the output field is in the displaced number state (25).

4 Summary

We have studied several properties of the displaced number states of the harmonic oscillator. In particular, the analysis for the photon number distribution showed that this function exhibits oscillations (these oscillations can be explained in terms of phase-space interference [11]). As we have mentioned earlier the oscillations of the photon number distribution could be experimentally observed even with a highly imperfect detection system. This is rather encouraging, recalling that recently one-photon state has been produced in the laboratory. We have shown that a coherent contribution to the amplitude of this state should, in principle, pose no insurmountable practical barier using the linear directional coupler or the ordinary homodyning scheme. Such experiments would also be of general importance in demonstrating that the coherent component added to the amplitude of a one-photon state is capable of making this state macroscopically different from its neighbors which differ from it only microscopically. This property of the displaced number state can be used in optical communication networks in which one bit of information may correspond to the presence or absence of one photon (i.e., one-photon state or vacuum state). By launching the signal field into port 1 and a strong coherent field into port 2 one can easily distinguish at the output of port 1 which bit of information was sent.

Acknowledgements

This work was supported in part by the UK Science and Engineering Research Council.

5 References

[1] R.LOUDON: *Rep. Prog. Phys.* **43** (1980) 913.

[2] D.MESCHEDE, H.WALTHER, and G.MÜLLER: *Phys. Rev. Lett.* **54** (1985) 551; M.BRUNE, J.M.RAIMOND, P.GOY, L.DAVIDOVICH, and S.HAROCHE: *Phys. Rev. Lett.* **59** (1987) 1899.

[3] P.FILIPOWICZ, J.JAVANAINEN, and P.MEYSTRE: *J. Opt. Soc. Am. B* **3** (1986) 906; F.CUMMINGS, and A.K.RAJAGOPAL: *Phys. Rev. A* **39** (1989) 3414.

[4] L.A.LUGIATO, M.O.SCULLY, and H.WALTHER: *Phys. Rev. A* **36** (1987) 740.

[5] G.REMPE, F.SCHMIDT-KALER, and H.WALTHER: *Phys. Rev. Lett.* **64** (1990) 2783.

[6] .K.HONG and L.MANDEL: *Phys. Rev. Lett.* **56** (1986) 58.

[7] C.M.Holmes, G.J.Milburn, and D.F.Walls: unpublished.

[8] R.J.Glauber: *Phys. Rev.* **131** (1963) 2766.

[9] L.Mandel: *Opt. Lett.* **4** (1979) 205.

[10] C.Cohen-Tannoudji, B.Diu, and F.Laloë: *"Quantum Mechanics"* (Wiley, New York, 1977).

[11] F.A.M.De Oliveira, M.S.Kim, P.L.Knight, and V.Bužek: *Phys. Rev. A* **41** (1990) 2645 and references therein.

[12] D.Marcuse: *"Light Transmission Optics"* (Van Nostrand Reinhold Comp., Dordrecht, 1982).

[13] J.Janszky, S.Sibilia, and M.Bertolotti: *J. Mod. Opt.* **35** (1988) 1757.

[14] S.Prasad, M.O.Scully, and W.Martienssen: *Opt. Commun.* **62** (1987) 139; H.Fearn and R.Loudon: *Opt. Commun.* **64** (1987) 485; Z.Ou, C.K.Hong, and L.Mandel: *Opt. Commun.* **63** (1987) 118. B.Yurke, S.L.McCall, and J.R.Klauder: *Phys. Rev. A* **33** (1986) 4033. R.A.Campos, B..A.Saleh, and M.Teich: *Phys. Rev. A* **40** (1989) 1371.

GENERALIZED QUADRATIC GENERATOR OF SQUEEZED STATES

Anil K. Roy and C.L. Mehta
Physics Department, Indian Institute of Technology
New Delhi 110 016 India

Abstract: In this paper we consider a unitary operator $U(\sigma,\tau,\alpha)$ which is the exponential of a general quadratic in $\hat{a}$ and $\hat{a}^{+}$ and discuss properties of states which are generated by such a unitary transformation of the vacuum viz. $|\sigma,\tau,\alpha\rangle = U(\sigma,\tau,\alpha)\,|0\rangle$. The state $|\sigma,\tau,\alpha\rangle$ is an eigenstate of the operator $U\hat{a}U^{+}$ with the eigenvalue 0. We evaluate the uncertainties in position and its canonical conjugate momentum variables. Conditions for the state being a minimum uncertainty product state and also conditions for squeezing in either of the two conjugate variables are then discussed. Number state representation and also the coordinate representation are obtained.

1. Introduction

In quantum optics the quantized electromagnetic field is described in terms of the harmonic oscillator modes. For simplicity we shall consider only one degree of freedom. The annihilation and creation operators $\hat{a},\hat{a}^{+}$ satisfy the commutation relation:

$$[\hat{a},\ \hat{a}^{+}] = 1. \tag{1.1}$$

The number operator $\hat{a}^{+}\hat{a}$, being hermitian, provides a useful Fock state representation. The basis states in this representation are the number states $|n\rangle$, which can be constructed from the vacuum

Lecture Notes in Physics, Vol. 378
C. Bendjaballah, O. Hirota, S. Reynaud (Eds.),
Quantum Aspects of Optical Communications

state $|0\rangle$ by repeated application of $\hat{a}^+$:

$$\hat{a}^+\hat{a}\ |n\rangle = n|n\rangle, \tag{1.2}$$

$$|n\rangle = (n!)^{-1/2}\hat{a}^{+n}|0\rangle. \tag{1.3}$$

The eigenstates $|\alpha\rangle$ of the annihilation operator $\hat{a}$ satisfying the eigenvalue equation:

$$\hat{a}|\alpha\rangle = \alpha|\alpha\rangle \tag{1.4}$$

provide another basis of representation. These states, called the coherent states, have been found extremely useful while considering quantum theory of coherence, operator ordering, phase space representation etc.

In recent years considerable interest has been generated in studying the properties and applications of related states, called under different names as 'pulsating wave packets'[1], 'minimum uncertainty states'[2,3], 'new coherent states'[4,5], 'two photon coherent states'[6] and 'ideal squeezed states'[7]. These squeezed states $|\sigma,\alpha\rangle$ are the eigenstates of the unitarily transformed annihilation operator $\hat{A} = S(\sigma)\hat{a}S^+(\sigma)$ and are obtained by the action of the displacement operator $D(\alpha)$ followed by the squeezed operator $S(\sigma)$ on to the vacuum:

$$|\sigma,\alpha\rangle = S(\sigma)\ D(\alpha)|0\rangle, \tag{1.5}$$

$$\hat{A}|\sigma,\alpha\rangle = \alpha|\sigma,\alpha\rangle, \tag{1.6}$$

where squeeze operator $S(\sigma)$ and displacement operator $D(\alpha)$ both are unitary operators and are generally taken as:

$$S(\sigma) = \exp[\tfrac{1}{2}(\sigma\hat{a}^{+2} - \sigma^*\hat{a}^2)], \tag{1.7}$$

$$D(\alpha) = \exp(\alpha\hat{a}^+ - \alpha^*\hat{a}), \tag{1.8}$$

in which case $\hat{A}$ is given by:

$$\hat{A} = \hat{a}\cosh r - \hat{a}^+\ e^{i\theta}\sinh r, \tag{1.9}$$

where we have written

$$\sigma = r\, e^{i\theta}; \; 0 \le r < \infty, \quad 0 \le \theta \le 2\pi. \tag{1.10}$$

In the present paper we consider a unitary operator $U(\sigma,\tau,\alpha)$ which is the exponential of a general quadratic in a and a^{+}:

$$U(\sigma,\tau,\alpha) = \exp[\tfrac{1}{2}(\sigma\hat{a}^{+2} - \sigma^{*}\hat{a}^{2}) + i\tau\hat{a}^{+}\hat{a} + \alpha\hat{a}^{+} - \alpha^{*}\hat{a}] \tag{1.11}$$

with τ real and σ or α any complex number. It is readily seen that a state $|\sigma,\tau,\alpha\rangle$ generated by such a unitary transformation of the vacuum

$$|\sigma,\tau,\alpha\rangle = U(\sigma,\tau,\alpha)\,|0\rangle \tag{1.12}$$

is an eigenstate of the transformed annihilation operator $U\hat{a}U^{+}$ with 0 eigenvalue. We study the properties of these states and derive conditions under which they exhibit squeezing.

In Sec. 2, we give transformations for the various operators by the unitary operator U. The expectation values of these operators are then derived in Sec. 3. In Sec. 4 we consider the conditions under which these states will exhibit properties such as being the minimum uncertainty product states or being the squeezed states. Different wave function representations corresponding to the state $|\sigma,\tau,\alpha\rangle$ are also considered. In Sec. 5 we obtain the eigenvalue equation first and then arrive at the number state representation. The coordinate or q-representation of the state $|\sigma,\tau,\alpha\rangle$ is discussed in Sec. 6. Similar calculations will lead to the momentum representation. These representations are found to be displaced Gaussian wave functions.

2. Unitary Transformations

In this section we obtain the effect of the transformation on the various operators such as $\hat{a}$, $\hat{a}^{+}$ etc. A straight forward application of the formula

$$e^{\hat{A}}\ \hat{B}\ e^{-\hat{A}} = \hat{B} + [\hat{A},\hat{B}] + (1/2!)[\hat{A},[\hat{A},\hat{B}]] + \ldots \qquad (2.1)$$

gives the following expression for the transformed operators

$$U\hat{a}U^{+} = k^{*}\hat{a} - \sigma S\hat{a}^{+} + m(1-C) - \alpha S, \qquad (2.2)$$

$$U\hat{a}^{+}U^{+} = k\hat{a}^{+} - \sigma^{*}S\hat{a} + m^{*}(1-C) - \alpha^{*}S, \qquad (2.3)$$

where we have written

$$C = \cosh \Omega, \qquad (2.4)$$

$$S = (\sinh \Omega)/\Omega, \qquad (2.5)$$

$$k = C + i\tau S, \qquad (2.6)$$

$$m = -(\sigma\alpha^{*} + i\tau\alpha)/\Omega^{2}, \qquad (2.7)$$

$$\Omega = [r^{2} - \tau^{2}]^{1/2}, \quad (\sigma = re^{i\theta}). \qquad (2.8)$$

The transformation of the coordinate and its canonical conjugate momentum operators are obtained readily from the relations

$$q = (\hat{a} + \hat{a}^{+})/\sqrt{2}, \qquad p = (\hat{a} - \hat{a}^{+})/\sqrt{2}i. \qquad (2.9)$$

We thus find that

$$U\hat{q}U^{+} = [(k^{*}-\sigma^{*}S)\hat{a} + (k-\sigma S)\hat{a}^{+} + (m+m^{*})(1-C) - (\alpha+\alpha^{*})S]/\sqrt{2}, \qquad (2.10)$$

$$U\hat{p}U^{+} = [(k^{*}+\sigma^{*}S)\hat{a} - (k+\sigma S)\hat{a}^{+} + (m-m^{*})(1-C) - (\alpha-\alpha^{*})S]/\sqrt{2}i. \qquad (2.11)$$

We also need expressions for the inverse transformations such as $U^{+}\hat{a}U$ etc. Since U^{+} is obtained by changing the sign of σ,τ and α in U, we immediately find from Eq (2.2) that

$$U^{+}\hat{a}U = k\hat{a} + \sigma S\hat{a}^{+} + m(1-C) + \alpha S, \qquad (2.12)$$

$$U^{+}\hat{a}^{+}U = k^{*}\hat{a}^{+} + \sigma^{*}S\hat{a} + m^{*}(1-C) + \alpha^{*}S. \qquad (2.13)$$

3. Expectation and Uncertainty Values

We next consider the expectation values of the various operators in the state $|\sigma,\tau,\alpha\rangle$. Using Eq (1.12) we obtain

$$\langle\hat{a}\rangle = \langle\sigma,\tau,\alpha|\ \hat{a}\ |\sigma,\tau,\alpha\rangle = \langle 0|U^{+}\hat{a}U|0\rangle. \tag{3.1}$$

We now use Eq (2.12) and the fact that $\hat{a}|0\rangle = \langle 0|\hat{a}^{+} = 0$ and obtain

$$\langle\hat{a}\rangle = m(1-C) + \alpha\ S. \tag{3.2}$$

Proceeding in this manner we obtain the following expressions for the expectation values

$$\langle\hat{a}^{+}\rangle = m^{*}(1-C) + \alpha^{*}S, \tag{3.3}$$

$$\langle\hat{n}\rangle = \langle\hat{a}^{+}\hat{a}\rangle$$

$$= r^{2}S^{2} + mm^{*}(1-C)^{2} + S(1-C)(m\alpha^{*} + m^{*}\alpha) + |\alpha|^{2}S^{2}, \tag{3.4}$$

$$\langle\hat{a}^{2}\rangle = k\sigma S + m^{2}(1-C)^{2} + 2m\alpha S(1-C) + \alpha^{2}S^{2}, \tag{3.5}$$

$$\langle\hat{a}^{+2}\rangle = k^{*}\sigma^{*}S + m^{*2}(1-C)^{2} + 2m^{*}\alpha^{*}S(1-C) + \alpha^{*2}S^{2}, \tag{3.6}$$

$$\langle\hat{q}\rangle = [(m + m^{*})(1 - C) + (\alpha + \alpha^{*})S]/\sqrt{2}, \tag{3.7}$$

$$\langle\hat{p}\rangle = [(m - m^{*})(1 - C) + (\alpha - \alpha^{*})S]/\sqrt{2}i, \tag{3.8}$$

$$\langle\hat{q}^{2}\rangle = \tfrac{1}{2}[(k\sigma + k^{*}\sigma^{*})S + (\alpha^{2} + \alpha^{*2})S^{2} + (m^{2} + m^{*2})(1-C)^{2}$$

$$+ 2(r^{2}S^{2} + mm^{*}(1-C)^{2} + S(1-C)(m\alpha^{*} + m^{*}\alpha) + |\alpha|^{2}S^{2})$$

$$+ 2S(1-C)(m\alpha + m^{*}\alpha^{*}) + 1], \tag{3.9}$$

$$\langle\hat{p}^2\rangle = -\tfrac{1}{2}[(k\sigma + k^*\sigma^*)S + (\alpha^2 + \alpha^{*2})S^2 + (m^2 + m^{*2})(1-C)^2$$

$$- 2(r^2S^2 + mm^*(1-C)^2 + S(1-C)(m\alpha^* + m^*\alpha + |\alpha|^2S^2)$$

$$+ 2S(1-C)(m\alpha + m^*\alpha^*) - 1]. \qquad (3.10)$$

The uncertainties in q and p are defined as

$$(\Delta q)^2 = \langle\hat{q}^2\rangle - \langle\hat{q}\rangle^2, \qquad (3.11)$$

$$(\Delta p)^2 = \langle\hat{p}^2\rangle - \langle\hat{p}\rangle^2. \qquad (3.12)$$

Using equations (3.7)-(3.10) we obtain

$$(\Delta q)^2 = \tfrac{1}{2}[1 + 2rS^2(r - \tau\sin\theta) + 2rCS\cos\theta], \qquad (3.13)$$

$$(\Delta p)^2 = \tfrac{1}{2}[1 + 2rS^2(r + \tau\sin\theta) - 2rCS\cos\theta]. \qquad (3.14)$$

We find that the uncertainties in q and p are independent of the parameter α. It may be noted that the result is consistent with the case $\tau = 0$ obtained earlier by Stoler[2]:

$$(\Delta q)^2_{\tau=0} = \tfrac{1}{2}[1 + 2\sinh^2 r + 2\sinh r\cosh r\cos\theta], \qquad (3.15)$$

$$(\Delta p)^2_{\tau=0} = \tfrac{1}{2}[1 + 2\sinh^2 r - 2\sinh r\cosh r\cos\theta]. \qquad (3.16)$$

4. Conditions for Minimum Uncertainty Product and Squeezing:

In the last section we derived expressions for the uncertainties in q and p. From these expressions [cf. Eq (3.13), (3.14)] we find that the uncertainty product is given by

$$(\Delta q)^2(\Delta p)^2 = \tfrac{1}{4} + r^2S^2[C\sin\theta + \tau S\cos\theta]^2. \qquad (4.1)$$

It is readily seen that state $|\sigma,\tau,\alpha\rangle$ is minimum uncertainty product state (MUS) if and only if

$$C \sin \theta + \tau S \cos \theta = 0, \tag{4.2}$$

or

$$\tan \theta = - (\tau \tanh \Omega) / \Omega. \tag{4.3}$$

Thus if τ were zero then σ must be real and vice versa. For any other value of τ we can choose θ from Eq (4.3) so that the given state is MUS.

We next consider conditions under which we can get squeezing either in q or its conjugate parameter p. For squeezing in q we must have, from eq (3.13),

$$r < \tau \sin \theta - (C/S) \cos \theta \tag{4.4}$$

On the other hand if we require squeezing in p, we must have

$$r < (C/S) \cos \theta - \tau \sin \theta \tag{4.5}$$

We thus conclude that if $\tau \sin \theta - (C/S) \cos \theta < 0$ (>0) then there can not be any squeezing in the parameter q (p). Further if $\tau \sin \theta - (C/S) \cos \theta < 0$ then there may be squeezing in the parameter p provided we also have $r < |(C/S) \cos \theta - \tau \sin \theta|$. If we further require that the state is also MUS [Eq (4.2)] then the condition for squeezing in the parameter p is

$$1/r > [\tanh \Omega)/\Omega] \cos \theta \tag{4.6}$$

or

$$r < -\tau / \sin \theta. \tag{4.7}$$

Analogous remarks hold for squeezing in the parameter q. Thus we see that the state $|\sigma,\tau,\alpha\rangle$ may be MUS with squeezing either in the parameter q or in p.

5. Eigenvalue Equation and Number State Representation

It has been observed that the state $|\sigma,\tau,\alpha\rangle$ is an eigenstate of $U\hat{a}U^{+}$ with 0 eigenvalue. Using Eq (2.2) we may rewrite this statement in the form :

$$\hat{G}\ |\sigma,\tau,\alpha\rangle = g\ |\sigma,\tau,\alpha\rangle, \tag{5.1}$$

where

$$\hat{G} = k^{*}\hat{a} - \sigma S\ \hat{a}^{+}, \tag{5.2}$$

and

$$g = \alpha S - m(1-C). \tag{5.3}$$

The operator G also satisfies

$$[\hat{G},\ \hat{G}^{+}] = 1. \tag{5.4}$$

In this section we derive the number state representation of the state $|\sigma,\tau,\alpha\rangle$. Writing

$$|\sigma,\tau,\alpha\rangle = \Sigma|n\rangle\langle n|\sigma,\tau,\alpha\rangle \tag{5.5}$$

and using Eq (5.1) we obtain after some substitutions and simplifications the following expressions :

$$\langle n|\sigma,\tau,\alpha\rangle = \Sigma N\ \{H_n(x)\ (\delta/2)^{n/2}\}/(n!)^{1/2}, \tag{5.6}$$

where

$$\delta = -\ \sigma S/k^{*}, \tag{5.7}$$

$$x = [\alpha S - m(1-C)]/(k^{*}(2\delta)^{1/2}), \tag{5.8}$$

and N is the normalisation constant chosen such that

$$\langle\sigma,\tau,\alpha|\sigma,\tau,\alpha\rangle = 1.$$

$H_n(x)$ is the Hermite polynomial of the nth degree. Using the orthonormality of $|n\rangle$ states we obtain

$$\langle n|\sigma,\tau,\alpha\rangle = (n!)^{1/2} \; (1-|\delta|^2)^{1/4}(\delta/2)^{n/2}H_n(x)$$

$$\times \exp[-\tfrac{1}{2}(|\beta|^2-\beta^2\delta^*)/(1-|\delta|^2)], \qquad (5.9)$$

where

$$\beta = [\alpha S - m(1-C)]/k^*. \qquad (5.10)$$

6. Coordinate Representation

In this section we express $|\sigma,\tau,\alpha\rangle$ in terms of the eigenstates of $\hat{q}$. We write

$$|\sigma,\tau,\alpha\rangle = \int |q\rangle\langle q|\sigma,\tau,\alpha\rangle \, dq, \qquad (6.1)$$

where the basis states of the q representation satisfy

$$\hat{q}\; |q\rangle = q\; |q\rangle, \qquad (6.2)$$

and

$$\hat{p}\; |q\rangle = -i(d/dq)|q\rangle. \qquad (6.3)$$

By making the operator $\hat{G}$ operate on $|\sigma,\tau,\alpha\rangle$ and using Eqs (6.2), (6.3) and the orthogonality of the states $|q\rangle$,we obtain the following expression for $\langle q|\sigma,\tau,\alpha\rangle$:

$$\langle q|\sigma,\tau,\alpha\rangle = B \exp[-\{q(k^*-\sigma S)^{1/2}/\surd 2 - g/(k^*-\sigma S)^{1/2}\}^2/(k^*+\sigma S)]. \qquad (6.4)$$

The normalisation constant B is given by

$$B = \left[\frac{\pi}{\mathrm{Re}\left(\frac{k^*-\sigma S}{k^*+\sigma S}\right)}\right]^{-1/4} \exp\left[\mathrm{Re}\,\frac{g^2}{(k^{*2}+\sigma^2 S^2)} - \frac{\left\{\mathrm{Re}\,\frac{g}{(k^*+\sigma S)}\right\}^2}{\mathrm{Re}\left(\frac{k^*-\sigma S}{k^*+\sigma S}\right)} + i\phi\right] \qquad (6.5)$$

where Ø is an arbitrary phase factor. As expected the q-representation is the exponential of a quadratic in q. Analogous expression may also be obtained for the momentum representation.

References:

1. H. Takahasi, Adv. commun. Syst., 1, 227 (1965).
2. D. Stoler, Phys. Rev. D, 1,3217 (1970).
3. D. Stoler, Phys. Rev. D, 4, 1925 (1971).
4. E.Y.C. Lu, Lett. Nuovo. Cim., 2,1241 (1971).
5. E.Y.C. Lu, Lett. Nuovo. Cim., 3,585 (1972).
6. H.P. Yuen, Phys. Rev. A, 13, 2226 (1976).
7. C.M. Caves, Phys. Rev. D, 23, 1693 (1981).

Thermal states with variances smaller than vacuum variances

M. Chiranov

ISEH Focsani, CTC Laboratory
Bd. Bucuresti 16, RO–5300 Focsani, Romania

We present a method to obtain thermal states with variances TSV smaller than vacuum variances in the formalism of non equilibrium thermo field dynamics $NETFD$ and discuss the possibility to change these states in an isoentropic manner.

1 Introduction

If we intend to study one quantum system and consider its interaction with a heat bath, it is possible to use Thermo Field Dynamics TFD. This method can be used for non equilibrum process, being known the fact that if we choose one good initial state for a quantum system, a net decrease in entropy is possible, so an increase in the quantity of information. Also, we are interested if it is possible to change the state of the system in an isoentropic, but energy non conserving, evolution. The second part of the work shows the connections between Information Entropy and Uncertainty Principle. These help us to prove the entropy's changes. Section 3 offers the fundamental notions from TFD, defines the Thermal Coherent States TCS and Thermal Squeezed States TSS. Also it is shown the existence of the possibility of decreasing the variances for a thermal state, under the zero and non zero temperature vacuum variances. Section 4 constructs a general method for obtain TSV, using the Heisenberg picture and the coherent state representation for the state vectors. The isoentropic transformations which are non conserving energy are presented in Section 5.

2 Informational Entropy and Uncertainty principle

In many cases, to express the uncertainty in one measurement process, the well-known Heisenberg Principle is used, which express the minimum acceptable value for the product between the variances of two cannonical conjugated observables Q and P :

$$(\Delta Q)(\Delta P) \geq \frac{1}{2} \, |< [Q, P] >| \tag{2.1}$$

Lecture Notes in Physics, Vol. 378
C. Bendjaballah, O. Hirota, S. Reynaud (Eds.),
Quantum Aspects of Optical Communications

where Q and P represent the *position* and *momentum*

$$(\Delta Q)^2 =< Q^2 > - < Q >^2$$

$$(\Delta P)^2 =< P^2 > - < P >^2$$

the braket $< . >$ represents the mean value and : $[Q, P] = QP - PQ = i$ is the commutator. The relation (2.1) has been discussed in [1] on the grounds that the right side of (2.1) is not a fixed lower bound and depends on the state used to compute the mean values used in (2.1). Also, it has been proposed the *entropic* uncertainty relation :

$$H(P) + H(Q) \geq -2ln\frac{1}{2}(1 + C) \tag{2.2}$$

where $c = \max\limits_{j,k} |< q_j \mid p_k >|$; $\mid q_j >$, $\mid p_k >$ $(j, k = 1, \ldots, N)$ are the eigenvectors of the operators Q and P and $H(p) = -\sum_i^N p_i ln p_i$ represents the informational entropy computed using the probability distribution $p_i \geq 0; \sum_i^N p_i = 1$. For the non equilibrum processes [2], H. Kandrup proved that thanks to the correlations between the system's components, a supplementary entropy appears which displays information about interaction between the different degrees of freedom. It is shown that for judicious choice of the initial conditions, an initial decrease in entropy can be induced. An example of state for which is possible to obtain the decrease in entropy is contained in [3]. In this paper, the relation between entropy generation and particle creation is shown and it is proved that the entropy can be decreased for the states which phase information is well known. Of course this process is a non equilibrum one.

3 Notions from thermo field dynamics

Thermo field dynamics is formulated in [4], taking into account the thermal degrees of freedom. A space is used which is the direct product between the zero temperature Hibert space H and the tildian space $\tilde{H}$ so that every state in H has one tildian correspondening state in $\tilde{H}$. The same thing is valid for operators : every operator $A \in H$ has its tildian conjugate $\tilde{A} \in \tilde{H}$. The thermal vacuum is generated [5] from the zero temperature vacuum $\mid 0, \tilde{0} >$

$$|0(\beta) >= \exp(-iG)|0, \tilde{0} >= U_\beta|0, \tilde{0} > \tag{3.1}$$

β is the inverse temperature, and

$$G = -i \sum_k \theta_k(\beta)[a_k^\dagger \tilde{a}_k^\dagger - \tilde{a}_k a_k] \tag{3.2}$$

The non zero temperature operators $a_k(\beta)$ and $\tilde{a_k}(\beta)$ are :

$$a_k(\beta) = U_\beta a_k U_\beta^\dagger = a_k cosh(\theta_k) + \tilde{a}_k^\dagger sinh(\theta_k) \tag{3.3a}$$

$$\tilde{a}_k(\beta) = U_\beta \tilde{a}_k U_\beta^\dagger = \tilde{a}_k cosh(\theta_k) + a_k^\dagger sinh(\theta_k) \tag{3.3b}$$

where

$$cosh[\theta_k(\beta)] = [1 - \exp(-\beta E_k)]^{-1/2} \equiv u_k(\beta) \tag{3.4a}$$

$$sinh[\theta_k(\beta)] = [\exp(\beta E_k) - 1]^{-1/2} \equiv v_k(\beta) \tag{3.4b}$$

The operators of $H \bigotimes \tilde{H}$ obey the tilde conjugation rules [6]:

$$\widetilde{AB} = \tilde{A}\tilde{B} \tag{3.5a}$$

$$(C_1 \widetilde{A + C_2} B) = C_1^* \tilde{A} + C_2^* \tilde{B} \tag{3.5b}$$

$$(\widetilde{A^\dagger}) = \tilde{A}^\dagger \tag{3.5c}$$

$$\tilde{(\tilde{A})} = \sigma A \tag{3.5d}$$

$$\widetilde{| 0 >} = | 0 > \tag{3.5e}$$

$$\widetilde{< 0 |} = < 0 | \tag{3.5f}$$

where $\sigma = +1(-1)$ for bosonic (fermionic) A. In the next we are working only with bosonic operators, so $\sigma = 1$ and we will neglect it. The tildian and non tildian operators obey the commutation rules:

$$[a, a^\dagger] = [\tilde{a}, \tilde{a}^\dagger] = 1 \tag{3.6a}$$

$$[a, \tilde{a}] = [a, \tilde{a}^\dagger] = 0 \tag{3.6b}$$

TCS are defined [5]to be the eigenstates of the annihilation operators $a(\beta)$,$\tilde{a}(\beta)$:

$$a(\beta) \mid \alpha, \tilde{\gamma}; \beta >= \alpha \mid \alpha, \tilde{\gamma}; \beta > \tag{3.7a}$$

$$\tilde{a}(\beta) \mid \alpha, \tilde{\gamma}; \beta >= \tilde{\gamma} \mid \alpha, \tilde{\gamma}; \beta > \tag{3.7b}$$

The state vector is obtained under the action of the translation operator:

$$\mid TCS >= \mid \alpha, \tilde{\gamma}; \beta >= \exp[\alpha a^\dagger(\beta) + \tilde{\gamma} \tilde{a}^\dagger(\beta) - \alpha^* a(\beta) - \tilde{\gamma}^* \tilde{a}(\beta)] \mid 0(\beta) > \tag{3.8}$$

The canonical variables Q, P are defined as combinations of $a, a^\dagger, \tilde{a}, \tilde{a}^\dagger$:

$$Q = \frac{1}{\sqrt{2\omega}}[a(\beta) + a^\dagger(\beta)] \ ; \ P = \frac{i\sqrt{\omega}}{2}[a^\dagger(\beta) - a(\beta)] \tag{3.9a}$$

$$\tilde{Q} = \frac{1}{\sqrt{2\omega}}[\tilde{a}(\beta) + \tilde{a}^\dagger(\beta)] \ ; \tilde{P} = \frac{i\sqrt{\omega}}{2}[\tilde{a}^\dagger(\beta) - \tilde{a}(\beta)] \tag{3.9b}$$

we have assumed $\hbar = 1$.

In [7],the TSS are defined:

$$\mid TSS >= \mid \eta, \tilde{\eta}, \alpha, \tilde{\gamma}; \beta >= U_\beta S_\eta \tilde{S}_{\tilde{\eta}} D_\alpha \tilde{D}_{\tilde{\gamma}} \mid 0, \tilde{0} > \tag{3.10}$$

D_α and $\tilde{D}_{\tilde{\gamma}}$ being the displacement operators in the non tildian and tildian spaces, with the quantity α and $\tilde{\gamma}$ respectively. S_η and $\tilde{S}_{\tilde{\eta}}$ are the squeezing operators :

$$S_\eta = \exp[\frac{1}{2}(\eta a^{\dagger 2} - \eta^* a^2)] \tag{3.11a}$$

$$\tilde{S}_{\tilde{\eta}} = \exp[\frac{1}{2}(\tilde{\eta}\tilde{a}^{\dagger 2} - \tilde{\eta}^* \tilde{a}^2)] \tag{3.11b}$$

where $\eta = r\exp(i\phi)$ and $\tilde{\eta} = \tilde{r}\exp(i\tilde{\phi})$ The TSS are eigenstates of the operators:

$$b = S_\eta a S_\eta^\dagger = a cosh(r) - a^\dagger \exp(i\phi) sinh(r) \tag{3.12a}$$

$$\tilde{b} = \tilde{S}_{\tilde{\eta}} \tilde{a} \tilde{S}_{\tilde{\eta}}^\dagger = \tilde{a} cosh(\tilde{r}) - \tilde{a}^\dagger \exp(i\tilde{\phi}) sinh(\tilde{r}) \tag{3.12b}$$

In [7], Kireev et al. computed the variances:

$$<(\Delta Q)^2>= \frac{1}{2\omega}[u_\beta^2(cosh(2r) + sinh(2r)cos(\phi)) + v_\beta^2(cosh(2\tilde{r}) + sinh(2\tilde{r})cos(\tilde{\phi}))] \tag{3.13a}$$

$$<(\Delta P)^2>= \frac{\omega}{2}[u_\beta^2(cosh(2r) - sinh(2r)cos(\phi)) + v_\beta^2(cosh(2\tilde{r}) - sinh(2\tilde{r})cos(\tilde{\phi}))] \tag{3.13b}$$

From $(3-13)$, it is very clear that if we give up the tilde conjugation rules given by $(3-5)$, then $(3-13)$ becomes smaller than the zero temperature vacuum variances [7]. So far, we showed that it is possible to obtain more information about the non tilde observables acting in the tildian space. So if we intend to obtain states with variances smaller than the zero temperature variances, it needs to solve the inequalities:

$$<(\Delta Q)^2> \leq \frac{1}{2\omega} \tag{3.14a}$$

$$<(\Delta P)^2> \leq \frac{\omega}{2} \tag{3.14b}$$

And if we look for states with variance smaller than thermal vacuum variances, it must to resolve:

$$<(\Delta Q)^2> \leq \frac{1}{2\omega}(u_\beta^2 + v_\beta^2) \tag{3.15a}$$

$$<(\Delta P)^2> \leq \frac{\omega}{2}(u_\beta^2 + v_\beta^2) \tag{3.15b}$$

The left hand side of these inequalities being given by $(3-13)$ and the right hand side represent the TCS variances, that are equal with finite temperature vacuum variances.

4 Thermal states with variances smaller than vacuum variances

Our next intention is to present a general method that can be used to obtain TSV. This fact is in direct connection with obtaining the decrease in entropy [1], that is possible in non equilibrum process [3]. $NETFD$ was formulated by Arimitsu and Umezawa [9] and used to treat the laser from a microscopic point of view (see [9] and references herein). In $NETFD$ the non tilde and tilde annihilation and creation operators $(3-3)$ vary in accord with (see [9]):

$$a(t) = \hat{S}^{-1}(t)a\hat{S}(t) \tag{4.1a}$$

$$a^{\ddagger}(t) = \hat{S}^{-1}(t)a^{\dagger}\hat{S}(t) \tag{4.1b}$$

$$\tilde{a}(t) = \hat{S}^{-1}(t)\tilde{a}\hat{S}(t) \tag{4.2a}$$

$$\tilde{a}^{\ddagger}(t) = \hat{S}^{-1}(t)\tilde{a}^{\dagger}\hat{S}(t) \tag{4.2b}$$

Here the index β is implied that $\beta = \infty$. (non zero temperature). Generally $\hat{S}$ is **not an unitary operator** so that $a(t)$ and $\tilde{a}(t)$ are not Hermitian conjugates of $a^{\ddagger}(t)$ and $\tilde{a}^{\ddagger}(t)$ respectively. But the canonical commutation is verified:

$$[a(t), a^{\ddagger}(t)] = [\tilde{a}(t), \tilde{a}^{\ddagger}(t)] = 1 \tag{4.3}$$

$$\hat{S}(t) = \exp(-i\hat{H}t); \hat{H} = H - \tilde{H} \tag{4.4}$$

We express $(4-1)$ as

$$a(t) \equiv f_1(a, a^{\dagger}; t) \tag{4.6a}$$

$$a^{\ddagger}(t) \equiv f_2(a, a^{\dagger}; t) \tag{4.6b}$$

Now the canonical variables Q(t) and P(t) become :

$$Q(t) = \frac{1}{\sqrt{2\omega}}[a(t) + a^{\ddagger}(t)] = \frac{1}{\sqrt{2\omega}}(f_1 + f_2) \tag{4.7a}$$

$$Q(t) = i\sqrt{\frac{\omega}{2}}[a^{\ddagger}(t) - a(t)] = i\sqrt{\frac{\omega}{2}}(f_2 - f_1) \tag{4.7b}$$

We compute the variances for the variables $Q(t)$ and $P(t)$ in the coherent state representation:

$$|>= \frac{1}{\pi}\int | \alpha >< \alpha |> d^2\alpha \tag{4.8}$$

where $d^2\alpha = d(\Re\alpha)d(\Im\alpha)$.

For this purpose a useful theorem [8] shows how to write the normal ordered mean values for the function $F(a, a^{\dagger})$:

$$F^{(n)}(a, a^{\dagger}) = N\{< \alpha | F(a, a^{\dagger}) | \alpha >\} = N\{\bar{F}(\alpha + \frac{\delta}{\delta\alpha^*}, \alpha^*)\} \tag{4.9}$$

where the superscript (n) means normal ordered form (all the creation operators are standing to the right of the annihilation operators). For example if $F(a,a^\dagger) = aa^\dagger$, $F^{(n)} = a^\dagger a + 1$. N denotes the normalizing operator : it puts the eigenvalues α^* before the eigenvalues α. In fact this theorem replaces the application of multiple commutation relations with differentiations. The overbar $\bar{F}$ means the mean value of the operator F. So the variances of the canonical observables $Q(t)$ and $P(t)$ are:

$$< (\Delta Q)^2 >= \frac{1}{2\omega}[(f_1 f_2)^{(n)} + (f_2 f_1)^{(n)} - 2f_1^{(n)} f_2^{(n)}] \qquad (4.10a)$$

$$< (\Delta P)^2 >= \frac{\omega}{2}[(f_1 f_2)^{(n)} + (f_2 f_1)^{(n)} - 2f_1^{(n)} f_2^{(n)}] \qquad (4.10b)$$

where we have droped the time dependence. Since we are looking for states with variances smaller than thermal vacuum variances, or zero temperature vacuum variances, it is necessary to resolve the inequalities $(3-15)$ or $(3-14)$ with the variances given by $(4-10)$. But the result $(4-10)$ is a particular one, because in general$\hat{S}$ is not unitary operator (see $(4-1)$, $(4-3)$ for justification). In this situation $Q(t)$ and $P(t)$ are not Hermitian operators, we use the results obtained by C. Caves and B.L. Schumaker [17]. Denoting by Q_1 and Q_2 the real and imaginary parts of Q :

$$Q_1 = \frac{1}{2}(Q + Q^\dagger) = \frac{1}{2\sqrt{2\omega}}[f_2^\dagger + f_1^\dagger + f_1 + f_2] \qquad (4.11)$$

$$Q_2 = \frac{-i}{2}(Q - Q^\dagger) = \frac{-i}{2\sqrt{2\omega}}[f_1 - f_1^\dagger - f_2^\dagger + f_2] \qquad (4.12)$$

the mean square uncertainty of Q is the sum of the variances of Q_1 and Q_2

$$<|\, \Delta Q \,|^2>=< (\Delta Q_1)^2 > + < (\Delta Q_2)^2 > \qquad (4.13)$$

Using the same definition for P, we obtain

$$P_1 = \frac{1}{2}(P + P^\dagger) = \frac{i}{2}\sqrt{\frac{\omega}{2}}[f_2 + f_2^\dagger - f_1^\dagger - f_1] \qquad (4.14)$$

$$P_2 = \frac{-i}{2}(P - P^\dagger) = \frac{1}{2}\sqrt{\frac{\omega}{2}}[f_2 - f_2^\dagger + f_1^\dagger - f_1] \qquad (4.15)$$

$$<|\, \Delta P \,|^2>=< (\Delta P_1)^2 > + < (\Delta P_2)^2 > \qquad (4.16)$$

Finally we will find for (4-13) and (4-16)

$$<|\, \Delta Q \,|^2>= \frac{1}{4\omega}[(f_1 f_1^\dagger)^{(n)} + (f_1^\dagger f_1)^{(n)} + (f_1 f_2^\dagger)^{(n)} + (f_2^\dagger f_1)^{(n)} + (f_2^\dagger f_2)^{(n)}$$
$$+(f_2 f_2^\dagger)^{(n)} + (f_1^\dagger f_2)^{(n)} + (f_2 f_1^\dagger)^{(n)} - 2f_1^{(n)} f_1^{\dagger(n)}$$
$$-2f_1^{(n)} f_2^{\dagger(n)} - 2f_2^{(n)} f_1^{\dagger(n)} - 2f_2^{(n)} f_2^{\dagger(n)}] \qquad (4.17)$$

$$<| \Delta P |^2>= \frac{\omega}{4}[-(f_1 f_1^\dagger)^{(n)} - (f_1^\dagger f_1)^{(n)} + (f_1 f_2^\dagger)^{(n)} + (f_2^\dagger f_1)^{(n)} + (f_2 f_1^\dagger)^{(n)}$$

$$+(f_1^\dagger f_2)^{(n)} - (f_2 f_2^\dagger)^{(n)} - (f_2^\dagger f_2)^{(n)} + 2f_2^{(n)} f_2^{\dagger(n)}$$

$$-2f_1^{\dagger(n)} f_2^{(n)} - 2f_1^{(n)} f_2^{\dagger(n)} + 2f_1^{(n)} f_1^{\dagger(n)}] \tag{4.18}$$

To conclude, in the case when $\hat{S}$ is non Hermitian, Q and P are also non Hermitian and the mean square uncertainty for the operators Q and P are smaller than the same quantities given by the zero and non zero temperature vacuum states, so it is necessary to resolve the inequalities $(3-14)$ and $(3-15)$ with the left hand side given by $(4-17)$ and $(4-18)$.

5 Isoentropic evolution

Because we have derived the situation when is possible to obtain TSV now, let us look for the way to preserve these variances. So, for our mixed state, composed from the tilde and non tilde variables, we intend to preserve, for finite time, the informations about the non tilde observables. In other words, we desire to act on the TSV in an isoentropic manner changing only the energy of the state. This problem is treated in [11] and more detailed in [12]. Our purpose is to find a Hamiltonian H that obeys the equation:

$$\frac{d\rho}{dt} = i[\rho, H] + \frac{\delta\rho}{\delta t} \tag{5.1}$$

where

$$\rho(Q_1, Q_2) = \sum_n p_n \psi_n(Q_1)\psi_n^*(Q_2) \tag{5.2}$$

and $\sum_n p_n = 1$. In equation $(5-1)$

$$\frac{d\rho}{dt} = \sum_n \psi_n \frac{dp_n}{dt} \psi_n^* \tag{5.3}$$

But, due to the fact that our purpose is to keep the entropy unchanged, we need:

$$\frac{dp_n}{dt} = 0 \tag{5.4}$$

In this case $(5-1)$ becomes:

$$\frac{d\rho}{dt} = i[\rho, H] \tag{5.5}$$

Generally, the Hamiltonian H is not quadratic, so the evolution is nonlinear and $(5-5)$ cannot be solved exactly. Using a variational ansatz, Eboli et al. [12] obtained tractable equations.

They choosed a Gaussian density matrix and make the ansatz

$$\rho(Q_1,Q_2)=\exp(-\gamma)\exp\{\frac{-1}{2}[\;Q_1^i(\frac{G^{-1}}{2}-2i\pi)_{ij}Q_1^j$$

$$+Q_2^i(\frac{G^{-1}}{2}+2i\pi)_{ij}Q_2^j-Q_1^i(G^{\frac{-1}{2}}\eta G^{-1})_{ij}Q_2^j\;]\} \quad (5.6)$$

The matrices G and π are real symmetric and η is Hermitian $\eta=\eta_R+4i\eta_i$; η_R is real symmetric and η_i is real antisymmetric. The density matrix is normalized

$$\gamma=\frac{1}{2}\;ln[det\{2\pi\sqrt{G}(1-\eta_R)^{-1}\sqrt{G}\}] \quad (5.7)$$

The vectors Q_1 and Q_2 are n component vectors ; η is the degree of mixing. For $\eta=0$ we have a pure state and $\rho(Q_1,Q_2)=\psi(Q_1)\psi^*(Q_2)$. The Gaussian density matrix formalism can be used to describe the coherent and squeezed states for one or multimode representations [13]. This formalism is useful in our problem. The density matrix at $\beta=\infty$ (**zero temperature**) being one characteristic for a pure state. When temperature becomes non zero, in the NETFD approximation, we use (4-1) and (4-2) to describe the temporal evolution of the canonical conjugated observables $a(t)$,$\tilde{a}(t)$,$a^{\ddagger}(t)$ and $\tilde{a}^{\ddagger}(t)$. We will observe the appearance of the mixed terms containing tilde and non tilde variables, so that η from $(5-6)$ will be non zero. For the density matrix $(5-6)$, the mean values of Q and P vanish:

$$<Q>=\int Q\rho(Q,Q)dQ=0 \quad (5.8)$$

$$<P>=\frac{1}{i}\int\nabla_Q\;\rho(Q,Q)dQ=0 \quad (5.9)$$

and the bilinear averages are :

$$<Q^iQ^j>=[\sqrt{G}(1-\eta_R)^{-1}\sqrt{G}]_{ij} \quad (5.10)$$

$$<P^iP^j>=\frac{1}{4}[G^{-1/2}(1+\eta_R)G^{-1/2}]_{ij}+4[(\pi+\eta_i)\sqrt{G}(1-\eta_R)^{-1}\sqrt{G}(\pi-\eta_i)]_{ij} \quad (5.11)$$

Following [12], if the choice for the Hamiltonian H from (5-5) is

$$H=\frac{1}{2}P^2+\frac{1}{2}Q^2\omega^2(t) \quad (5.12)$$

with

$$\omega^2(t)=\omega_0^2+\frac{n(n+1)\mu^2}{cosh^2(\mu t)} \quad (5.13)$$

where μ is a parameter and n any positive integer. The initial and final times are $-\infty$ and ∞ and $\omega_i=\omega_f=\omega_0$. For $n=1$ it is obtained

$$<Q^2>=\frac{1}{2\omega_0}coth(\beta\frac{\omega_0}{2})[1-\frac{\mu^2}{(\omega_0^2+\mu^2)cosh^2(\mu t)}] \quad (5.14)$$

$$<P^2>=\frac{1}{2\omega_0}coth(\frac{\beta\omega_0}{2})[\frac{\nu}{\Delta}] \quad (5.15a)$$

$$\nu = \mu^6 sinh^2(\mu t) + \omega_0^2(\omega_0^2 + \mu^2)^2 cosh^6(\mu t) \tag{5.15b}$$

$$\Delta = (\omega_0^2 + \mu^2)(\omega_0^2 cosh^2 \mu t + \mu^2 sinh^2 \mu t) cosh^4(\mu t) \tag{5.15c}$$

For finding the TSV we do solve the inequalities (3-14) and (3-15) with $<(\Delta Q)^2>$ and $<(\Delta P)^2>$ given by (5-14), (5-15) and recall that $<Q>$ and $<P>$ are zero (eq.(5-8) and (5-9)).

6 Conclusions

So, it is possible to obtain thermal states that have variances smaller than zero and non zero temperature vacuum variances, due to the fact in a mixed state is possible to obtain a decrease in entropy [2]and [3]. Also TFD formalism can be used to treat the dissipative systems [6], [14]. In this case it is possible to obtain a decrease in number of particles, so in entropy. This problem can be included in the thermal treatment of the microscopic world [15], the quantum thermodynamics [16].

Acknowledgments

Many thanks to the organizers of **QOC '90** for the grant offered for participation and to Prof. T. Arimitsu for the papers sent to me.

References

1. D. Deutsch : Phys. Rev. Lett. **50**, 631 (1987).
2. H. Kandrup : Jour. Math. Phys., **28**, 1398 (1987).
3. H. Kandrup : Phys. Rev.**D 37**, 3505 (1988).
4. H. Umezawa, H. Matsumoto and H. Tachiki: *Thermo Field Dynamics and Condensed States* (North–Holland, Amsterdam 1982).
5. A. Mann and M. Revzen : Phys. Lett. **A 134**, 273,(1989);
A. Mann, M. Revzen, K. Nakamura, H. Umezawa, and Y. Yamanaka, Jour. Math. Phys. **30**, 2883, (1989).
6. T. Arimitsu, H. Umezawa and Y. Yamanaka : J. Math. Phys. **28**, 2741 (1987);
I. Hardman, H. Umezawa and Y. Yamanaka : J. Math. Phys. **28**,2925, (1987).
7. A. Kireev, A. Mann, M. Revzen and H. Umezawa: Phys. Lett. **A 142**, 215, (1989).
8. W.H. Louisell : *Quantum Statistical Properties of Radiation* (John Wiley and Sons, New York 1973).
9. T. Arimitsu, H. Umezawa : Prog. Theor. Phys. **74**, 429, (1985);
T. Arimitsu, H. Umezawa : Prog. Theor. Phys. **77**, 32 and 53, (1987).

10. T. Arimitsu : *Proc. of the Workshop on Thermo–Field Dynamics and their applications*, Physica **A 158**, 317 (1989).
11. R. Jackiw : see 10. p. 269.
12. O. Eboli, R. Jackiw and So-Young Pi : Phys. Rev. **D 37**, 3557 (1988).
13. B.L. Schumaker : Phys. Rep. **135**, 317, (1986).
14. T. Arimitsu,M. Guida and H. Umezawa: Physica **A 148**, 1,(1988);
T. Arimitsu, H. Umezawa, Y. Yamanaka, and N.J. Papastamatiou: Physica **A 148** 27, (1988).
15. H. Umezawa: Phys. Essays **2**, 101,(1989).
16. I. Hardman, H. Umezawa, and Y. Yamanaka : Phys. Lett. **A 146**, 293, (1990).
17. C. Caves and B.L. Schumaker: Phys. Rev. **A 31**, 3068, (1985).

SQUEEZING-SYMMETRY OF THE BALANCED HOMODYNE DETECTOR

G. M. D'Ariano

Dipartimento di Fisica 'Alessandro Volta', via Bassi 6, I-27100 Pavia, Italy

1. INTRODUCTION

Detection of squeezed light requires high quantum efficiency, phase sensitivity, and freedom from amplifier noise or any other kind of disturbance. At present, the homodyne detector is the optimum device for detection of a quadrature component of the field [1]. Its phase sensitivity comes from combining the signal beam through a beam splitter (BS) with an intense 'local oscillator' (LO) field operating at the same frequency. The combined field is then directed to a photodetector and the amplitude component of the field is revealed as the beating between the two input fields. Noise from intensity fluctuations of the LO is canceled by means of the balanced configuration of Fig.1, involving two photodetectors with equal responsivity and a 50-50 BS: the difference photocurrent $\hat{I}_D$ between the two photodetectors D1 and D2 measures the interference between the signal beam and the LO, the interference being constructive at one photodetector and destructive at the second one.

In this paper, a novel symmetry of the balanced homodyne detector is presented, which relates different input pairs of beams to the same output photocurrent $\hat{I}_D$. I call the symmetry 'squeezing symmetry' because the input pairs of beams—which are equivalent in producing the same current $\hat{I}_D$—are related through a squeezing transformation of the fields. The symmetry is presented in Sec. 2, where the physical meaning of the squeezing transformation of the fields is also discussed. In Sec. 3 the extension of the symmetry to four-port linear devices is briefly sketched, and an application to a simple interferometer, build up as a cascade of linear devices, is illustrated.

2. SYMMETRY OF THE DIFFERENCE PHOTOCURRENT IN THE BALANCED HOMODYNE DETECTOR

In the following, a single-mode analysis is given, in the assumption of lossless BS and ideal photodetectors (having unit quantum efficiency). The input fields a and b combine at the 50-50 BS giving the sum and the difference fields c and d in the output arms. After tuning the overall phases (by adjusting the path lengths), one has

$$c = \frac{a+b}{\sqrt{2}}\,, \qquad d = \frac{a-b}{\sqrt{2}}\,. \tag{1}$$

The output photocurrents $\hat{I}_1$ and $\hat{I}_2$ are proportional to the number operators $c^\dagger c$ and

Lecture Notes in Physics, Vol. 378
C. Bendjaballah, O. Hirota, S. Reynaud (Eds.),
Quantum Aspects of Optical Communications

$d^\dagger d$ and the difference photocurrent $\hat{I}_D$ has the form

$$\hat{I}_D = \hat{I}_1 - \hat{I}_2 \propto a^\dagger b + b^\dagger a \ . \tag{2}$$

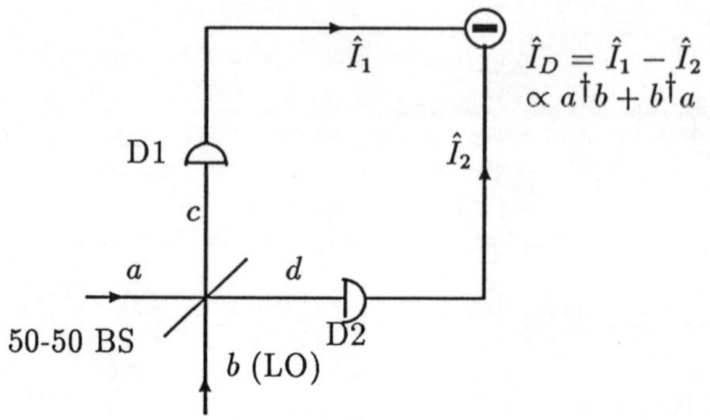

Figure 1: Scheme of the balanced homodyne detector.

The particular form of $\hat{I}_D$ in Eq.(2) is highly symmetrical, as a consequence of the balanced scheme. Besides the trivial symmetry under permutation of the input fields, $\hat{I}_D$ commutes with the unitary operator $\hat{U}(\mu,\nu)$

$$[\hat{I}_D, \hat{U}(\mu,\nu)] = 0 \ , \tag{3}$$

where

$$\hat{U}(\mu,\nu) = \hat{S}_a^\dagger(\overline{\mu},\nu)\hat{S}_b(\mu,\nu) \ . \tag{4}$$

Here $\hat{S}_{a,b}(\mu,\nu)$ are the squeezing operators of Yuen [2] acting on the input fields as follows

$$\hat{S}_a(\mu,\nu)a\hat{S}_a^\dagger(\mu,\nu) = \mu a + \nu a^\dagger \ , \qquad \hat{S}_b(\mu,\nu)b\hat{S}_b^\dagger(\mu,\nu) = \mu b + \nu b^\dagger \ . \tag{5}$$

and the complex numbers μ and ν satisfy the relation $|\mu|^2 - |\nu|^2 = 1$. Invariance (3) can be verifed by using the identity $\hat{S}_{a,b}^\dagger(\mu,\nu) = \hat{S}_{a,b}(\overline{\mu},-\nu)$. For a real parameter μ Eqs.(3) and (4) state that *the difference photocurrent $\hat{I}_D$ is invariant under inverse squeezing of the two input beams.* Due to the form of Eq.(4) the symmetry transformation $\hat{U}(\mu,\nu)$ does not affect the correlation between the input beams, as opposed to the $SU(2)$ symmetries of the beam splitter derived in Ref.[3].
In the Schrödinger picture the invariance (3) means that the quantum statistics of the difference photocurrent $\hat{I}_D$ does not change if the input state is symmetry-transformed as

$$\hat{R}_{in} \longrightarrow \hat{U}^\dagger(\mu,\nu)\hat{R}_{in}\hat{U}(\mu,\nu) \ . \tag{6}$$

where $\hat{R}_{in}$ denotes the density matrix of the input (with the two beams which are in general quantum-correlated). For the particular case of uncorrelated beams, namely

$\hat{R}_{in} = \hat{\rho}_a \hat{\rho}_b$, the symmetry transformation (6) is equivalent to the following pair of single-mode transformations

$$\hat{\rho}_a \longrightarrow \hat{S}_a(\overline{\mu}, \nu)\hat{\rho}_a \hat{S}_a^\dagger(\overline{\mu}, \nu) , \qquad \hat{\rho}_b \longrightarrow \hat{S}_b^\dagger(\mu, \nu)\hat{\rho}_b \hat{S}_b(\mu, \nu) . \tag{7}$$

Here, some remarks regarding the physical meaning of the squeezing transformations (7) are in order. The present squeezing transformation corresponds to a simultaneous squeezing of both the noise and the signal, whereas in the usual squeezed states, the unsqueezed signal is superimposed to the squeezed fluctuations (see also Ref.[4]). In practice, the present squeezing is equivalent to an ideal noisless phase-sensitive amplification [5], which enhances a phase component of the field, while reduces the conjugated one. As a consequence of the squeezing symmetry, one infers that, for example, the use of a squeezed LO is equivalent to a phase-sensitive amplification of the remote signal. However, no improvement of the signal-to-noise ratio is achieved, and the homodyne with squeezed LO is essentially equivalent to the homodyne with coherent LO, where the homodyne gain is shared by both the signal and the squeezing power components of the LO.

3. GENERALIZATION TO OTHER FOUR-PORT LINEAR DEVICES

In this section I briefly present the derivation of the squeezing symmetries of a linear four-port device with output photocurrent quadratic in the output fields. In particular, I suggest an application to a four-port device, built up as a cascade of elementary four-port devices.

By a linear device I mean that the Heisenberg equations relating the output to the input fields are linear. The beam splitter provides an example of linear four-port device, with Heisenberg equations

$$\begin{pmatrix} c \\ d \end{pmatrix} = \begin{pmatrix} ma + nb \\ -\overline{n}a + \overline{m}b \end{pmatrix} , \qquad |m|^2 + |n|^2 = 1 . \tag{8}$$

The degenerate four wave mixer (FWM) with classical (non depleted) pump waves is an other example of four-port linear device, with evolution equations for the signal and idler waves given by [6]

$$\begin{pmatrix} c \\ d \end{pmatrix} = \begin{pmatrix} ma + nb^\dagger \\ na^\dagger + mb \end{pmatrix} , \qquad |m|^2 - |n|^2 = 1 . \tag{9}$$

$|m|^2$ being the signal gain. If one of the waves is assumed to have a constant field amplitude, the parametric amplifier (PA) can be regarded as a four-port device and Eqs.(9) apply to this case as well. (The symmetrical form of Eqs.(8) and (9) pertains to a suited choice of the field phases related to the path lengths).

For a general four-port linear device the Heisenberg equations can be written in the form

$$\mathbf{M}\begin{pmatrix} a \\ b \end{pmatrix} = \hat{M}\begin{pmatrix} a \\ b \end{pmatrix}\hat{M}^\dagger , \tag{10}$$

where $\mathbf{M}$ denotes the linear transformation corresponding to the unitary operator $\hat{M}$.

Table 1: Some examples of linear transformations and conserved currents.

Symbol	*Linear transformations of the fields a, b*	*Constraints*	*Some conserved currents*
$\mathbf{V}(\mu,\nu,e^{i\phi})$	$\mu a+\nu b^\dagger$ $\mu b+e^{i\phi}\nu a^\dagger$	$\lvert\mu\rvert^2-\lvert\nu\rvert^2=1$	$a^\dagger a-b^\dagger b\,,\quad(\phi=0)$
$\mathbf{W}(\mu,\nu,e^{i\phi})$	$\mu a+\nu b$ $\overline{\mu}b+e^{i\phi}\overline{\nu}a$	$\lvert\mu\rvert^2+\lvert\nu\rvert^2=1$	$a^\dagger a+b^\dagger b\,,\quad(\phi=\pi)$
$\mathbf{U}(\mu,\nu,e^{i\phi})$	$\mu a+\nu a^\dagger$ $\mu b+e^{i\phi}\nu b^\dagger$	$\lvert\mu\rvert^2-\lvert\nu\rvert^2=1$	$a^\dagger b+b^\dagger a\,,\quad(\phi=\pi)$
$\mathbf{Z}(\mu,\nu,e^{i\phi})$	$\mu a+\nu a^\dagger$ $\overline{\mu}b+e^{i\phi}\overline{\nu}b^\dagger$	$\lvert\mu\rvert^2-\lvert\nu\rvert^2=1$	$a^\dagger b^\dagger+ab\,,\quad(\phi=\pi)$

Table 2: Symmetry transformations of the four beams leaving the Heisenberg equations invariant. (The phases ϕ and ψ are given by: $\phi=\arg(m)$, $\psi=\arg(n)$).

Device	*Heisenberg evolutions of the input fields*	*Symmetry transformations*	
		Input fields	*Output fields*
BS	$\mathbf{W}(m,n,-1)$	$\mathbf{V}(\mu,\nu,-1)$	$\mathbf{V}(\mu,\nu,-1)$
		$\mathbf{U}(\mu,\nu e^{-2i\phi},e^{2i(\phi-\psi)})$	$\mathbf{U}(\mu,\nu,e^{-2i(\phi+\psi)})$
50-50 BS		$\mathbf{U}(\mu,-\nu e^{-i(\phi-\psi)},-e^{2i(\phi-\psi)})$	$\mathbf{V}(\mu,\nu,1)$
FWM/PA	$\mathbf{V}(m,n,1)$	$\mathbf{V}(\mu,\nu e^{-2i\phi},e^{2i(\psi-\phi)})$	$\mathbf{V}(\mu,\nu,e^{2i(\phi+\psi)})$
		$\mathbf{W}(\mu,\nu,1)$	$\mathbf{W}(\mu,\nu,1)$
		$\mathbf{Z}(\mu,\nu e^{-2i\phi},e^{2i(\psi-\phi)})$	$\mathbf{Z}(\mu,\nu,e^{2i(\phi+\psi)})$

Some examples of linear transformations, which are relevant in the present context, are defined in Table 1, where the transformations (8) and (9) are represented by the linear operators $\mathbf{W}(m,n,-1)$ and $\mathbf{V}(m,n,1)$.
The derivation of the squeezing symmetries of a four-port linear device with measured photocurrent $\hat{I}_o$ can be divided into three steps:
i) evaluation of the symmetry transformations (ST) of the output beams which preserve the current $\hat{I}_o$ (some examples can be found in Table 1);
ii) evaluation of the ST of the four beams which preserve the Heisenberg equations of the device; in other words: evaluation of the ST of the output beams corresponding to a given ST of the input beams;
iii) matching between the current-preserving ST of step i) and the Heisenberg-preserving ST of step ii).
Step i) through step iii) allow one to obtain the squeezing transformation of the input beams leaving the output photocurrent invariant. In Table 2, some ST leaving the Heisenberg equations invariant are given, for both the BS and FWM/PA devices. They can be simply derived by a linear analysis, or by means of the group theoretical approaches of Refs.[3],[7],[8].
Using the above method with the help of Tables 1 and 2, the squeezing symmetry of the 50-50 BS ($\phi=\psi=0$) can be obtained as follows: a) the output photocurrent $\hat{I}_D=c^\dagger c-d^\dagger d$ is preserved by $\mathbf{V}(\mu,\nu,1)$; b) the ST $\mathbf{V}(\mu,\nu,1)$ of the output is related

to the ST $\mathbf{U}(\mu, -\nu, -1)$ of the input: the second transformation corresponds to the unitary operator of Eq.(4).
If a cascade of linear four-port devices is considered (where, for example, the output beams of an element coincide with the input beams of the following), it is not necessary to derive its Heisenberg equations in order to apply the above procedure. In fact, the evaluation of the Heisenberg-preserving ST in step ii) for every element of the cascade allows one to connect the ST of the inputs of two consecutive devices, and the process can be iterated, until the input beams of the whole cascade are reached. Eventually, the compatibility of the ST of two consecutive elements may restrict the class of the symmetry or impose phase-matching between the devices. This mechanism actually corresponds to a smaller symmetry class of the whole cascade, and it occurs when the dynamical groups of the contiguous devices are different, as, for example, when a BS is followed by a FWM/PA [3],[8].

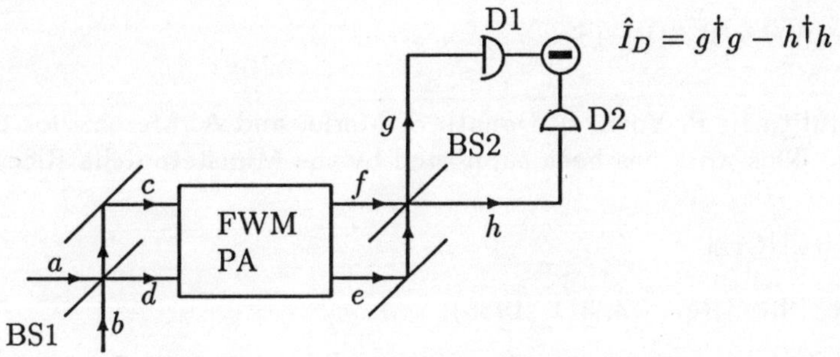

As an example illustrating the above concepts I consider the interferometer in Fig. 2, which is built up as a cascade of a BS followed by a FWM/PA and finally by a balanced homodyne detector. I denote by $\phi_{B1}, \psi_{B1}, \phi_{B2}, \psi_{B2}$ the phases of the beam splitters and by ϕ_F, ψ_F those of the four-wave mixer (for the conventions, see Table 2). Choosing, for simplicity, $\phi_{B2} = \psi_{B2} = 0$, as in the previous example, the ST of the BS2 input is $\mathbf{U}(\mu, -\nu, -1)$. The transformation $\mathbf{U}$ is not in the Table 2 at the entry of the FWM: as a consequence, the symmetry matching between the BS2 and the FWM requires a restriction of the symmetry class. This is attained through the identity

$$\mathbf{U}(\mu, \nu, e^{i\theta}) = \mathbf{Z}(\mu, \nu, e^{i(\theta + 2\arg(\nu))}) , \qquad \mu \text{ real}, \tag{11}$$

which implies that the ST for f and e coincides with $\mathbf{Z}(\mu, -\nu, -e^{2i\arg(\nu)})$. From Table 2 one can see that the phase of ν is constrained to the values $\arg(\nu) = \phi_F + \psi_F + \pi/2 + k\pi$ (k integer), namely the restricted symmetry for e and f is $\mathbf{Z}(\mu, i\nu e^{i(\phi_F+\psi_F)}, e^{2i(\phi_F+\psi_F)})$, where now both μ and ν are real. The corresponding ST for the input beams c and d is $\mathbf{Z}(\mu, i\nu e^{i(\psi_F-\phi_F)}, e^{2i(\psi_F-\phi_F)})$, which is the global squeezing symmetry of the partial cascade FWM/PA-BS2.
In order to obtain the squeezing symmetry of the whole cascade one has to match the ST of BS1 and FWM. Eq.(11) leads to $\mathbf{Z}(\mu, i\nu e^{i(\psi_F-\phi_F)}, e^{2i(\psi_F-\phi_F)}) = \mathbf{U}(\mu, i\nu e^{i(\psi_F-\phi_F)}, -1)$, and the complete cascade has a squeezing symmetry only if the phases of BS1 are constrained by $\phi_{B1} + \psi_{B1} = \pi/2 + k\pi$. In this case the ST of a and b is $\mathbf{U}(\mu, i\nu e^{i(\psi_F-\phi_F-2\phi_{B1})}, -e^{4i\phi_{B1}})$.

In conclusion, the interferometer in Fig.2 has a squeezing symmetry only if the phases of the beam splitter BS1 are related by the equation

$$\phi_{B1} + \psi_{B1} = \pi/2 + k\pi \ . \tag{12}$$

The output photocurrent $\hat{I}_D$ is invariant under the squeezing transformation of the input beams

$$a \longrightarrow \mu a + i\nu e^{(\psi_F - \phi_F - 2\phi_{B1})} a^\dagger \ , \qquad b \longrightarrow \mu b - i\nu e^{(\psi_F - \phi_F + 2\phi_{B1})} a^\dagger \ , \tag{13}$$

where μ and ν are real. The same result can also be obtained by evaluating the Heisenberg equations of the cascade and deriving the related ST: this direct way, however, is lengthy, and the symmetry breaking mechanism due to the matching of different devices is not evident.

4. ACKNOWLEDGMENTS

I am grateful to H. P. Yuen, S. Donati, S. Merlo, and A. Mecozzi for discussions and suggestions. This work has been supported by the Ministero della Ricerca Scientifica.

5. REFERENCES

1 B. Yurke, Phys. Rev. **32**, 311 (1985)

2 H. P. Yuen, Phys. Rev. A **13**, 2226 (1976) ($\hat{S}_a(\mu,\nu)$ is the evolution operator of a Hamiltonian $\hat{H}$ which is a quadratic function of a and $a^\dagger$. In the simplest case of μ real and $\hat{H}$ constant as a function of time $\hat{S}_a(\mu,\nu)$ is the usual squeezing operator $\hat{S}_a(\mu,\nu) = \exp(\frac{1}{2}\zeta a^{\dagger 2} - \frac{1}{2}\overline{\zeta} a^2)$, where $\mu = \cosh|\zeta|$, $\nu = (\zeta/|\zeta|)\sinh|\zeta|$).

3 R. A. Campos, B. E. Saleh, And M. C. Teich, Phys. Rev. A **40**, 1371 (1989)

4 O. Hirota, *Optical communication with coherent squeezed state. Realization of received quantum state control*, in this volume.

5 A. Mecozzi and P. Tombesi, Opt. Commun. **75**, 256 (1990)

6 H. P. Yuen and J. H. Shapiro, Opt. Lett. **4**, 334 (1989)

7 B. Yurke, S.L. McCall, and J. R. Klauder, Phys. Rev. A **33**, 4033 (1986)

8 F. Singer, R. A. Campos, M. C. Teich and B. E. A. Saleh, Quantum Optics **2**, 307 (1990)

Session VI

– Devices for Quantum Communication –

Chairman : P. Chavel

OPTICAL COMMUNICATION WITH NOVEL QUANTUM DEVICES

Horace P. Yuen
Department of Electrical Engineering and Computer Science
Department of Physics and Astronomy
Northwestern University, Evanston, IL 60208

ABSTRACT

The nature of optical quantum communication is briefly reviewed. The important roles of the different quantum amplifiers are discussed. The concepts of quantum duplicators for various observables are introduced. These devices may have significant applications in optical systems including local area networks. Their possible realization via nonlinear optical effects will be indicated.

The fundamental performance of a communication system is determined through its statistical description. For optical systems the statistical description is ultimately a quantum mechanical one - it is specified, via the laws of quantum physics, by the quantum *state* of the field and the quantum *measurement* performed on the field. At the receiver, the field quantum state is determined by the nature of the transmission medium and the state of the transmitted field (which is determined by the nature of the source), whereas the quantum measurement corresponds to the particular optical detection scheme employed. In all the existing optical systems, the source quantum states are coherent states (CS) or their random superpositions, and the detection scheme is either direct detection, or homodyne, or heterodyne detection. Coherent states or their random mixtures are obtained from conventional light sources including conventional lasers and light-emitting diodes, and the above three detection schemes correspond to the quantum measurements of photon number, single field quadrature and both field quadratures respectively [1]. It turns out that the unavoidable quantum noise in such conventional systems is the order $\hbar\omega$, which is thousands of degrees Kelvin at optical frequencies. Form 1965-1975, considerable effort was spent on mathematically characterizing the optimum quantum measurements for various performance criteria and fixed quantum states, for single-link communication. A brief history is outlined in Ref. [2]. For CS sources such work has not led to the discovery of any greatly improved optical communication. Since the early 1970's, the study of optimal quantum states [3,4] led to the investigation of two-photon coherent state (TCS) or

Lecture Notes in Physics, Vol. 378
C. Bendjaballah, O. Hirota, S. Reynaud (Eds.),
Quantum Aspects of Optical Communications

squeezed state light [5] which is a subject of active current study, and of near number state (NNS) light [6] which has just begun to be explored. Both TCS and NNS could lead to greatly improved optical communications. Their advantages can be briefly explained as follows.

Consider a field mode with annihilation operator a in a number state $|n\rangle$,

$$N_a|n\rangle \equiv a^\dagger a|n\rangle = n|n\rangle. \tag{1}$$

If the photon number operator N_a is measured, the outcome is certain to be n, assuming ideal devices. For a coherent state $|\alpha\rangle$ with $\langle N_a\rangle = |\alpha|^2 = n$, the probability that m photons are obtained in ideal direct detection is the Poisson distribution with mean n. For the purpose of illustration we use signal-to-noise ratios to compare the performance capabilities of different states. Let $\langle M\rangle = \mathrm{tr}\rho M$ be the quantum average of an operator M of a mode in state ρ. For CS we have

$$SNR_{N_a} \equiv \frac{\langle N_a\rangle^2}{\langle \Delta N_a^2\rangle} = |\alpha|^2 = n \tag{2}$$

whereas $SNR_{N_a} \to \infty$ for $|n\rangle$. It should not be surprising that even after digitization (from the discrete nature of $|n\rangle$ versus the continuum $|\alpha\rangle$) is taken into account, states $|n\rangle$ still outperform CS $|\alpha\rangle$ in a variety of situations for the same average energy.

Similar to $|n\rangle$ that matches the measurement of N_a, the field quadrature eigenstates would match homodyne detection. However, a quadrature eigenstate has infinite energy and can only be approximately approached in reality by TCS, which are basically minimum uncertainty states for the field quadratures. Let a_1, a_2 be the quadratures of the field mode, $a \equiv a_1 + ia_2, a_1^\dagger = a_1, a_2^\dagger = a_2$, and let

$$a_\phi \equiv a_1 cos\phi - a_2 sin\phi \tag{3}$$

be an arbitrary quadrature. The uncertainty principle

$$\langle \Delta a_\phi^2\rangle\langle \Delta a_{\phi+\frac{\pi}{2}}^2\rangle \geq 1/16 \tag{4}$$

puts no lower limit on $\langle \Delta a_\phi^2\rangle$ for any state, although a small $\langle \Delta a_\phi^2\rangle$ would necessitate a large $\langle \Delta a_{\phi+\frac{\pi}{2}}^2\rangle$ that enters as noise energy in the state. A *squeezed state* in the general sense is a state, pure or mixed, for which $\langle \Delta a_\phi^2\rangle < \frac{1}{4}$ (the coherent state level) for some ϕ. A TCS or squeezed state in the restricted sense is a pure state that satisfies (4) with equality. For a properly chosen TCS with energy S, the lowest noise quadrature, say a_1, has homodyne signal-to-noise ratio [4].

$$SNR_{a1} \equiv \frac{\langle a_1\rangle^2}{\langle \Delta a_1^2\rangle} = 4S(S+1), \tag{5}$$

Therefore, the quadrature noise in the b-mode cannot be reduced below a floor level independently of the a-mode state. Thus, the effect of loss on TCS is not limited to signal attenuation; it also leads to a degradation in the achievable low noise level. If the a-mode is in a NNS, the corresponding partition noise introduced by v is similarly detrimental.

This effect especially affects two components of an optical communication system - the transmission medium because of the propagation loss, and the receiver because of nonideal devices such as nonunity photodetectore quantum efficiency. It turns out that the nonideal detection problem can be overcome by appropriate optical preamplification. For short haul fiber optic communication, the loss problem may not be particularly serious. For the long haul case, there are several possibilities of creating special fibers with distributive gain and of incorporating amplifiers along fibers. Thus, the development of amplifiers that match the nonclassical sources of NNS and TCS is important for solving both these problems and making communication with nonclassical states practical. It also happens that such novel quantum amplifiers are very useful even for communication with classical sources, as explained in Ref. [9,10]. Briefly, a parametric amplifiers is a phase-sensitive linear amplifier (PSA) which preserves the homodyne signal-to-noise ratio between the input a_ϕ and output b_ϕ, for arbitrary state of the input field,

$$SNR_{b\phi} \equiv \frac{\langle b_\phi \rangle^2}{\langle \Delta b_\phi^2 \rangle} = \frac{\langle a_\phi \rangle^2}{\langle \Delta a_\phi^2 \rangle} \equiv SNR_{a_\phi}. \tag{8}$$

With an ordinary phase-insensitive linear amplifier (PIA) one obtains, for strongly squeezed inputs,

$$SNR_{b_\phi} \gg SNR_{a_\phi}; \tag{9}$$

and for coherent states inputs,

$$SNR_{b_\phi} = \frac{1}{2} SNR_{a_\phi}. \tag{10}$$

Thus, PSA are useful even in long-haul fiber communication with conventional sources. Similar advantages are obtained with the photon number amplifier (PNA) [6],[9]-[11] which, ideally, would transform $|n\rangle$ into $|Gn\rangle$ for an integer G and thus preserves the direct detection signal-to-noise ratio at the output mode [6],

$$SNR_{N_b} = SNR_{N_a}. \tag{11}$$

With a PIA for strongly sub-Poissonian sources, one only obtains

$$SNR_{N_b} \gg SNR_{N_a} \tag{12}$$

which is to be compared with $4S$ obtained for CS with the same energy S. Typically at the transmitter $S \gg 1$; thus TCS are vastly superior to CS if the signal information is in the low noise quadrature. Even after the loss of bandwidth by a factor of 1/2 is taken into account, TCS are still superior to CS in many communication spplications [1,3,4,7,8] with various standard modulation scheme.

On the other hand, the ultimate capacity, i.e., the channel capacity constrainted only by power and bandwidth but not by any signaling or modulation restriction, is never increased by more than a factor of two with TCS compared to CS [3]. In fact, the ultimate quantum capacity, i.e., the one optimized over all possible quantum states and quantum measurements, differs just insignificantly from CS with one of the three standard detection schemes chosen according to the power bandwidth ratio [3]. This ultimate quantum capacity of the electromagnetic field has been widely assumed to be achieved by number states and photon counting. A complete proof via a theorem of Linbald was recently obtained by Ozawa and myself. Since the ultimate capacity of a noisy channel has not so far turned out to be a relevant parameter in actual communications due to a variety of reasons, TCS and NNS may still be technologically significant even though their ultimate capacity are not much improved compared to that of CS as in the case of coherent versus incoherent fiber optic communications. In particular, the use of number states can be viewed as a physical alternative to coding - its capacity can be achieved without coding because the number state channel is noise free in the ideal limit we are discussing.

Let us call any quantum state which is a CS or a random superposition of CS a *classical* state and the rest *nonclassical* states. Both NNS and TCS are nonclassical. A major problem on the application of nonclassical states is that they are very sensitive to loss [2,3,5,7] - any kind of linear loss including absorption, scattering, diffraction, coupling, deletion, etc., that enters into the system from the source to the nonideal optical detection. Basically, this arises from the fact that the output annihilation operator b of a linear lossy mode can be represented by

$$b = \eta^{\frac{1}{2}} a + (1-\eta)^{\frac{1}{2}} v, \tag{6}$$

where η is the fractional transmission and v is the annihilation operator of a vacuum mode. The state of b is determined from that of a and the vacuum state $|0\rangle$ of the v-mode, with appropriate weighting through η and $1-\eta$. Since $|0\rangle$ is a CS, a great deal of the nonclassical nature of the a-mode state is lost if $1-\eta$ is not very small. In particular, since $\langle \Delta v_\phi^2 \rangle = \frac{1}{4}$ for a CS, it follows immediately from (6) that

$$\begin{aligned} \langle \Delta b_\phi^2 \rangle &= \eta \langle \Delta a_\phi^2 \rangle + (1-\eta)\frac{1}{4} \\ &\geq (1-\eta)\frac{1}{4} \end{aligned} \tag{7}$$

and for CS sources,

$$SNR_{N_b} = \frac{1}{2} SNR_{N_a}. \tag{13}$$

On the basis of these properties we have suggested [9] the utilization of PSA or PNA to form *a lossless tap which, in a local area network (LAN) environment, would enable a very large number of users to obtain the same performance as the first user.* We now describe a simpler, more readily realizable class of quantum devices, which we call "duplicators", that would perform the same function.

Functionally, a duplicator has one "input" mode a and two "output" modes, b, c in addition to other possible auxiliary modes, such that the probabilities for measuring certain observables are equal for the three input-output modes for arbitrary input state ρ_a. For example, in a *photon number duplicator* (PND) one would have

$$\langle n|\rho_a|n\rangle = \langle n|\rho_b|n\rangle = \langle n|\rho_c|n\rangle, \tag{14}$$

whereas in a *single quadrature duplicator* (SQD) one has

$$\langle \alpha_\phi|\rho_a|\alpha_\phi\rangle = \langle \beta_{\phi'}|\rho_b|\beta_{\phi'}\rangle = \langle \gamma_{\phi''}|\rho_c|\gamma_{\phi''}\rangle. \tag{15}$$

The $|\alpha_\phi\rangle, |\beta_{\phi'}\rangle, |\gamma_{\phi''}\rangle$ in (15) are the eigenstate of a_ϕ, $b_{\phi'}$, and $c_{\phi''}$ respectively for fixed phase angles ϕ, ϕ', and ϕ''. A photon number duplicator is similar to a power gain $G = 2$ photon number amplifier. Nore that state duplication, viz., $\rho_a = \rho_b = \rho_c$ is impossible when ρ_a varies in general because unitarity would then be violated [11]. It can be seen that a duplicator for a class of observables provides a way to perform quantum nondemolition detection (QND) of such observables if the output b, c modes are perfectly correlated in the eigenstates of those observables. In fact, in the abstract mathematical specification of a duplicator in ref.[11], the output modes are indeed perfectly correlated, as is also the case with the experimental realization to be described in the following. The application of QND to local area network was first suggested by Yamamoto [12]. Note however, that a realization of QND does *not* provide a realization of a duplicator.

Duplicators can obviously be used in series to provide a corresponding amplifier or larger gain $G = 2^m$ with m stages. However, a single duplicator already leads to a lossless tap in a communication system: mode a is the signal mode, b the mode transmitting down the line and c the readout mode. In general, the b and c modes are correlated at least to some degree. The duplicator tap is superior to our previous amplifier tap in avoiding the need of providing large power, and in maintaining the same power down the line. In addition, photon number duplicators appear to be easier to realize than photon number amplifiers, while the realizations of SQD and PSA are comparable.

Similar to the quantum amplifiers, *quantum duplicators are useful even in optical communications with conventional sources.* Table 1 (below) summarizes the difference source states and their natural matching detection schemes and amplifiers/duplicators.

TABLE 1

STATES	DETECTION	AMPLIFIER	DUPLICATOR
CS	heterodyne	PIA	BQD
TCS	homodyne	PSA	SQD
NS	direct	PNA	PND

ABBREVIATIONS

CS	coherent states
NS	number states
TCS	two-photon coherent states (or squeezed states in one sense of the word)
NNS	near number states
PIA	phase-insensitive linear amplifier
PNA	photon number amplifier
PSA	phase-sensitive linear amplifier
LAN	local area network
QND	quantum nondemolition detection
PND	photon number duplicator
SQD	single quadrature duplicator
BQD	both quadratures duplicator

A photon number duplicator can be physically spproximated by either of the following two processes as indicated in ref. [13], when the input signal power flux is not too strong. The first one is depicted schematically in Fig. 1. It involves a three-wave mixing process in which a signal

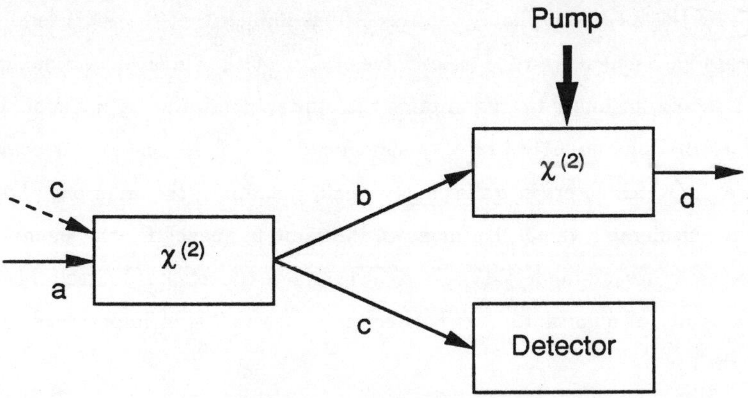

Figure 1: Photon number duplication via parametric down conversion; an input "signal" photon is split into an output photon and a readout photon. The output photon can be up-converted back into a "signal" photon for the d-mode down the line. A small probe input for the read mode c may ease the operation of the scheme.

photon is converted to an output photon plus a readout photon. Classically, it is well known that with a small input at the readout mode, the total energy of the a-mode would be completely converted at an interation length L smaller than but close to that determined by

$$\kappa L = \frac{\pi}{2}, \tag{16}$$

where κ, the gain coefficient per unit length, is proportional to $\chi^{(2)} I_a^{1/2}$, $\chi^{(2)}$ being the second order nonlinear susceptibility, and I_a the average power flux of the signal. Quantum mechanically the conversion process, which would start even from vacuum b, c initial states, is described by an effective Hamiltonian of the form

$$H_I \sim \lambda \left(a^\dagger bc + ab^\dagger c^\dagger \right) \tag{17}$$

with $\lambda \propto \chi^{(2)}$. It is an immediate consequence of (17) that in the paraxial approximation with z being the propagation direction,

$$N_b(z) - N_b(0) = N_c(z) - N_c(0) = N_a(0) - N_a(z), \tag{18}$$

which is the quantum mechanical form of the Manley-Rowe relation. It is a direct consequence of (18) that, as far as photon counting is concerned, at the point L of complete conversion, the number of photons at each of the initially vacuum b, c modles is equal to the input signal photon number, i.e.,

$$n_b(L) = n_c(L) = n_a(0), \tag{19}$$

where n_a, etc. are the measured photon numbers. This implies that the PND requirement (14) is satisfied. Classically, L depends on the exact value of I_a via (16). Quantum mechanically, the effect of quantization shows up: under the assumption that the response time of the nonlinear material is bigger than the signal photon arrival rate, L is independent of I_a so long as it is below the vacuum fluctuation level I_a^0, which enters into the effective gain κ and is of the order one photon per mode, which can be a considerable value. To maintain the same frequency for the signal down the line, one can up-convert b into a d-mode, i.e., $\rho_b = \rho_d$ at complete conversion. In itself this would provide a general transformer of quantum states between modes which is of importance in manipulating nonclassical light.

A second possible realization is depicted schematically in Fig. 2. It involves a four-wave mixing process in which an input signal photon at the a-mode is combined with a pump photon to produce the b, c output photons, which can now be of the same frequency as the a-mode photon. Under the

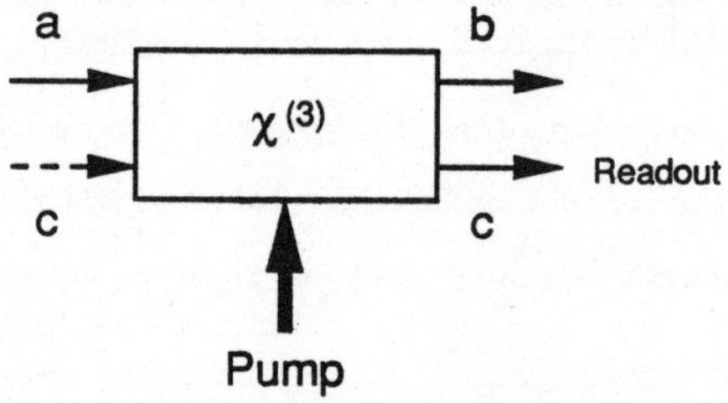

Figure 2: Photon duplication via four-wave mixing; an input signal photon combines with a pump photon to yield an output signal photon and a readout photon. Again a small c-input may be convenient.

undepleted classical pump approximation, this process goes exactly as the previous one except that the κ of (16) is now proportional to $\chi^{(3)}\sqrt{I_p I_a}$ where $\chi^{(3)}$ is the third order susceptibility and I_p the average power flux of the pump. The same effective H_I (17) applies with $\lambda \propto \chi^{(3)} I_p^{1/2}$. Again the vacuum fluctation value I_a^0 is to be used in κ for input signal level below I_a^0, our region of operation.

It is significant that our photon-number duplicator also provides a convenient realization of the quantum nondemolition measurement of photon number. This is because the b, c modes are perfectly correlated in photon numbers as indicated by (18). Compared to the Kerr-effect realization [14] which requires an infinite gain for an exact realization even in the ideal limit, our scheme requires only a finite gain determined by (16).

The two schemes described above also provide a single quadrature duplicator, i.e., Eq. (15) is satisfied for properly chosen phase angles after frequency normalization, if any. Physically, this may be expected from the photon number conversion and the fact that the nonlinear mixings are phase coherent processes. They also provide approximate realizations of single quadrature QND, approximate in this case because the b, c amplitudes are only approximately correlated for finite gains.

This work was supported by the Office of Naval Research.

References

[1] H. P. Yuen and J. H. Shapiro, "Optical Communication with Two-Photon Coherent States - Part III: Quantum Measurement Realizable with Photoemissive Detectors", IEEE Trans. Inf. Theory **IT-26**, 78-92 (1980).

[2] H. P. Yuen, "Quantum Communication, Quantum Measurement, TCS and QND," in *Quantum Optics, Experimental Gravitation and Measurement Theory*, P. Maystre and M. O. Scully, Eds., (Plenum, New York, 1983) pp. 249-268.

[3] H. P. Yuen, "Generalized Coherent States and Optical Communication," in Proc. 1975 Conf. Information Sciences and Systems, Johns Hopkins Univ., pp. 171-177, (1975).

[4] H. P. Yuen, "States that give the Maximum Signal-to-Noise Ratio for a Fixed Energy," Phys. Lett. A **56**, 105-106 (1976).

[5] H. P. Yuen, "Two-Photon Coherent States of the Radiation Field," Phys. Rev. A **13**, 2226-2243 (1976).

[6] H. P. Yuen, "Generation, Detection, and Application of High-Intensity Photon-Number-Eigenstate Fields," Phys. Rev. Lett. **56**, 2176-2179 (1986).

[7] H. P. Yuen and J. H. Shaprio, "Optical Communication with Two-Photon Coherent States - Part I: Quantum-State Propagation and Quantum-Noise Reduction," IEEE Trans. Inf. Theory **IT-24**, 657-668 (1978).

[8] J. H. Shapiro, H. P. Yuen, and J. A. Machado Mata, "Optical Communication with Two-Photon Coherent States - Part II: Photoemissive Detection and Structured Receiver Performance," IEEE Trans. Inf. Theory **IT-25**, 179-192 (1979).

[9] H. P. Yuen "Design of Transparent Optical Networks by using Novel Quantum Amplifiers and Sources", Opt. Lett. **12**, 789-791 (1987).

[10] H. P. Yuen, "Nonclassical Light", in *Photons and Quantum Fluctuations*, E.R. Pike and H. Walther, Eds., (Adam Hilger, Bristol, 1988) pp. 1-9.

[11] H. P. Yuen, "Amplification of Quantum States and Noiseless Photon Amplifiers," Phys. Lett. A **113**, 405-407 (1986).

[12] Y. Yamamoto, private communication, 1986.

[13] H.P. Yuen, "Efficient Photon Number Detection," in *Coherence and Quantum Optics VI*, ed. by J. H. Eberly, et al., Plenum, New York, 1990; pp. 1265-1268.

[14] N. Imoto, H. A. Hans, and Y. Yamamoto, "Quantum Nondemolition Measurement of the Photon Number via the Optical Kerr Effect," Phys. Rev. A32, 2287-2292 (1985).

Construction of Modulator for Quantum Orthogonal Signal by Bright Squeezer

Masao OSAKI† **Keizou INAGAKI**††

†Faculty of Science and Technology, Keio Univ.
††Advanced Telecommunication Research Inst.

1. Introduction

Recently, the squeezed states and their applications have received much attention with development of quantum communication theory, because it clarifies the existence of useful quantum communication systems [1~3]. The mathematical framework of the quantum communication theory was devised by Helstrom and others [4], and more the concept of quantum state control has been shown. However, the physical realizations of the fruitful results in the theory are, in general, very difficult, because the squeezed state with large amount of coherent component, so-called coherent squeezed state can not be controlled arbitrary in the sense of experiments [5,6]. But, recent progress in quantum optics verified that the coherent squeezed state is available. Especially, Oliveira suggested the injected-field optical bistability model to obtain "bright squeezed state" [7,8].

In this paper, we analyze numerical properties of quantum noise of optical bistable as the bright squeezer to study the possibility of the application to optical communications. Based on these results, we try to design several quantum communication systems using "bright squeezer".

2. Quantum Communications System

Here, we consider the quantum communication system using $M-ary$ orthogonal signals. In general, the optimization of this kind of systems is treated as the minimization of the error probability with respect to operator valued measure under the fixed signal state. The optimum error probability of the system with the coherent state was given by Yuen, and it can be improved by changing the signal quantum state from coherent state to squeezed state. This analysis was done by our group [9]. The final form of the error probability is

$$Pe = \frac{M-1}{M^2}\left[\{1+(M-1)\kappa_{12}\}^{1/2} - (1-\kappa_{12})^{1/2}\right]^2, \tag{1}$$

where κ_{12} is the inner product among orthogonal signals. When the squeezed state is used, the quantum states of the $M-ary$ orthogonal signals $|\psi_i>$ $(i = 1 \sim M)$ are represented as follows :

$$|\psi_i> = \prod_{k=1}^{i-1} |0;\mu,\nu>_k |\alpha;\mu,\nu>_i \prod_{k=i+1}^{M} |0;\mu,\nu>_k, \quad \text{for} \quad i = 1 \sim M, \tag{2}$$

Lecture Notes in Physics, Vol. 378
C. Bendjaballah, O. Hirota, S. Reynaud (Eds.),
Quantum Aspects of Optical Communications

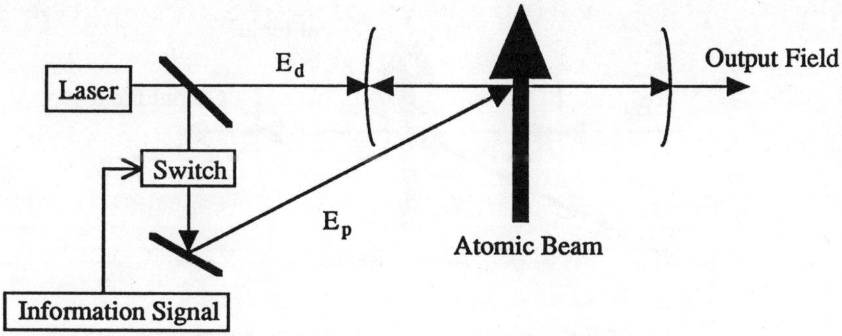

Figure 1: Schematic diagram of modulator using bright squeezer.

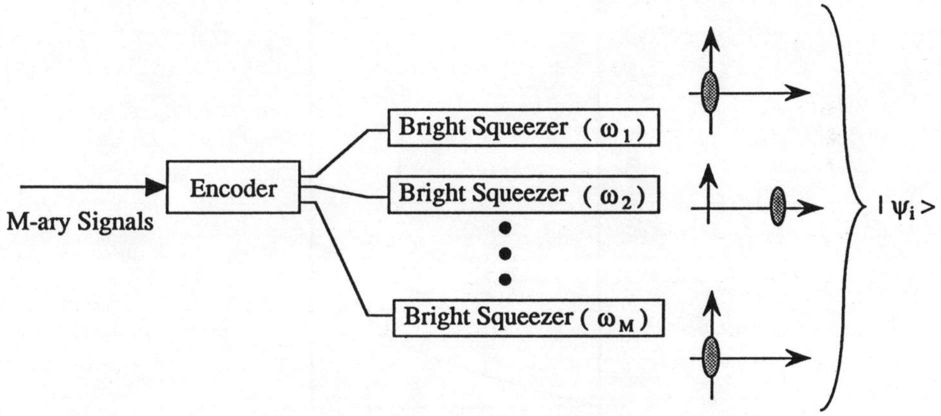

Figure 2: Transmitter of $M - ary$ orthogonal signals.

where μ, ν are squeezing parameters and $|0; \mu, \nu>_k$, $|\alpha; \mu, \nu>_i$ are the k_{th} mode squeezed vacuum state and the i_{th} mode coherent squeezed state, respectively. In this case, inner product κ_{12} is given as follows :

$$\kappa_{12} = exp\left[-(\mu+\nu)^2(<\hat{n}> -M|\nu|^2)\right], \tag{3}$$

where $<\hat{n}>$ is average signal power per signal.

Although the system with the coherent squeezed state indicates the superior property, the realization of this system had been thought to be impossible because of the difficulty of generation of the squeezed state with controlled coherent component. However, the bright squeezer shown in section 3 is applicable to the construction of the light source of this $M - ary$ orthogonal signals with squeezed state. In Fig.1 and Fig.2, we show the constructions of the bright squeezer with a modulator and the transmitter of $M - ary$ orthogonal signals. In this transmitter, if the signal is "i", pumping field E_p is injected into the "i"$_{th}$ bright squeezer, and other squeezers have driving field E_d only. So we can obtain the $M - ary$ orthogonal signals represented in Eq.(2).

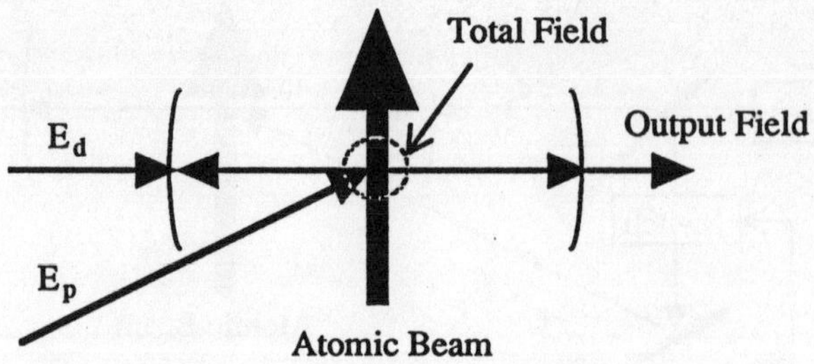

Figure 3: Constructon of bright squeezer.

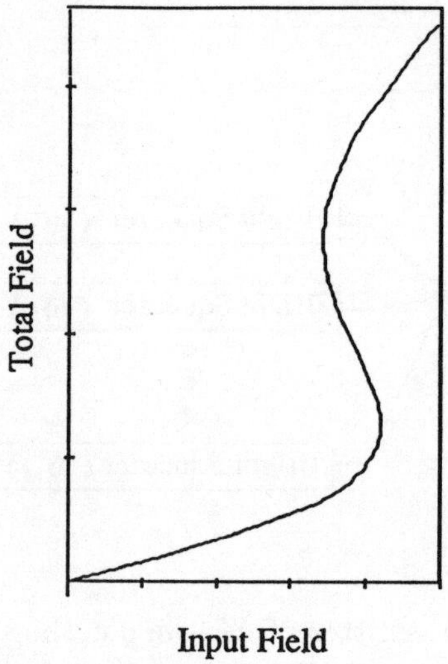

Figure 4: Phenomenon of optical bistability.

3.Principle of Bright Squeezer

To generate the bright squeezed state, we consider the bright squeezer suggested by Oliveira and Knight, using the optical bistability shown in Fig.3.

We first explain the system configuration. Driving field (E_d) is injected into the cavity with a nonlinear media passing through the mirror. It produces the optical bistability with squeezing phenomenon. But E_d doesn't have enough intensity to generate dipole moments at the atoms, because of the high damping rate of mirrors to get a large degree of squeezing. So the coherent component is nearly equal to zero. If pumping field (E_p) is driven to the cavity without passing through the mirror, it provides enough intensity to generate large dipole moments at the atoms. And it results in a squeezed state with a large amount of coherent component.

The hysteresis property between the input field and the total field is shown in Fig.4. Squeezing process is occurred in the lower brunch of the hysteresis.

3.1 Squeezing by Optical Bistability

We shall now define the relation between the phase space and the complex amplitude operators as follows :

$$\hat{X}_\theta = \hat{a}e^{-i\theta} + \hat{a}^\dagger e^{i\theta}. \tag{4}$$

To analyze the quantum noise spectrum density, we use the two-time correlation function $< \Delta\hat{X}_{\theta(t)}\Delta\hat{X}_{\theta(0)} >$ as follows :

$$\begin{aligned} n_s^{-1} <: \Delta\hat{X}_{\theta(t)}\Delta\hat{X}_{\theta(0)} :> =& < \xi_{1(\tau)}\xi_{1(0)} > e^{-2i\theta} + < \xi_{3(\tau)}\xi_{3(0)} > e^{2i\theta} \\ &+ < \xi_{1(\tau)}\xi_{3(0)} > + < \xi_{3(\tau)}\xi_{1(0)} > . \end{aligned} \tag{5}$$

where $\Delta\hat{X}_{\theta(t)} = \hat{X}_{\theta(t)} - < \Delta\hat{X}_{\theta(t)} >$, that is, $\Delta\hat{X}_{\theta(t)}$ is the deviation of $\hat{X}_{\theta(t)}$, ξ_1 , ξ_3 are also the deviations of the eigenvalues of total cavity field, n_s is a saturating intensity, and :: means normally-ordered. The quantum noise spectrum density is obtained by Fourier transform of the correlation function,

$$\begin{aligned} S_{\theta(\omega)} &= 1 + 2\kappa' \int e^{-i\omega t} <: \Delta\hat{X}_{\theta(t)}\Delta\hat{X}_{\theta(0)} :> dt \\ &= 1 + (2\kappa'/\gamma_{||}) \int e^{-i\Omega\tau} <: \Delta\hat{X}_{\theta(\tau)}\Delta\hat{X}_{\theta(0)} :> d\tau. \end{aligned} \tag{6}$$

where

κ' : damping rate of mirror,
$\gamma_{||}$: inversion decay rate of cavity,
Ω : $(= \omega/\gamma_{||})$, normalized frequency,
τ : $(= t\gamma_{||})$, normalized time.

By substituting Eq.(5) into Eq.(6), the quantum noise spectrum density is given as follows :

$$\begin{aligned} S_{\theta(\Omega)} =& 1 + 2\mu n_s[G_{11(\Omega)}exp(-2i\theta) \\ &+ G_{33(\Omega)}exp(2i\theta) + G_{13(\Omega)} + G_{31(\Omega)}]. \end{aligned} \tag{7}$$

where $\mu = \kappa'/\gamma_{||}$,

$$G_{ij(\Omega)} = \int e^{-i\Omega\tau} < \xi_{i(\tau)}\xi_{j(0)} > d\tau, (i,j = 1, ...5). \tag{8}$$

The phase depending terms are first two terms in the bracket in the right hand side of Eq.(7). For the optical bistability, these two terms $G_{11(\Omega)}$ and $G_{33(\Omega)}$ are not equal to zero. So the quantum noise has phase depending property, and we can minimize it for each Ω by controlling the phase θ.

In order to calculate the quantum noise spectrum density, we apply the Fourier transform to Ito statical deferential equation of quantum fluctuations, and use the solution of algebraic equation as follows :

$$\mathbf{G}_{(\Omega)} = (\mathbf{A} - \mathrm{i}\Omega\mathbf{I})^{-1}\mathbf{B}\mathbf{B}^{\mathrm{t}}(\mathbf{A}^{\mathrm{t}} + \mathrm{i}\Omega\mathbf{I})^{-1}. \tag{9}$$

where **I** is a 5×5 identity matrix, and **A** and **B** represent a drift and a diffusion matrixes, respectively.

3.2 Optimum Conditions of Squeezing

In this section, we define the desirable properties for the bright squeezed state to use as optical sources for quantum communication systems. The desirable properties are

1) signal to noise ratio (SNR) is maximized,

2) wide bandwidth of squeezing.

The first condition requires that coherent component and minimum quantum noise are occurred in the same phase. And second condition is necessary for high bit rate communication systems because the modulated signal has side band spectrums around the carrier frequency. The squeezed bandwidth is defined as the less quantum than the standard quantum noise level. From now, we analyze the squeezing property of the optical bistability based on these two points.

In order to clarify the direction of squeezing, Eq.(7) should be payed attention. The optimal phase θ_{opt} , which gives minimum quantum noise, is defined by

$$exp(2i\theta_{opt}) = -[G_{11(\Omega)}/G_{33(\Omega)}]^{1/2}. \tag{10}$$

Substituting Eq.(10) into Eq.(7), the quantum noise spectrums with maximum suppression are obtained as shown in Fig.5. Parameters are the same as used in Ref.[8]. $S_{\theta(\Omega)} = 1.0$ for the standard quantum level, and $S_{\theta(\Omega)} < 1$ indicates the squeezing.

The other hand, Eq.(10) shows that the optimal phase θ_{opt} has a possibility to change with the baseband frequency Ω. If θ_{opt} depends on Ω , the effective squeezed bandwidth will be reduced in the case of the coherent detection quantum communication systems, because it is impossible to give the optimal phase for each baseband frequency. So we need the condition that θ_{opt} is independent of Ω.

To obtain this condition, we calculate $G_{11(\Omega)}$ and $G_{33(\Omega)}$. And as a result, we obtained the cavity detuning and the atoms detuning must be equal to zero as the condition that θ_{opt} is independent of Ω . Under this condition, the optimal phase θ_{opt} is a constant $\pi/2$ for the cavity mode.

By substituting this condition into Eq.(7), however, the little squeezing occurs in the quantum nise spectrum density as Fig.6.

Then, we need the analysis of the phase depending property of quantum noise spectrum density in Fig.5, and clarify the optimal detection phase which gives the widest squeezed bandwidth under the requirement of the degree of quantum noise reduction. The phase depending property of quantum noise spectrum density, Fig.7, can be calculated with parameters used in Fig.5 (iii). For an quantum communication, if 20% reduction of quantum noise is required, one should choose about $2.14_{(rad)}$ as the optimal detection phase.

The optimal detection phase may change with the conditions and may be solved by mathematically. However, Eq.(9) becomes very complex, and we consider that its solution is very difficult. So the simulations are useful to obtain the optimal detection phase under the given conditions.

Then, if the pumping field E_p is injected at this optimal detection phase, the wider bandwidth and maximum SNR bright squeezed state is generated. Hence, controlling the intensity of E_p, one can modulate the bright squeezed state directly (Fig.8). Or

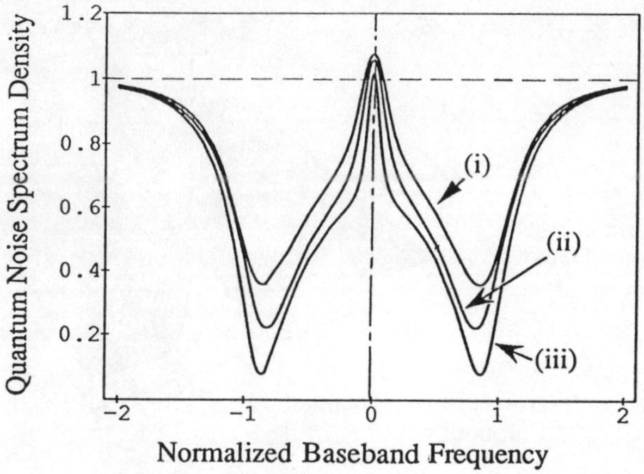

Figure 5: Most suppressed quantum noise spectrum densities.
(i);$\mu = 5.4, C = 28, \delta = 7.3, \phi = -0.9, |x| = 7.0$,
(ii);$\mu = 13.5, C = 52, \delta = -14.6, \phi = 1.0, |x| = 16.4$,
(iii);$\mu = 50, C = 200, \delta = 60, \phi = -1.0, |x| = 70$.

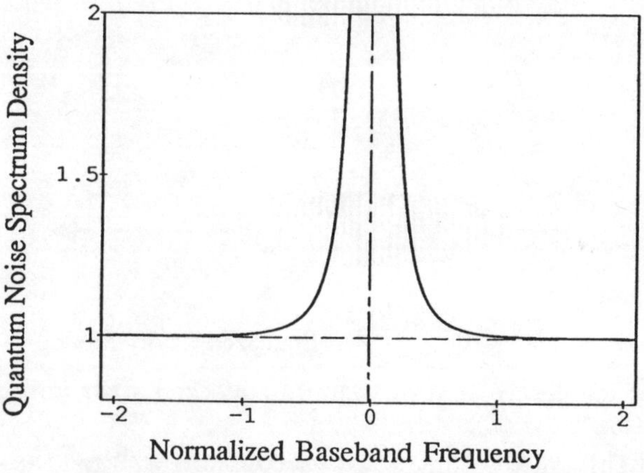

Figure 6: Quantum noise spectru density with constant θ_{opt}.
(iii);$\mu = 50, C = 200, \delta = 0, \phi = 0, |x| = 70$.

switching the phase of E_p "0" or "π", binary phase shift keying (BPSK) modulator can be realized.

4. Conclusions

In this paper, we analyzed the quantum noise property of the optical bistability, in order to apply it to the quantum communication systems. As a result, in general, the optimal phase θ_{opt} which gives minimum quantum noise, may depend on the baseband frequency. Hence the squeezed bandwidth may be reduced in the case of the coherent detection quantum communication systems. Thus we need the condition that θ_{opt} is independent of baseband frequency. The condition is that the cavity and the atomic detunings are

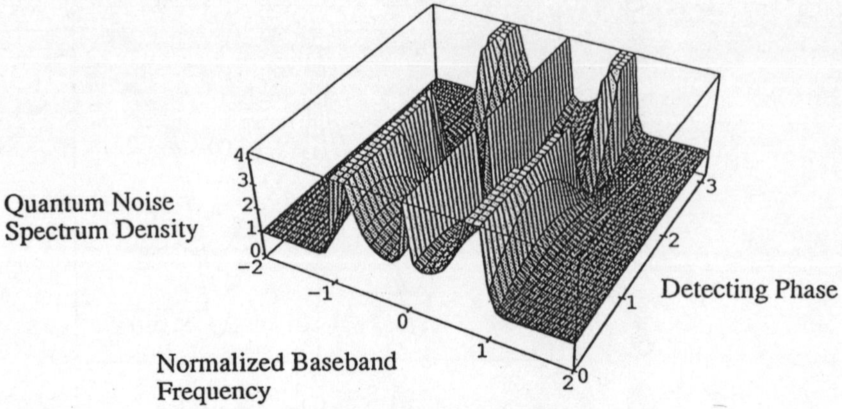

Figure 7: Phase depending property of quantum noise spectrum density. (iii);$\mu = 50, C = 200, \delta = 60, \phi = -1.0, |x| = 70$.

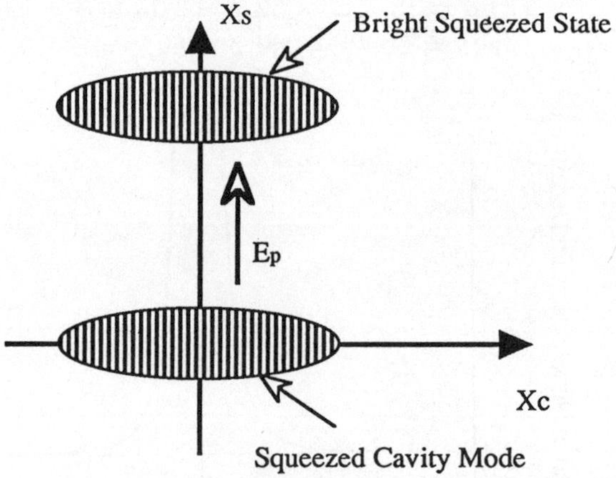

Figure 8: Phase-space description of vacuum squeezed state and bright squeezed state.

equal to zero, and this means that θ_{opt} is a constant $\pi/2$ for the cavity mode. Under this condition, however, the degree of squeezing is nearly equal to zero. Namely, it is impossible for a single phase to detect the largest degree of squeezing occurred in each baseband frequency. To the next, we analyzed the phase depending property of the quantum noise spectrum density, and showed the optimal detection phase which gives widest squeezed bandwidth under the conditions. So the widest squeezed bandwidth and maximum SNR bright squeezed state can be obtained by injecting the pumping field at this optimal detection phase. Then we showed the application of the bright squeezed state to a quantum communication system using the $M-ary$ orthogonal signals which provides a superior property to that using coherent state. The realization of the bright squeezer will contribute much to the development of the quantum communication systems.

Acknowledgments

We wish to thank Professor O. Hirota (Tamagawa Univ.) for his helpful discussions, and Dr. T. Uyematsu (Tokyo Ins. Tech.) for informative advices of this problem.

References

[1] O.Hirota :"Optical communication theory," Morikita Pub. Company,Tokyo, (1985).
[2] H.P.Yuen and J.H.Shapiro :"Optical communication with two-photon coherent state, part1," IEEE Trans. Inf. Theory, IT-24, pp.657-668, (1978).
[3] H.P.Yuen and J.H.Shapiro :"Optical communication with two-photon coherent state, part2," IEEE Trans. Inf. Theory, IT-25, pp.179-192, (1978).
[4] C.W.Helstrom :"Quantum detection and estimation theory," Academic Press, New York, N.Y. (1976).
[6] R.E.Slusher, L.W.Hollberg, B.Yurke, J.C.Mertz, and F.J.Valley :"Observation of squeezed states generated by four-wave mixing in an optical cavity," Phys. Rev. Lett. vol-55, pp.2409-2412, (1985).
[7] F.A.M. de Oliveira, and P.L.Knight :"Bright squeezing," Phys. Rev. Lett. vol-61, pp.830-833, (1988).
[8] F.A.M. de Oliveira, and P.L.Knight :"Injected-field optical bistability model of squeezed coherent light generation," Phys. Rev. A, vol-39, pp.3417-3431, (1989).
[9] K.Yamazaki :"On quantum optimum receiver for M-ary orthogonal squeezed state signal," National Convention Record IEICE, B-911, Mar. (1990).

Efficiency of Optical Modulations with Coherent State

Noboru WATANABE

Department of Information Sciences
Science University of Tokyo
Noda City, Chiba 278, Japan

Abstract

In information of communication by means of laser, quantum communication theory describing the quantum effect has been needed instead of conventional Shannon's theory. In this paper, (1) we briefly review the mathematical formulation of quantum communication theory and (2) consider the efficiency of modulations (ASK(OOK), FSK, PSK, PPM) with coherent state by using the Ohya's mutual entropy.

Introduction

In communication processes of information, an electric current or an electric wave is generally used as a medium transmitting information. We often call this medium a carrier wave. An optical communication using a laser as a carrier wave has been developed and studied. In information communication by means of laser, a new theory (optical communication theory) should directly include the quantum effect in itself.

The main purpose of this paper is to consider the efficiency of optical modulations (ASK(OOK), FSK, PSK, PPM) with the coherent states by using the mutual entropy and the quantum channel introduced by Ohya several years ago.

In §1, we briefly review the mathematical construction of a quantum mechanical channel and attenuation process based on [8,9].

In §2, we explain about the formulation of the mutual entropy (information) introduced in [9] for quantum input state through the quantum channel.

In §3, we consider the efficiency of the optical degital modulations (ASK(OOK), FSK, PSK) by using the mutual entropy in quantum communication theory.

§1. Quantum Mechanical Channel

1.1 Construction of quantum mechanical channel

Let $B(\mathcal{H})$ be the set of all bounded linear operators on a separable Hilbert space $\mathcal{H}$, $\mathfrak{S}(\mathcal{H})$ be the set of all density operators on $\mathcal{H}$. Then a quantum dynamical system is described by $(B(\mathcal{H}),\mathfrak{S}(\mathcal{H}))$. In order to discuss communication processes, we need two dynamical systems: an input system $(B(\mathcal{H}_1),\mathfrak{S}(\mathcal{H}_1))$ and an output system $(B(\mathcal{H}_2),\mathfrak{S}(\mathcal{H}_2))$.

Lecture Notes in Physics, Vol. 378
C. Bendjaballah, O. Hirota, S. Reynaud (Eds.),
Quantum Aspects of Optical Communications

The input state generally change to the output state under the influence of noise and loss associated with a channel. The quantum mechanical channel describing this change is defined as follows [8,9]: A mapping Λ^* from $\mathfrak{S}(\mathcal{H}_1)$ to $\mathfrak{S}(\mathcal{H}_2)$ is said to be a channel if its dual map Λ from $B(\mathcal{H}_2)$ to $B(\mathcal{H}_1)$ satisfies the following three conditions: (1) Λ is completely positive (i.e., for any $A_j \in B(\mathcal{H}_2)$, $B_i \in B(\mathcal{H}_1)$ and $n \in N$, $\sum_{i,j=1}^{n} B_i^* \Lambda(A_i^* A_j) B_j \geq 0$), (2) Λ is normal, (3) $\Lambda(I_2) = I_1$, where I_i is an identity operator on $\mathcal{H}_i$.

When an input state is given by a density operator $\rho \in \mathfrak{S}(\mathcal{H}_1)$, von Neumann introduced [28] the entropy of the input state ρ such as

$$S(\rho) = -tr\rho \log \rho.$$

We here briefly explain a general form of a quantum channel introduced by Ohya [9] for an optical communication process, and it includes the direct effects of noise and loss existing in the course of the communication processes. In addition to the Hilbert spaces $\mathcal{H}_1$ and $\mathcal{H}_2$, we need two more Hilbert spaces $\mathcal{K}_1$ and $\mathcal{K}_2$ describing the effect of noise and loss, respectively. Let $\xi \in \mathfrak{S}(\mathcal{K}_1)$ be a state representing the noise and Δ, Π, Γ be the following maps: (1) the map Δ is an amplification from $B(\mathcal{H}_2)$ to $B(\mathcal{H}_2 \otimes \mathcal{K}_2)$ given by $\Delta(A) = A \otimes I$ for any $A \in B(\mathcal{H}_2)$, (2) the map Π is a completely positive map from $B(\mathcal{H}_2 \otimes \mathcal{K}_2)$ to $B(\mathcal{H}_1 \otimes \mathcal{K}_1)$ with $\Pi(I) = I$, (3) the map Γ is given by $\Gamma(Q) = tr_{\mathcal{K}_1} \xi Q$ for any $Q \in B(\mathcal{H}_1 \otimes \mathcal{K}_1)$, where $tr_{\mathcal{K}_1}$ is the partial trace with respect to the Hilbert space $\mathcal{K}_1$. Then, we define a mapping Λ from $B(\mathcal{H}_2)$ to $B(\mathcal{H}_1)$ such as

$$\Lambda = \Gamma \circ \Pi \circ \Delta.$$

We next consider the dual maps of Δ, Π, Γ: (1*) the dual map Δ^* of Δ is a map from $\mathfrak{S}(\mathcal{H}_2 \otimes \mathcal{K}_2)$ to $\mathfrak{S}(\mathcal{H}_2)$ such that $\Delta^*(\sigma) = tr_{\mathcal{K}_2} \sigma$, (2*) the dual map Π^*: $\mathfrak{S}(\mathcal{H}_1 \otimes \mathcal{K}_1) \to \mathfrak{S}(\mathcal{H}_2 \otimes \mathcal{K}_2)$ is given by $tr\Pi^*(\sigma)W = tr\sigma\Pi(W)$ for any $\sigma \in \mathfrak{S}(\mathcal{H}_1 \otimes \mathcal{K}_1)$ and any $W \in B(\mathcal{H}_2 \otimes \mathcal{K}_2)$, (3*) the dual map Γ^*: $\mathfrak{S}(\mathcal{H}_1) \to \mathfrak{S}(\mathcal{H}_1 \otimes \mathcal{K}_1)$ is given by $\Gamma^*(\rho) = \rho \otimes \xi$. Therefore, we know the noise state ξ and the mechanism of the transformation Π, we can write down a channel explicitly such that

$$\Lambda^*(\rho) = \Delta^* \circ \Pi^* \circ \Gamma^*(\rho) = tr_{\mathcal{K}_2} \Pi^*(\rho \otimes \xi), \text{ for any } \rho \in \mathfrak{S}(\mathcal{H}_1).$$

1.2 Attenuation process

We briefly review a construction of the optical channel for an attenuation process.

Let α^* and α be the creation and annihilation operators of photon, respectively. The system composed of photons is described by a Hamiltonian H

$$H = \alpha^* \alpha + \frac{1}{2}.$$

The eigenvalue of Schrödinger equation

$$Hx(q) = \epsilon_n x(q),$$

is given by $\epsilon_n = n + \frac{1}{2}$, and the associated eigenvector is given by

$$x_n(q) = \frac{1}{\sqrt{\pi^{1/2} n!}} H_n(\sqrt{2}x)\exp(-\frac{x^2}{2}) \text{ for each } n,$$

where $H_n(x)$ is a Hermite polynomial of degree n.

Moreover we define the CONS $\{x_{n_1}^{(1)}\}$, $\{y_{m_1}^{(1)}\}$, $\{x_{n_2}^{(2)}\}$ and $\{y_{m_2}^{(2)}\}$ of Hilbert spaces $\mathcal{H}_1$, $\mathcal{K}_1$, $\mathcal{H}_2$ and $\mathcal{K}_2$ of the input, noise, output and loss systems by

$$x_{n_1}^{(1)}(q_1) = \pi^{-\frac{1}{4}} 2^{-\frac{n_1}{2}} (n_1!)^{-\frac{1}{2}} \exp\{\frac{-q_1^2}{2}\} H_{n_1}(q_1)$$

$$y_{m_1}^{(1)}(t_1) = \pi^{-\frac{1}{4}} 2^{-\frac{m_1}{2}} (m_1!)^{-\frac{1}{2}} \exp\{\frac{-t_1^2}{2}\} H_{m_1}(t_1)$$

$$x_{n_2}^{(2)}(q_2) = \pi^{-\frac{1}{4}} 2^{-\frac{n_2}{2}} (n_2!)^{-\frac{1}{2}} \exp\{\frac{-q_2^2}{2}\} H_{n_2}(q_2)$$

$$y_{m_2}^{(2)}(t_2) = \pi^{-\frac{1}{4}} 2^{-\frac{m_2}{2}} (m_2!)^{-\frac{1}{2}} \exp\{\frac{-t_2^2}{2}\} H_{m_2}(t_2).$$

We here suppose the following linear transformation satisfying the conservation of energy (i.e., $n_1 + m_1 = n_2 + m_2$, where n_1, m_1, n_2 and m_2 are the number of photons in the input, noise, output and loss systems) among the coodinates q_1, t_1, q_2, t_2 of the input, noise, output and loss systems, respectively:

$$\begin{cases} q_2 = \mu q_1 + \nu t_1, \\ t_2 = -\nu q_1 + \mu t_1. \end{cases} \qquad (\mu^2 + \nu^2 = 1)$$

For simplicity, we put a noise state $\xi = |y_0^{(1)}><y_0^{(1)}| \in \mathfrak{S}(\mathcal{K}_1)$, where $y_0^{(1)}$ is a vacuum state vector in $\mathcal{K}_1$.

By using this linear transformation, the mapping Π^* introduced in [22] is defined by $V(\cdot)V^*$, where

$$\begin{aligned} & V(x_n^{(1)} \otimes y_0^{(1)})(q_2, t_2) \\ & = x_n^{(2)} \otimes y_0^{(2)}(\mu q_2 - \nu t_2,\ \nu q_2 + \mu t_2) \\ & = \pi^{-\frac{1}{2}} 2^{-\frac{n}{2}} (n!)^{-\frac{1}{2}} \exp\{\frac{-(\mu q_2 - \nu t_2)^2}{2}\} \exp\{\frac{-(\nu q_2 + \mu t_2)^2}{2}\} \\ & \times H_n(\mu q_2 - \nu t_2) H_0(\nu q_2 + \mu t_2) \\ & = \sum_{j=0}^{n} C_j^n x_j^{(2)} \otimes y_{n-j}^{(2)}(q_2, t_2), \end{aligned}$$

$$C_j^n = \sqrt{\frac{n!}{(n-j)!j!}} \mu^j \nu^{n-j}$$

and μ^2 can be regarded as the transmition ratio η for the channel. Then the channel Λ^* is expressed by

$$\Lambda^*(\rho) = tr_{\mathcal{K}_2} V(\rho \otimes \xi) V^*.$$

1.3 Coherent states

We here briefly explain a coherent state. Let θ be the complex parameter represented by

$$\theta = |\theta| \exp(i\omega),$$

where $|\theta|$ is a parameter for amplitude and ω is a parameter for frequency. For this parameter θ, we denote the unitary operator D(θ) as

$$D(\theta) = \exp\{\theta\alpha^* - \bar{\theta}\alpha\}$$

By using D(θ), a coherent state vector $|\theta>$ is defined by

$$\begin{aligned} |\theta> &= D(\theta)|x_0^{(1)}> \\ &= \exp\{-\frac{1}{2}|\theta|^2\} \sum_{n=0}^{\infty} \frac{\theta^n}{\sqrt{n!}} |x_n^{(1)}>, \end{aligned}$$

and we called $|\theta><\theta|$ a coherent state.

More concrete channel is constructed in [9] for an attenuation process. This channel can be used to study the communication process, for instance, to calculate the error probability [14,15,17,18,19].

§2. Mutual Entropy

When an input state is given by a density operator ρ, the amount of information carried by the input state ρ is described by the von Neumann entropy S(ρ). The quantum mechanical channel Λ^* is mathematical investigated as a mapping sending a quantum state from an input system to an output system. In communication theory, it is important to know the amount of information correctly transmitted from the input system to the output system through the channel Λ^*. The mutual entropy (information) represents this amount of information transmitted from ρ to $\Lambda^*\rho$, and it is denoted by I(ρ;Λ^*). Ohya introduced the mutual entropy I(ρ;Λ^*) by using the compound state expressing the correlation between ρ and $\Lambda^*\rho$.

We here discuss about the quantum mutual entropy was introduced by Ohya [9] using by a compound state expressing the correlation between ρ and $\Lambda^*\rho$. We here review this definition.

Let us denote the Schatten decomposition [20,26] of ρ by

$$\rho = \sum_n \lambda_n E_n,$$

where λ_n is the eigenvalue of ρ with $\lambda_1 \geq \lambda_2 \geq, \cdots$ and E_n is the projection operator to the subspace generated by the eigenvector x_n of λ_n, that is, $E_n = |x_n><x_n|$. This

decomposition is unique if and only if no eigenvalue of ρ is degenerate. The compound state [9] expressing the correlation between ρ and $\Lambda^*\rho$ has been given by

$$\sigma_E = \sum_n \lambda_n E_n \otimes \Lambda^* E_n$$

for a decomposition $E = \{E_n\}$. This compound state does depend on the decomposition of ρ, that is, on the choice of $E = \{E_n\}$. The reference [1] explains the relation between this compound state and a lifting in quantum Markov chain.

The mutual entropy with respect to an input state ρ and a communication channel Λ^* is defined by

$$I(\rho; \Lambda^*) \equiv \sup\{S(\sigma_E|\sigma_0); E = \{E_n\}\},$$

where $\sigma_0 = \rho \otimes \Lambda^*\rho$ is a trivial compound state and $S(\sigma_E|\sigma_0)$ is relative entropy [25] of σ_E from σ_0 given by

$$S(\sigma_E|\sigma_0) \equiv tr\sigma_E(\log \sigma_E - \log \sigma_0).$$

Then this mutual entropy satisfies the similar properties of mutual information of Shannon [21,25], and the following fundamental inequality of Shannon's type is essential here.

Theorem 2.1[9]: *$0 \leq I(\rho; \Lambda^*) \leq \min\{ S(\rho), S(\Lambda^*\rho)\}$* .

§3. Efficiency of Modulations With Coherent State

The optical modulation is a method to produce an optical signal expressing a message generated in an input source. This optical signal is described by a quantum coherent state in the mathematical framework. A optical channel transmitting the optical signal from the input system to the output system is a kind of the quantum mechanical channel Λ^*. We obtain the mutual entropy $I(\rho_M;\Lambda^*)$ with respect to the modulated state ρ_M and the optical channel Λ^*. When the channel is fixed, it is natural for us to consider a transmission efficiency by the ratio

$$r(\rho_M) \equiv \frac{I(\rho_M;\ \Lambda^*)}{S(\rho_M)}.$$

By using this ratio, the efficiency of modulations for photon number state is discused in [29]. In this paper, we use a digital symbols {0,1} as the alphabets of the input source. The digital symbol 0 (resp. 1) is denoted by each quantum coherent state $E_{M(0)}$ (resp. $E_{M(1)}$) depending on each optical modulation M. When the probability distribution of the input source is given by P = {λ, 1-λ}, the modulated state ρ_M is expressed by

$$\rho_M = \lambda E_{M(0)} + (1 - \lambda) E_{M(1)}.$$

We here discuss about the efficiency of optical modulations (PPM, ASK(OOK), FSK, PSK)[3,13]. Optical modulators generate the coherent states corresponding to the input signals and it is generally used that the following methods of optical modulations:

(1) ASK (Amplitude-Shift-Keying)(OOK (On-Off-Keying))

ASK is the method to carry the informaion on the Amplitude of the coherent state corresponding to the inputs signals $\{0,1\}$.

(2) FSK (Frequency-Shift-Keying)

FSK is the method to carry the information on the frequency of the coherent state corresponding to the input signals $\{0,1\}$.

(3) PSK (Phase-Shift-Keying)

PSK is the method to carry the information on the Phase of the coherent state corresponding to the input signals $\{0,1\}$.

(4) PPM (Pulse Position Modulation)

PPM is a method to carry the information on the position generating the optical pulse corresponding to the value of an input signal.

In the sequel, we write each modulated state as follows: (1) ρ_{PPM}^{Photon} with photon number state for PPM, (2) ρ_{PPM} with coherent state for PPM, ρ_{OOK} with coherent state for OOK, ρ_{FSK} with coherent state for FSK, ρ_{PSK} with coherent state for PSK. Each input signal is represented as

(1) PPM with photon number state

Input signal 0 $\rightarrow E_{PPM(0)}^{Photon} = |x_p^{(1)}><x_p^{(1)}| \otimes |x_0^{(1)}><x_0^{(1)}|$

Input signal 1 $\rightarrow E_{PPM(1)}^{Photon} = |x_0^{(1)}><x_0^{(1)}| \otimes |x_p^{(1)}><x_p^{(1)}|$

$$\rho_{PPM}^{Photon} = \lambda E_{PPM(0)}^{Photon} + (1-\lambda) E_{PPM(1)}^{Photon}.$$

(2) PPM with coherent state

Input signal 0 $\rightarrow E_{PPM(0)} = |\theta_{PPM}><\theta_{PPM}| \otimes |x_0^{(1)}><x_0^{(1)}|$

Input signal 1 $\rightarrow E_{PPM(1)} = |x_0^{(1)}><x_0^{(1)}| \otimes |\theta_{PPM}><\theta_{PPM}|$

$$\rho_{PPM} = \lambda E_{PPM(0)} + (1-\lambda) E_{PPM(1)}.$$

$$\theta_{PPM} = |\theta| \exp(i\omega)$$

(3) OOK(ASK) with coherent state

Input signal 0 $\rightarrow E_{OOK(0)} = |x_0^{(1)}><x_0^{(1)}|$

Input signal 1 $\rightarrow E_{OOK(1)} = |\theta_{OOK}><\theta_{OOK}|$

$$\rho_{OOK} = \lambda E_{OOK(0)} + (1-\lambda) E_{OOK(1)}.$$

$$\theta_{OOK} = |\theta| \exp(i\omega)$$

(4) FSK with coherent state

Input signal 0 $\rightarrow E_{FSK(0)} = |\theta_{FSK(0)}><\theta_{FSK(0)}|$

Input signal 1 $\rightarrow E_{FSK(1)} = |\theta_{FSK(1)}><\theta_{FSK(1)}|$

$$\rho_{FSK} = \lambda E_{FSK(0)} + (1-\lambda) E_{FSK(1)}.$$

$$\theta_{FSK(0)} = |\theta| \exp(i\omega)$$
$$\theta_{FSK(1)} = |\theta| \exp(i\omega_{FSK(1)})$$

(5) PSK with coherent state

Input signal 0 $\rightarrow E_{PSK(0)} = |\theta_{PSK(0)} >< \theta_{PSK(0)}|$
Input signal 1 $\rightarrow E_{PSK(1)} = |\theta_{PSK(1)} >< \theta_{PSK(1)}|$

$$\rho_{PSK} = \lambda E_{PSK(0)} + (1-\lambda) E_{PSK(1)},$$

$$\theta_{PSK(0)} = |\theta| \exp(i\omega)$$
$$\theta_{PSK(1)} = |\theta| \exp(i\omega + i\kappa_{PSK(1)}).$$

For the above modulations, we obtain the following results:

(1) For the input state ρ_M (M $= \{PPM, OOK, FSK, PSK\}$), we have the spectral decomposition of ρ_M such that

$$\rho_M = \|\rho_M\| F_{M(0)} + (1 - \|\rho_M\|) F_{M(1)},$$

where $F_{M(0)}$ and $F_{M(1)}$ are given as follows: Let $f(t) = 1 - 4\lambda(1-\lambda)(1-\exp(t))$, $g(t) = \frac{1-\sqrt{f(t)}}{2(1-\exp(t))}$. Then for PPM, we put

$$\begin{aligned} F_{PPM(0)} &= (a_{PPM}|x_0^{(1)} \otimes \theta_{PPM} > + b_{PPM}|\theta_{PPM} \otimes x_0^{(1)} >) \\ &\quad \times (a_{PPM} < x_0^{(1)} \otimes \theta_{PPM}| + b_{PPM} < \theta_{PPM} \otimes x_0^{(1)}|) \\ F_{PPM(1)} &= (c_{PPM}|x_0^{(1)} \otimes \theta_{PPM} > + d_{PPM}|\theta_{PPM} \otimes x_0^{(1)} >) \\ &\quad \times (c_{PPM} < x_0^{(1)} \otimes \theta_{PPM}| + d_{PPM} < \theta_{PPM} \otimes x_0^{(1)}|) \end{aligned}$$

and for $\{OOK, FSK, PSK\}$, we put

$$F_{M(0)} = (a_M|\theta_{M(0)} > + b_M|\theta_{M(1)} >)(a_M < \theta_{M(0)}| + b_M < \theta_{M(1)}|)$$
$$F_{M(1)} = (c_M|\theta_{M(0)} > + d_M|\theta_{M(1)} >)(c_M < \theta_{M(0)}| + d_M < \theta_{M(1)}|),$$

where

$$a_M = a_{M1} + a_{M2}i, \; b_M = b_{M1} + b_{M2}i$$
$$c_M = c_{M1} + c_{M2}i, \; d_M = d_{M1} + d_{M2}i.$$

$\|\rho_M\|$ is given as follows:

$$\|\rho_M\| = \frac{1}{2}\{1 + \sqrt{f(-2|\theta_M|^2)}\}.$$

Moreover let

$$\begin{aligned} h_{PPM} &= -2|\theta|^2 \\ h_{OOK} &= -|\theta|^2 \\ h_{FSK} &= -|\theta_{FSK0} - \theta_{FSK1}|^2 \\ h_{PSK} &= -|\theta_{PSK0} - \theta_{PSK1}|^2 \\ p_{PPM} &= \exp(h_{PPM}) \\ p_{OOK} &= \exp(h_{OOK}) \end{aligned}$$

and let

$$\begin{aligned} p_{FSK} &= \exp\{-\frac{1}{2}(|\theta_{FSK(0)}|^2 + |\theta_{FSK(1)}|^2)\}Re(\exp(\theta_{FSK(0)}\overline{\theta_{FSK(1)}})) \\ p_{PSK} &= \exp\{-\frac{1}{2}(|\theta_{PSK(0)}|^2 + |\theta_{PSK(1)}|^2)\}Re(\exp(\theta_{PSK(0)}\overline{\theta_{PSK(1)}})) \\ q_{FSK} &= \exp\{-\frac{1}{2}(|\theta_{FSK(0)}|^2 + |\theta_{FSK(1)}|^2)\}Im(\exp(\theta_{FSK(0)}\overline{\theta_{FSK(1)}})) \\ q_{PSK} &= \exp\{-\frac{1}{2}(|\theta_{PSK(0)}|^2 + |\theta_{PSK(1)}|^2)\}Im(\exp(\theta_{PSK(0)}\overline{\theta_{PSK(1)}})). \end{aligned}$$

Here

$$\begin{aligned} |a_M|^2 &= \frac{\lambda - g(h_M)}{\sqrt{f(h_M)}} \\ |b_M|^2 &= \frac{1 - \lambda - g(h_M)}{\sqrt{f(h_M)}} \\ a_{M1}b_{M1} + a_{M2}b_{M2} &= \frac{\{2\lambda(1-\lambda) - g(h_M)\}p_M}{\sqrt{f(h_M)}} \\ a_{M1}b_{M1} - a_{M2}b_{M2} &= \frac{\{2\lambda(1-\lambda) - g(h_M)\}q_M}{\sqrt{f(h_M)}} \end{aligned}$$

$$\begin{aligned} |c_M|^2 &= \frac{\lambda - \|\rho_M\||a_M|^2}{1 - \|\rho_M\|} \\ |d_M|^2 &= \frac{\lambda - \|\rho_M\||b_M|^2}{1 - \|\rho_M\|} \\ c_{M1}d_{M1} + c_{M2}d_{M2} &= \frac{-\|\rho_M\|(a_{M1}b_{M1} + a_{M2}b_{M2})}{1 - \|\rho_M\|} \\ c_{M1}d_{M1} - c_{M2}d_{M2} &= \frac{-\|\rho_M\|(a_{M1}b_{M1} - a_{M2}b_{M2})}{1 - \|\rho_M\|}. \end{aligned}$$

For the input state ρ_M, we have the von Neumann entropy $S(\rho_M)$ and the output state $\Lambda^*\rho_M$ such that

$$S(\rho_M) = -\|\rho_M\|\log\|\rho_M\| - (1 - \|\rho_M\|)\log(1 - \|\rho_M\|).$$

For PPM, we have

$$\Lambda^* \rho_{PPM} = \lambda |x_0^{(2)} >< x_0^{(2)}| \otimes |\sqrt{\eta}\theta_{PPM} >< \sqrt{\eta}\theta_{PPM}|$$
$$+ (1-\lambda)|\sqrt{\eta}\theta_{PPM} >< \sqrt{\eta}\theta_{PPM}| \otimes |x_0^{(2)} >< x_0^{(2)}|$$

and we obtain for $\{OOK, FSK, PSK\}$

$$\Lambda^* \rho_M = \lambda |\sqrt{\eta}\theta_M >< \sqrt{\eta}\theta_M|$$
$$+ (1-\lambda)|\sqrt{\eta}\theta_M >< \sqrt{\eta}\theta_M|.$$

For the output state $\Lambda^* \rho_M$, we get the von Neumann entropy $S(\Lambda^* \rho_M)$ of $\Lambda^* \rho_M$ such that

$$S(\Lambda^* \rho_M) = -\|\Lambda^* \rho_M\| \log \|\Lambda^* \rho_M\| - (1 - \|\Lambda^* \rho_M\|) \log(1 - \|\Lambda^* \rho_M\|),$$

$$\|\Lambda^* \rho_M\| = \frac{1}{2}\{1 + \sqrt{f(\eta h_M)}\}.$$

For the output states $\Lambda^* F_{M(0)}$ and $\Lambda^* F_{M(1)}$, we have the entropies of $\Lambda^* F_{M(0)}$ and $\Lambda^* F_{M(1)}$ such as

$$S(\Lambda^* F_{M(0)}) = -\|\Lambda^* F_{M(0)}\| \log \|\Lambda^* F_{M(0)}\| - (1 - \|\Lambda^* F_{M(0)}\|) log(1 - \|\Lambda^* F_{M(0)}\|),$$
$$S(\Lambda^* F_{M(1)}) = -\|\Lambda^* F_{M(1)}\| log \|\Lambda^* F_{M(1)}\| - (1 - \|\Lambda^* F_{M(1)}\|) log(1 - \|\Lambda^* F_{M(1)}\|),$$

$$\|\Lambda^* F_{M(0)}\| = \frac{1}{2}\{1 + \sqrt{1 - 4|a_M|^2|b_M|^2(1 - \exp(-\frac{1}{2}(1-\eta)h_M))(1 - \exp(\eta h_M))}\}$$

$$\|\Lambda^* F_{M(1)}\| = \frac{1}{2}\{1 + \sqrt{1 - 4|c_M|^2|d_M|^2(1 - \exp(-\frac{1}{2}(1-\eta)h_M))(1 - \exp(\eta h_M))}\}$$

Then we obtain the Ohya's mutual entropy $I(\rho_M; \Lambda^*)$ with respect to ρ_M and Λ^* such that

$$I(\rho_M; \Lambda^*) = S(\Lambda^* \rho_M) - \|\rho_M\| S(\Lambda^* F_{M(0)}) + (1 - \|\rho_M\|) S(\Lambda^* F_{M(1)}).$$

For the input state ρ_M, we get the ratio $r(\rho_M)$ such as

$$r(\rho_M) = \frac{I(\rho_M; \Lambda^*)}{S(\rho_M)}$$
$$= \frac{S(\Lambda^* \rho_M) - \|\rho_M\| S(\Lambda^* F_{M(0)}) + (1 - \|\rho_M\|) S(\Lambda^* F_{M(1)})}{S(\rho_M)}.$$

For the above notations, we obtain the followings:

Lemma 3.1

$$S(\rho_{PPM}^{Photon}) \geq S(\rho_{PPM}) \geq S(\rho_{OOK}) \geq S(\rho_{FSK}) \geq S(\rho_{PSK}).$$

Lemma 3.2

$$I(\rho_{PPM}^{Photon};\Lambda^*) \geq I(\rho_{PPM};\Lambda^*) \geq I(\rho_{OOK};\Lambda^*) \geq I(\rho_{FSK};\Lambda^*) \geq I(\rho_{PSK};\Lambda^*).$$

By using the above lemmas, we get the following theorem.

Theorem 3.3

$$r(\rho_{PPM}^{Photon}) \geq r(\rho_{PPM}) \geq r(\rho_{OOK}) \geq r(\rho_{FSK}) \geq r(\rho_{PSK}).$$

Acknowledgments

The author wishes to express his hearty gratitude to Professor Masanori Ohya for his valuable suggestions and advice.

References

[1] L. Accardi and M. Ohya: "Compound channels, transition expectations and liftings, in preparation.
[2] C.W. Helstrom, J.W.S. Liu and J.P. Gordon: "Quantum mechanical communication theory", Proc. IEEE 58, 1578- 1598 (1970).
[3] O. Hirota: "Introduction to Optical Communication Theory (in Japanese)", Morikita Pub. Comp., Tokyo (1985).
[4] A.S. Holevo: "Information theoretical aspects of quantum measurment", Problemy Peredaci Informacii 9, 31-42 (1973).
[5] R.S. Ingarden and K. Urbanik: "Quantum information thermodynamics", Acta Phys. Polon. 21, 281-304 (1962).
[6] R.S. Ingarden: "Information theory and thermodynamics of light, Part I and Part II", Fortschritte der Physik, 12, 567-594 (1964) and 13, 755-805 (1965).
[7] G. Lindblad: "Completely positive maps and entropy inequalities", Commun. Math. Phys. 40, 147-151 (1975).
[8] M. Ohya: "Quantum ergodic channels in operator algebras", J. Math. Anal. Appl. 84, 318-328 (1981).
[9] M. Ohya: "On compound state and mutual information in quantum information theory", IEEE Trans. Infor. Theory, 29 770-774 (1983).
[10] M. Ohya: "Note on quantum probability", L. Nuovo Cimento 38, 402-404 (1983).
[11] M. Ohya: "Entropy transmission in C*-dynamical systems", J. Math. Anal. Appl. 100, 222-235 (1984).
[12] M. Ohya: "State change and entropies in quantum dynamical syatems", Springer Lecture Note in Math., 1136, 397-408 (1985).
[13] M. Ohya: "Optical Communication Theory and Their Applications (in Japanese)",

Edited by Researcher on Optical Communication Theory, Morikita Pub. Comp., Tokyo (1986).

[14] M. Ohya: "Some aspects of quantum information theory and their applications to irreversible processes", Reports on Mathematical Physics, 27, 19-47 (1989).

[15] M. Ohya: "Information dynamics and its applications to optical communication processes", in this volume.

[16] M. Ohya and N. Watanabe: "Construction and analysis of a mathematical model in quantum communication processes (in Japanese)", IECE of Japan, J67-A, No.6, 548-552 (1984).

[17] M. Ohya, H. Yoshimi and O. Hirota: "Rigorous derivation of error probability in quantum control communication processes (in Japanese)", IEICE of Japan, J71-B, No4, 533-539 (1986).

[18] M. Ohya and H. Suyari: "Optimization of error probability in uantum control communication processes (in Japanese)", IEICE of Japan, J73-B-I, No.3, 200-207, (1990).

[19] M. Ohya and H. Suyari: "Rigorous derivation of error probability in coherent optical communication", in this volume.

[20] R. Schatten: "Norm Ideals of Completely Continuous Operators", Springer-Verlag (1970).

[21] C.E. Shannon: "A mathematical theory of communication", Bell Syatem Tech. J. 27, 379-423 and 623-656 (1948).

[22] H. Takahashi: "Information theory of quantum mechanical channels", Advances in Communication Systems 1, Academic Press 227-310 (1966).

[23] H. Uhlmann: "Relative entropy and the Wigner-Yanase- Lieb concavity in interpolation theory", Commun. Math. Phys. 54, 21-32 (1977).

[24] H. Umegaki: "Conditional expectation in an operator algebra, IV (entropy and information)", Kodai Math. Sem. Rep. 14, 59-85 (1962).

[25] H. Umegaki and M. Ohya: "Entropies in Probabilistic Systems (in Japanese)", Kyoritsu Pub. Comp. (1983).

[26] H. Umegaki and M. Ohya: "Quantum Mechanical Entropies (in Japanese)", Kyoritsu Pub. Comp. (1984).

[27] H. Umegaki, M. Ohya and F. Hiai: "Introduction to Operator Algebras (in Japanese)", Kyoritsu Pub. Comp. (1985).

[28] J. von Neumann: "Die Mathematischen Grundlargen der Quantenmechanik", Springer-Berlin (1932).

[29] N. Watanabe: "Efficiency of optical modulations for photon number states (edited by L. Accardi and W. von Waldenfels) ", KLUWER Pub. Comp. (1991).

AN OPTICAL SYSTEM AT THE QUANTUM LIMIT OF COHERENT DETECTION : THE WIND LIDAR

M. OUHAYOUN
S.A.T
19, Rue de la Gare
94230 CACHAN

I - INTRODUCTION

There is now a large interest in the scientific community for improved understanding of the atmospheric wind profiles, especially in the troposphere : this should help weather forecast, lead to better knowledge of large scale atmospheric circulation as well as climate dynamics.

The major probing system now under preparation for the next decades is a space borne CO_2 Doppler lidar that should provide precise data on wind velocities, 500 to 700 km away from the ground.

This is by now the only technically mature device available for that purpose. But, as it will be accounted for below, this lidar is expected to work at the quantum limit of coherent (heterodyne) detection, due to very faint aerosol backscattering.

II - SYSTEM DESCRIPTION [1, 2]

The wind lidar consists of a pulsed CO_2 laser emitter, transmit and receive telescope, heterodyne detection and signal processing units (fig. 1).

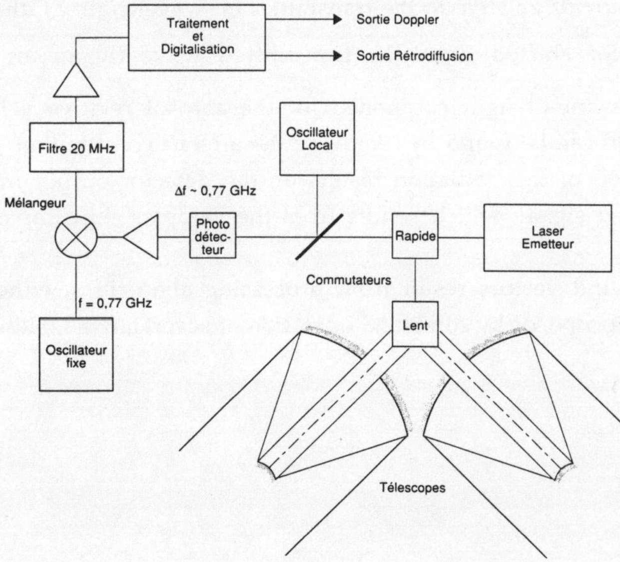

FIG. 1

Lecture Notes in Physics, Vol. 378
C. Bendjaballah, O. Hirota, S. Reynaud (Eds.),
Quantum Aspects of Optical Communications

Operation of the System is shown on fig. 2.

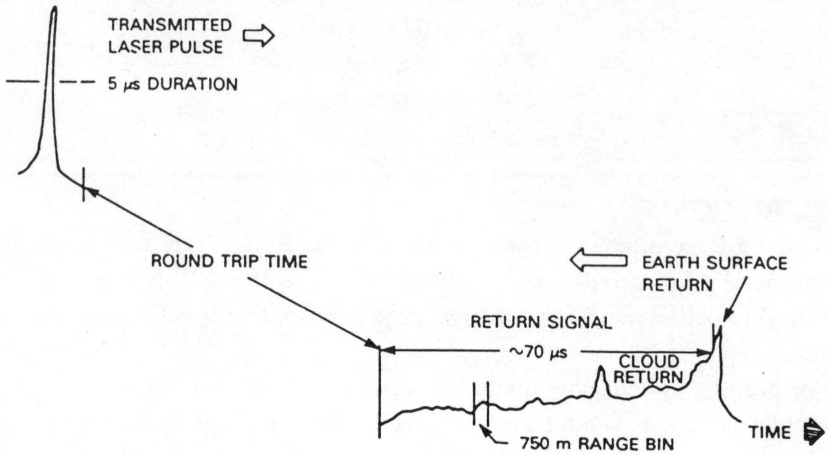

FIG.2

A laser pulse is directed through the telescopes in the atmosphere, and a small fraction of the incident radiation is backscattered by naturally moving aerosols. This radiation is collected by the same telescopes that were used to transmit the laser pulse and combined with the radiation from a local oscillator laser for heterodyne detection.

Since aerosols are moving relative to the transmitter (at wavelength λ), the backscattered radiation is Doppler shifted from the transmitted wave by an amount $\Delta\gamma = 2\,\frac{v}{\lambda}$ proportional to the line-of-sight component of the aerosol relative velocity (v) from which the air motion can be found by removing the aircraft contribution.

Heterodyne detection of this radiation results on the detector output in the form of a frequency modulated signal, which results from the Doppler shift imparted by the air motion.

True (horizontal) wind vectors result from processing algorithms, either by scanning through a single telescope, or by combined operation of several, fixed telescopes (fig.3).

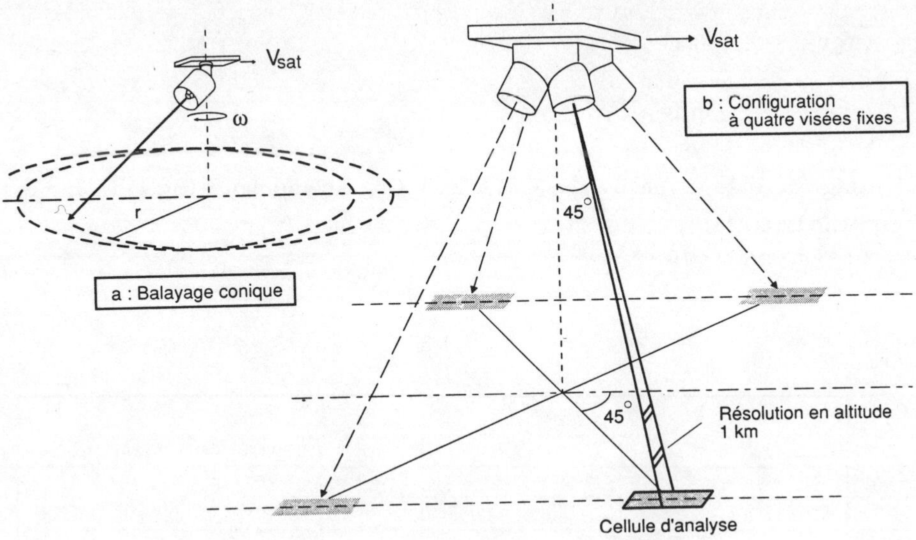

FIG.3

Some wind lidars at shorter infrared wavelengths (e.g. $\lambda \sim 2\ \mu m$) including solid-state lasers are now also considered, but we will focus here on CO_2 lidars ($\lambda \sim 10\ \mu m$) where the technology is most mature.

III - REQUIREMENTS ON THE LASER TRANSMITTER

Prediction of system requirements can be first assessed with some simple arguments :

A - Evaluation of the required transmitter energy [3]

The radar equation relevant for the wind lidar is :

$$\frac{P_R}{P_E} = A_a\, A_f\, \eta_o\, \beta LS/R^2$$

where :

P_E	= emitted optical power
P_R	= received optical power
A_a	= atmospheric attenuation
$(A_f\, \eta_o)$	= lidar efficiency
$\beta(R)$	= aerosol backscatter coefficient
L	= atmospheric sample thickness
S	= telescope collecting area
R	= range

with :

$A_a \sim 1$

$(A_f \eta_o)$ = 0.63

L = 450 m

S = 0.38 m^2

R = 750 km

And using the date given by the CROUZET CO_2 anemometer (fig.4) in the 3-13 km atmospheric layers [4].

$$10^{-19} \lesssim P_R/P_E \lesssim 10^{-18}$$

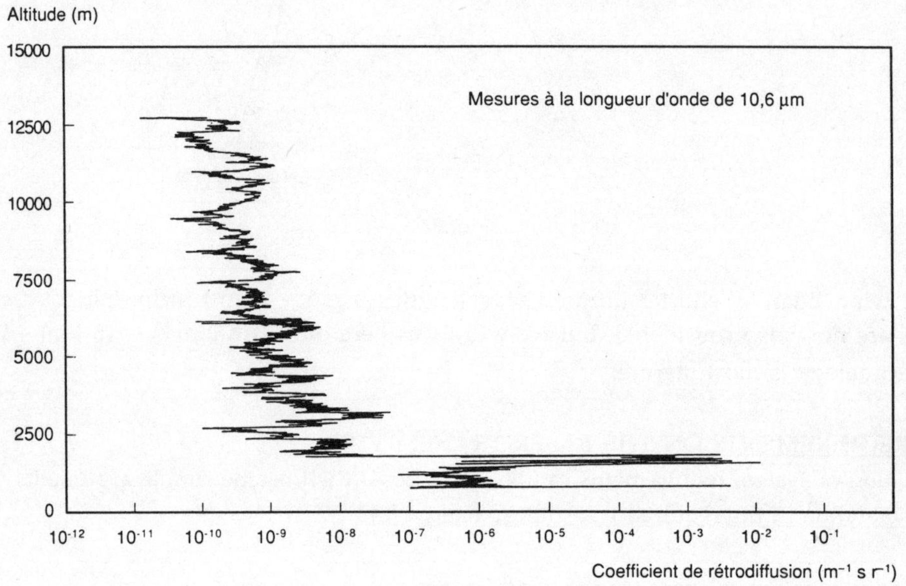

FIG.4

Taking also into account the signal to noise ratio in heterodyne detection as :

$$(SNR) = \frac{\eta_h \eta_q P_R}{F(h\nu)B} = \frac{1}{B}\frac{P_R}{(NEP)}$$

where :

η_h = heterodyne mixing efficiency

$\bar{\eta}_q$ = detector efficiency

F = preamplifier noise figure

$h\nu$ = photon energy

(NEP) = Noise Equivalent Power

B = detection bandwidth

Using :

η_h = 0.5

η_q = 0.6 (HgCdTe/PV)

F = 1.5

B = 360 kHz

Gives : (NEP) = 10^{-19} W/Hz

For a B = 360 kHz bandwidth, a (SNR) = 10 is equivalent to :

$P_R = 3{,}6\ 10^{-13}$ W

The required : laser transmitter power must then be :

$0.4\ \text{MW} < P_E < 4\ \text{MW}$

Or for a ~2 µS wide pulse :

$$\boxed{0.8\ \text{J} \leq W_E \leq 8\ \text{J}}$$

B - Other requirements

. Measuring wind velocities with sufficient accuracy (~1 m/s) requires that both lasers, transmitter and local oscillator, are frequency stable (± 100 kHz) according to the Doppler shift formula during and between consecutive measurement periods.

This is currently achievable for a CW, CO_2 local oscillator of moderate (P ~1 W) power by active control of the resonator length, but much more difficult in the case of the pulsed, high peak power transmitter.

. The lidar will be spaceborne for at least 3 years operation periods. This is equivalent to a 10^8 - 10^9 shots lifetime for the pulsed transmitter, which needs a continuous control of the lasing gas chemistry at high electrical injection level (~200 J/l atm).

C - Status of laser transmitter concepts

To fulfill the energy/single frequency requirements two solution are pursued. They both involve rapid discharges (~100 ns) on large optical beams (40 - 100 mm wide) for low frequency chirps :

- A high energy, CW injection locked, single TEA oscillator,
- An oscillator + amplifier sequence.

The last solution allows to achieve separately the frequency and energy specification. The oscillator could either be a conventional TEA CO_2 laser, with means for shot-to-shot frequency control, or preferably a TEA laser with internal frequency locking by four-wave mixing [5].

Meeting the three requirement of energy, spectral purity and lifetime is a very difficult challenge, beyond the status of present CO_2 laser technology. Many laboratories in the world are working now to achieve that goal in the next decades.

IV - FUTURE TRENDS

Much effort should be paid to release the power requirement on the laser transmitter to turn to more realistic objectives, in term of frequency stability and lifetime :

. "Classical" optical preamplification before detection, which is equivalent in theory to a detecting efficiency $\eta_h \eta_q = 1$ should be studied.

. Any proposal to overcome the classical limitation of coherent detection, i.e. the shot noise limit of the local oscillator, including measurements with squeezed quadratures of the electromagnetic field will be welcome.

REFERENCES

[1] Laser Atmospheric Wind Sounder (LAWS)
Instrument Panel Report, NASA-TM-101204 (1987)

[2] Bilan Energetique du Système Tropical (BEST), CNES Report (1988)

[3] SAT Report (CNES/89/5878/00)

[4] B.MORBIEU, J. MANDLE
3d International Conference on Advanced Infrared Detectors and Systems, London, UK (1986)

[5] M. OUHAYOUN, Y.GUERN
"Laser mirror by degenerate four wave mixing in a saturable amplifier"
IEEE JQE 22, 11, 2150-3 (1986)

Simulation of Optimum Quantum Receiver

Kouichi YAMAZAKI

Dept. of Information & Communication Science,
Faculty of Engineering, Tamagawa Univ.
Tokyo, 194 JAPAN

1. Introduction

Recent progress in quantum communication has led the reliability of optical communication systems to so called standard quantum limit, which is determined by the quantum noise of the vacuum state [1]. There are two schemes to overcome the standard quantum limit. One is an application of signals in non-standard quantum states such as squeezed state and sub-Poisson state, and the other is an improvement of detection scheme of quantum signals [2]. The "quantum communication theory" as the basic theory was developed by Helstrom and others [3]. The quantum communication theory, however, has no information about physical quantities by which the signals are to be observed. So far, only a few implementable receivers have been clarified.

Dolinar proposed an implementable quantum receiver for the detection of binary coherent signals with the minimum error probability [4]. However, it seems to be impossible for the receiver to be operated from a technical viewpoint.

Kennedy proposed one whose performance is exponentially optimal [5]. Its error probability is worse than that of the optimal receiver by a half.

In this article, we modify these quantum receivers in order to realize the quantum communication overcoming the standard quantum limit. First, we relax the local-oscillator-controlling conditions of Dolinar receiver to make it more implementable. Effects of the relaxation on the error probability are clarified by computer simulation. Then, the quasi-optimal receiver proposed by Kennedy is compared with the optimal one by their likelihood ratios. As the result, it is shown that difference of their error probabilities arises from the difference of their efficiencies of energy-utilization for deciding signals. Then, it is proposed to apply sequential testing to the quantum receivers to improve the efficiency. Properties of the receivers are clarified.

2. Quantum Receivers

2-1. Minimum error prorbability

The minimum error probability is given for detection of binary signals with a priori probabilities q_0 and q_1, respectively, as follows [3]:

$$P_e = \frac{1}{2}[1 - \sqrt{1 - 4q_0q_1| < S_0|S_1 > |^2}]. \tag{1}$$

Here, $< S_0|S_1 >$ is the inner product of $|S_0 >$ with $|S_1 >$.

2-2. Super Homodyne receiver [5]

Kennedy proposed a modified homodyne detection scheme in which the requirements on the local laser are more strict. That is, complex amplitude of the local laser is the

Lecture Notes in Physics, Vol. 378
C. Bendjaballah, O. Hirota, S. Reynaud (Eds.),
Quantum Aspects of Optical Communications

same as that corresponding to one of two signals. This is referred as "super homodyne receiver". In Kennedy receiver, the homodyne detector adds the local field of amplitude equal to negative of one of two signals to the receiver field via a beam splitter. Then, the error probability is given by means of energy of a signal $< n >$ as follows:

$$P_e = \frac{1}{2} e^{-4<n>}. \tag{2}$$

This shows that, Kennedy receiver improves 3 dB over the standard quantum limit.

2-3. Dolinar receiver [4]

Dolinar proposed the receiver (detection and decision scheme) to achieve the minimum error probability for detecting binary coherent signals. The receiver can be considered as a modification of the super homodyne receiver by applying feedback control of local lasers. Dolinar receiver consists of a photon counter, two local lasers and their controller. Complex amplitude of each local laser $\ell_j(t)$ $(j = 0, 1)$ is controlled as follows:

$$\ell_1(t) = -s(t)[\frac{f_\gamma(t)+1}{f_\gamma(t)-1}], \; \ell_0(t) = s(t)[\frac{f_\gamma(t)+1}{f_\gamma(t)-1}], \tag{3}$$

$$f_\gamma(t) = \frac{(1+\gamma)^2}{2\gamma} e^{m(t)} - 1 + \frac{1+\gamma}{2\gamma}\sqrt{(1+\gamma)^2 e^{2m(t)} - 4\gamma e^{m(t)}}, \tag{4}$$

$$m(t) = C \int_0^t 4|s(\sigma)|^2 d\sigma, \tag{5}$$

where $\gamma (= q_1/q_0)$ is assumed to be equal to a unit here.

The operation of Dolinar receiver is explained as follows. First, one of the two local lasers is used. The signal field is detected after combined with the local field via the beam splitter. When one photon is counted at the photodetector, the local laser alternates with the other immediately. Then actual local field, which is to be combined with the signal, depends very trivially on the set of times $t_1 < t_2 < ...$ at which count occurs,

$$\hat{\ell}(t : \underline{t}) = \begin{cases} \ell_1(t), \underline{t} = (t_1, t_2, ..., t_n), & n \; even \\ \ell_0(t), \underline{t} = (t_1, t_2, ..., t_n), & n \; odd \end{cases}$$

Finally, a received signal is decided at the end of the bit duration according to whether the photon number counted in the duration is even or odd.

3. Relaxation of Locla-Lalser Control Conditions

In Dolinar receiver, strict conditions for controlling the local lasers as well as the unit quantum efficiency of photodetector are essentially significant to achieve the minimum error probability. They are

1) continuous control of amplitude of local lasers
2) immediate change of two local lasers.

It may be impossible to realize these conditions from a technical viewpoint. In this section, we consider more implementable receivers by relaxing the conditions of Dolinar receiver. First, we divide the bit duration T into M time intervals with duration t $(T = Mt)$. If one or more photons are counted in a time interval, the local laser

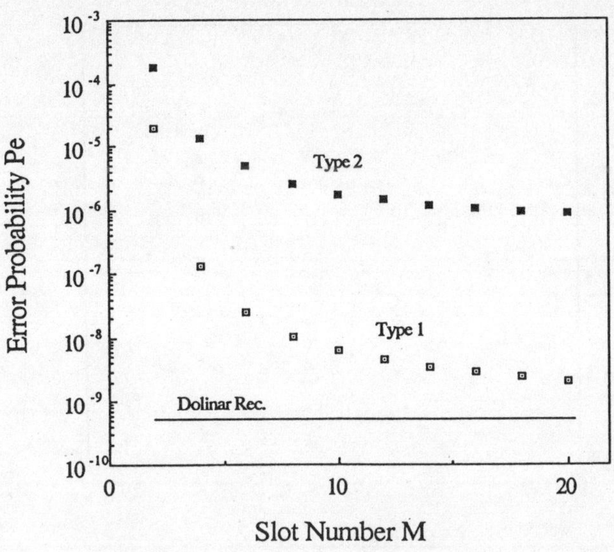

Figure 1: Effect of relaxation of local laser control conditions.

alternates with the other at the end of the time interval. This receiver does not require the immediate change of the local lasers ($Type1$). In this case, the amplitude of the local lasers is ideally controlled like Dolinar receiver. Next, we relax the condition to control the amplitude of local lasers. The amplitude of each laser is constant during each time interval, and it changes at the end of time interval discontinuously. The amplitude at each time interval is the value given by Eq.(3) at the mid-time of each time interval ($Type2$). Dolinar receiver corresponds to both $Type1$ and $Type2$ when division number M is infinity.

The properties of error probability of $Type1$ and $Type2$ receivers are shown for average photon number $<n>= 5$ in Fig.1. The vertical and the horizontal axes are error probability and division number, respectively. In $Type1$ receiver, the error probability approaches to that of Dolinar receiver rapidly as division number M increases. This is more remarkable for small number of average photon, because the probability which more than two photons are counted during each time interval is less than that for larger number of $<n>$. On the other hand, the error probability of $Type2$ receiver approaches to that of Dolinar receiver gradually. For the error probability of Dolinar receiver is characterized by the integral of squared sum of the signal and the local fields, and this cannot accords with that of Dolinar receiver for both S_0 and S_1 at once if the amplitude is changed by step-shaped function. It is found that the amplitude control of local lasers is essentially significant to achieve the minimum error probability.

In the practical system, the local lasers can not be changed immediately, and some non-zero time delay is unavoidable. That is, if two photons are counted in a short time interval, the controller may not respond to the later photon. We provide each time interval of $Type1$ receiver with the dead time in which controller does not change the local lasers even if photons are counted. Then, we consider the effect of the feedback delay time to change the local lasers on the error probability.

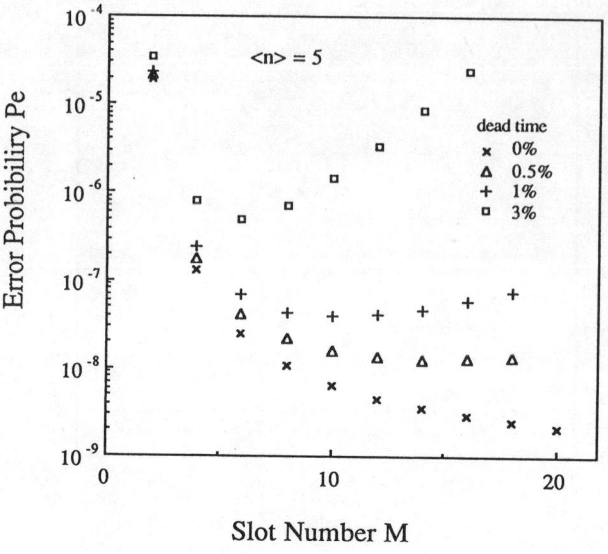

Figure 2: Effect of dead time of feedback.

The effect of dead time on the error probability is shown for average photon number $< n >= 5$ in Fig.2. Length of dead time is defined as the ratio of a dead time duration with a bit duration. In each length of dead time, there exists the optimum value of division number with which error probability is minimum, and this value increases as dead time decreases. Because the total length of dead time becomes larger compared with the bit duration as M increases. This result shows that bit duration of communication must be long compared with the dead time of the feedback control loop in order to achieve the minimum error probability.

4. Sequential Likelihood Ratio Testing for Quantum Receivers

4-1 Comparison between Dolinar and Kennedy Receivers

In order to analyze properties of error probability of the quantum receivers, we will show the cause of difference between error probabilities of Dolinar and the super homodyne receivers. Both of these two receivers use the maximum likelihood ratio method to decide the received signals. Then, we consider likelihood ratios of these receivers. In case of Dolinar receiver, the likelihood ratio of detecting signals is expressed by means of the local-laser weighting function $f_\gamma(t)$ in Eq.(4) as follows:

$$\gamma\Lambda(t:\underline{t}) = [f_\gamma(t)]^{(-1)^n}, \; \underline{t} = (t_1, t_2, ..., t_n), \tag{6}$$

where n is the photon number detected within $[0, t]$, and $\underline{t}$ is the set of photon counting time $(t_1, t_2, ..., t_n)$. Assuming that γ is equal to a unit, $ln\,\Lambda(t:\underline{t})$ changes sign at each count, but its magnitude is a deterministically increasing function of time. It is schematically shown in Fig.(3-a).

In super homodyne receiver, probability which k photons are counted within [0,t], conditioned on a signal m_j being sent, is expressed as follows:

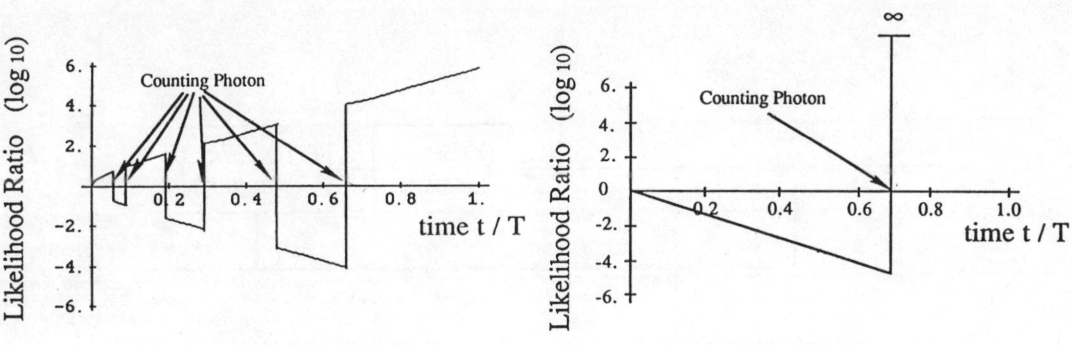

(a) Dolinar Receiver (b) Super Homodyne Receiver

Figure 3: Likelihood ratio of quantum receiver.

$$P(k:t|m_j) = \begin{cases} \left.\begin{matrix} 1, & k=0 \\ 0, & k>0 \end{matrix}\right\} & for\ j\ =0 \\ \\ \dfrac{[<n> \frac{t}{T}]^k}{k!} exp[-[<n> \dfrac{t}{T}] & for\ j=1\,. \end{cases} \tag{7}$$

Then, the likelihood ratio is given by

$$\Lambda(t|k) = \begin{cases} exp[-<n> \dfrac{t}{T}], & k=0\ in\ [0,t] \\ \infty, & k>0\ in\ [0,t]\,. \end{cases} \tag{8}$$

When m_0 is sent, the logarithm of likelihood ratio decreases from 0 monotonously as t increases. On the other hand, when m_1 is sent, the ratio jumps to infinity at the instant when the first photon is detected in the bit duration, and photons detected after that have no influence on the ratio. It is shown in Fig.(3-b).

Comparing Figs.(3-a) and (3-b), it is found that the logarithm of likelihood ratio of Dolinar receiver parts from a unit as t increases in spite of n, and its magnitude is the most at the end of the bit duration, while in the super homodyne receiver the ratio jumps to infinity if a photon is counted as mentioned before. This means that Dolinar receiver uses the energy detected in the bit duration efficiently to decide the signals, while the super homodyne receiver does not. This causes the different error probabilities between Dolinar and super homodyne receivers

It is the sequential testing that may improve the efficiency of energy utilization for deciding signals. Then, we propose to apply the sequential testing as a signal decision scheme to quantum receivers.

4.2 Sequential Testing for Quantum Receivers

The concept of sequential testing is the following [6],[7]. In the testing hypotheses, the number of observation , i.e. the size of the samples on which the test is based is not predetermined, and is a random valuable. That is, the number of observations required by the test depends on the outcome of the observations.

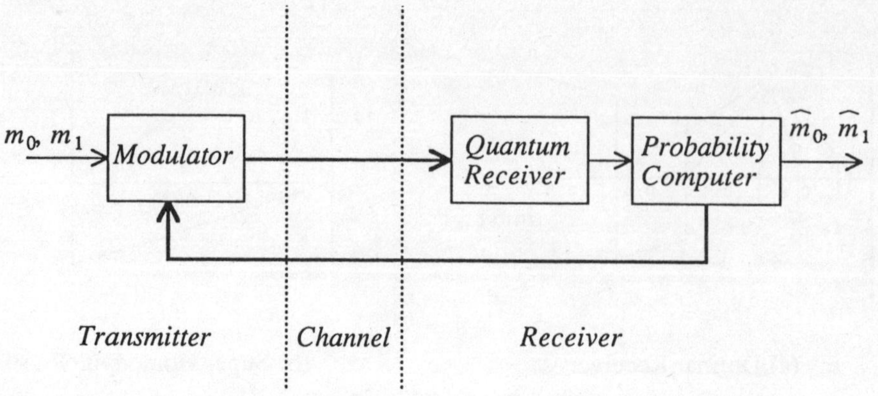

Figure 4: Schematic diagram of sequential quantum receiver.

In optical communication, the time within which outcome of photo detecter is integrated corresponds to the number of samples. Let $\Lambda(t)(= P(k : t|m_1)/P(k : t|m_0))$ be the likelihood ratio of a optical receiver, where, $P(k : t|m_j)$ is the probability which k photons are detected within $[0, t]$, conditioned on m_j being sent. For parameters A and B predetermined according to the required error probability, if

$$A < \Lambda(t) < B \tag{9}$$

the testing is continued by testing an additional observation. If

$$\Lambda(\mathrm{t}) \leq \mathrm{A} \tag{10}$$

the process is terminated with the decision that m_0 is sent. If

$$\Lambda(t) \geq B \tag{11}$$

the process is terminated with the decision that m_1 is sent.

At the instance that the testing is terminated, the receiver sends massage to tell the transmitter to stop sending signal during the corresponding bit duration. This may improve the efficiency to use optical energy for deciding signals. The schematic diagram of a optical communication system using the sequential testing is shown in Fig.(4).

(a) Dolinar receiver
As shown in Fig.(3), the value of the weighting function of Dolinar receiver, determining the probability ratio, depends only on the time t. Thus, the time when the testing is terminated is determined exactly by one time for predetermined parameters A and B. Therefore, the sequential testing is not available for Dolinar receiver.

(b) Super homodyne receiver
In the super homodyne receiver, the time terminating the test may be a random valuable since the likelihood ratio jumps to infinity at the time counting the first-photon, which occurs unexpectedly. Thus, the sequential testing is available for the super homodyne receiver. In the testing, what is important is to decide the decision parameters A and B. It is found in Fig.(3) that the receiver requires one photon to increase the ratio more

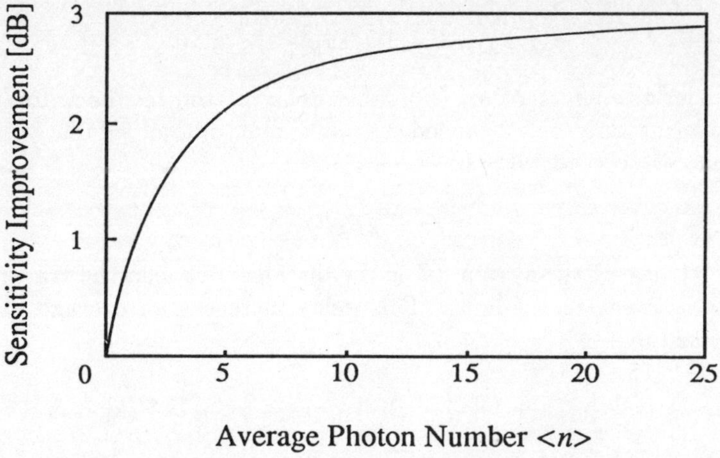

Figure 5: Sensitivity improvement of super homodyne receiver.

than a unit, and the photon results in the ratio to be infinity. This value continues until the end of the bit duration. So, A can be set to any positive value. To decide the signal m_0, however, we observe during a certain time interval in which no photon is counted. So, we have to determine the maximum interval $[0, T_{max}]$ within which the testing is done. The value of T_{max} depends on the required error probability P_e. Then we can determine the decision parameters A and B as follows:

$$\begin{aligned} \mathrm{A} &= C \quad (> 0), \\ B &= \ell n(2P_e)\,. \end{aligned} \tag{12}$$

We consider how much energy used to decide signals is saved by the sequential likelihood ratio testing for the super homodyne receiver. Assuming that the feedback for the testing is an ideal one without delay, the average photon number detected during the testing are $< n >$ and $1 - e^{-4<n>}$ for the signals m_0 and m_1, respectively. So the average photon number for the receiver is given by

$$< n >_{seq} = \frac{1}{2}(< n > +1 - exp[-4 < n >]). \tag{13}$$

This shows that the sequential testing improves the sensitivity of super homodyne receiver about 3dB for large number of $< n >$. The sensitivity improvement is shown in Fig.(5).

(c) Direct detection receiver
It is obvious from the result of the super homodyne receiver that it may be possible to improve the sensitivity more if one uses on-off keying (OOK) instead of BPSK as a modulation scheme. For no photons are sent when signal m_0 is sent. The OOK signal is detected by a direct detection receiver. For this system, the average photon number is described by rewriting Eq.(13) as follows:

$$< n >_{seq} = \frac{1}{2}(1 - exp[-2 < n >]) \,. \tag{14}$$

Here, we consider the effect of an avoidable delay of the feedback for the sequential testing. Considering only delay caused by a traveling optlcal signal, we can define the normalized delay time τ as follows:

$$\tau = 2R_b Z/c, \tag{15}$$

where R_b is a bit rate of the system, Z is the distance between the transmitter and the receiver, and c is the speed of light. This delay increases an average photon number, which is expressed by

$$\begin{aligned} < \mathrm{n} >_{\mathrm{seq}} = \tfrac{1}{2} \sum_{\mathrm{k}=0}^{\infty} (\mathrm{k}+1)[\textstyle\int_0^{1-\tau} \mathrm{P}(0| < \mathrm{n} > \mathrm{t}) < \mathrm{n} > \mathrm{P}(\mathrm{k}| < \mathrm{n} > \tau)\mathrm{dt} \\ + \textstyle\int_{1-\tau}^{1} P(0| < n > t) < n > P(k| < n > (1-t))dt] \end{aligned} \tag{16}$$

Fig.(6) shows improvement of the sensitivity for the direct detection receiver for $\tau = 0.01$, 0.2, 0.5, and 0.7. Noticing that an ordinary super homodyne receiver is about 3 dB higher sensitive than the direct detection receiver, it is found that this has higher sensitivity than that when the sequential testing is employed for the decision of the signals.

5. Conclusions

We have considered more implementable receivers than Dolinar receiver by relaxing the conditions to control local lasers of the receiver. The effect of these relaxation on the error probability was clarified. It was found that the bit duration of communication must be long compared with the time to change the local lasers to achieve the minimum error probability.

Next we proposed to apply the sequential testing to the optical receivers. We showed that this improves the sensitivity of quantum receivers except for Dolinar receiver, and that the direct detection receiver using the testing is the most sensitive receiver when the delay of the feedback can be negligible.

References

[1] B. F. Levine and C. G. Bethea, "Error Rates Measurement for Single photon Detection at 1.3μm," Appl. Phys. Lett. , vol.44, pp.649-650, Apr., 1984.

[2] O. Hirota, Optical Communication Theory, 1985, Tokyo:Morikita Pub. (in Japanese).

[3] C. W. Helstrom, Quantum Detection and Estimation Theory," 1976, New York, N.Y:Academic Press.

[4] S. J. Dolinar, "An Optimum Receiver for the Binary Coherent State Quantum Channel," M.I.T. Rec. Lav. Elec. Quard. Prog. Report, No.108, pp.219-225, Jan. 1973.

[5] R. S. Kennedy, "A Near-Optimum Receiver for the Binary Coherent State Quantum Channel," M.I.T. Rec. Lav. Elec. Quard. Prog. Report, No.111, pp.115-120, Oct. 1973.

[6] O. Hirota, "Quantum Sequential Probability Ratio Testing," IECE of Japan, Group Meeting Technical Report, vol.IT-83, pp.25-30, May, 1983 (in Japanese).

[7] A. Wald, Sequential Analysis, 1947, New York:Wiley.

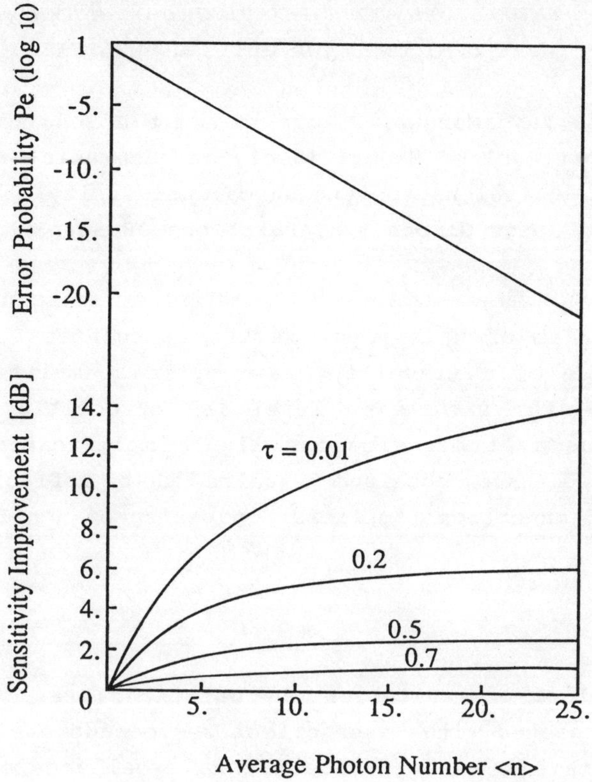

Figure 6: Sensitivity improvement of direct detection receiver.

GAIN, SATURATION AND NOISE FIGURE OF A TRAVELLING WAVE SEMICONDUCTOR OPTICAL AMPLIFIER

S.Ruiz-Moreno, J.Guitart and M.J.Soneira
Department of Signal Theory and Communications
E.T.S.I.T. Barcelona, Polytechnical University of Catalonia
c/ Jorge Girona s/n, Barcelona 08034, Spain

Abstract. The aim of this paper is the calculation of the gain and the noise figure of a travelling wave optical amplifier (TWOA) in a situation where the saturation level is appreciable. Once the non-linear equations that describe the light amplification process have been solved, we have obtained results that reflect the peculiar performances of an optical amplifier operating on non-linear regime.

1. Introduction

In the field of communications by optical fibre, the high technological level reached with laser effect semiconductors (GaAs) has made possible that this kind of device offers excellent performances whether as transmitter element (Fabry-Perot laser diode), as repeater element (optical amplifier) or even as preamplifier (optical preamplifier) placed just in front of the photodetector.

When the lateral facet reflectivities of the amplifier have an appreciable value, i.e., in the case of a Fabry-Perot optical cavity (FPOA), the spectral response shows an important ripple and the bandwidth remains limited by the frequency separation between two consecutive resonant modes. In an ideal situation, in which the reflectivities would be zero, the ripple desappears and then one deals with a travelling wave optical amplifier (TWOA). We will refer to a semiconductor laser amplifier which has been realized with very low residual facet reflectivities (lower than 10^{-4}). Such an optical device [1], [2] is called near travelling wave optical amplifier (NTWOA). Under these conditions the spectral response of the device shows a very small ripple (lower than ±0.5dB) and then the bandwidth depends on the spectral response of the semiconductor material [1], [4]. Due to that, the bandwidth offered by a NTWOA (some THz) is three orders of magnitude higher than the FPOA bandwidth (tens of GHz).

Lecture Notes in Physics, Vol. 378
C. Bendjaballah, O. Hirota, S. Reynaud (Eds.),
Quantum Aspects of Optical Communications

These high bandwidths that can be reached at the present time with NTWOA are much greater than the maximum transmission speeds (bit rate) achieved by modern optical fibre digital transmission systems (see, e.g.,[8]). Nevertheless, in spite of the fact that the optical bandwidth is sufficient, the problem appears when the incident optical power is enough to attain an appreciable saturation level of the amplifier. In these conditions the amplifier performances (gain and noise figure) are actually different from those predicted in linear conditions (absence of saturation).

The main goal of this communication is focused on the analysis of both the gain and the noise figure of a semiconductor optical amplifier with appreciable saturation levels.

In the following sections the equations chosen to characterize its non-linear behaviour are described. Concrete results are presented about gain and noise figure which have been obtained by means of the computer simulation of the 3691#14 optical amplifier used within the RACE Project No.1027, [6], [7]. In particular, it is shown how a gain reduction is accompanied by a considerable improvement in the amplifier noise figure.

2. Model of a saturated TWOA

Due to the statistical nature of the light as well as its amplification process, the tool chosen to characterize the non-linear behaviour of a semiconductor optical amplifier has been the photon density matrix equation (PDME), since its diagonal elements describe the variations of the statistical moments of the light during the amplification process [3], [9]. It becomes adequate to express the general equation for the rth order moment as

$$\frac{d\langle m^r\rangle}{dz} = \frac{1}{v}\sum_{j=1}^{r}\binom{r}{j-1}\Big[\Big\langle\frac{(m+1)m^{j-1}}{1+\ (m+1)s}\Big\rangle A + (-1)^{r+j-1}\Big\langle\frac{m^j}{1+ms}\Big\rangle B + (-1)^{r+j-1}\langle m^j\rangle C\Big] \qquad (1)$$

where z represents the amplification coordinate, i.e., $z=0$ (or $\mathcal{L}$) corresponds to the input (output) of an amplifier of length $\mathcal{L}$. In this equation $m(z)$ is the random variable expressing the photon number during the amplification process and $\langle\ \rangle$ represents the statistical mean value. In (1) A and B are the stimulated emission and absorption coefficients, respectively, C is a coefficient representing other loss mechanisms, independent of the saturation phenomenon, s is the saturation parameter and v is the speed of light in the amplifier medium . The photon number is counted per second and per unit of amplified spontaneous emission bandwidth (optical bandwidth), such that

$$m(z) = \frac{P(z)}{hf \cdot \Delta f} \tag{2}$$

being P(z) the optical power level and hf the photon energy. Since the work carried out in this paper has been realized for a wavelength corresponding to $\lambda \simeq 1.5 \mu m$ (third window), the photon energy is $hf \simeq 1.32 \cdot 10^{-19}$J. There exist several criteria in order to estimate the bandwidth of a NTWOA [1], [4]. Acting as in [4], one can demonstrate that a good estimation of the optical bandwidth is

$$\Delta f \simeq \frac{4}{3} \cdot \frac{c}{\lambda_p^2} \sqrt{\frac{g_p}{\gamma \Gamma}} \tag{3}$$

where c is the speed of light in the vaccum and λ_p is the wavelength at which the semiconductor presents the maximum non saturated net gain g_p per unit length. Constants γ and Γ are typical of the semiconductor selected and they represent, respectively, the parabolic coefficient and the confinement factor of the structure. The values of all these parameters, as well as those of A, B, C, s, $\mathcal{L}$ and v, are given in Table 1.

The relationship between the parameters A, B and C in (1) and the maximum non saturated net gain per unit length is

$$g_p = \frac{A-B-C}{v} \tag{4}$$

which allows us to define the unsaturated single pass gain, G_s^o, of the amplifier as

$$G_s^o = e^{g_p \mathcal{L}} = e^{(A-B-C)\frac{\mathcal{L}}{v}} \tag{5}$$

This last expression would correspond to the amplifier gain when its behaviour would be absolutely linear (s=0).

Equation (1) gives a coupled non-linear equations system for different values of r. In order to calculate the mean value and the variance of the photon number, two equations corresponding to the first and second order moments must be solved. Taking $r=1$ and $r=2$ in (1) one can obtain

$$\frac{d\langle m\rangle}{dz} = \frac{A}{v}\left\langle\frac{m+1}{1+(m+1)s}\right\rangle - \frac{B}{v}\left\langle\frac{m}{1+ms}\right\rangle - \frac{C}{v}\langle m\rangle \qquad (6)$$

$$\frac{d\langle m^2\rangle}{dz} = \frac{A}{v}\left\langle\frac{(m+1)(2m+1)}{1+(m+1)s}\right\rangle - \frac{B}{v}\left\langle\frac{(2m-1)m}{1+ms}\right\rangle - \frac{C}{v}\langle(2m-1)m\rangle \qquad (7)$$

The variance of the photon number can be written as

$$\sigma_m^2 = \langle m^2\rangle - \langle m\rangle^2 \qquad (8)$$

Then, the differential equation for the variance will be

$$\frac{d\sigma_m^2}{dz} = \frac{d\langle m^2\rangle}{dz} - 2\langle m\rangle\frac{d\langle m\rangle}{dz} \qquad (9)$$

We can use the set of equations (6) and (9) in order to calculate the mean value (power) and the variance (fluctuation), respectively, of the photon number. These equations are difficult to solve due to their double nature, statistical and non-linear, but they have the advantage that they take into account the saturation during the amplification process. Using the method developed in [5], we have calculated the mean and the variance at the amplifier output as a function of the incident optical power. Equations (6) and (9) have also been numerically solved by other authors [3].

3. Gain and saturation

Taking into account expression (2) we can relate the mean value of the photon number to the optical power. Then we define the photon mean number at the input and at the output of the amplifier as

$$\langle m(0)\rangle = \frac{P(0)}{hf\cdot\Delta f} = \frac{P_{IN}}{hf\cdot\Delta f} \qquad (10a)$$

and

$$\langle m(\mathcal{L})\rangle = \frac{P(\mathcal{L})}{hf\cdot\Delta f} = \frac{P_{OUT}}{hf\cdot\Delta f} \qquad (10b)$$

where P_{IN} and P_{OUT} are the average powers at the input and at the output of the amplifier, respectively. Then, we can calculate the amplifier gain G_s by means of

$$G_s = \frac{P_{OUT}}{P_{IN}} = \frac{\langle m(\mathcal{L})\rangle}{\langle m(0)\rangle} \tag{11}$$

In a situation where the saturation phenomenon is appreciable, the result obtained for G_s will depend, obviously, on both the incident optical power P_{IN} and the saturation parameter s. In this work we assume that the incident light is coherent and, therefore, its statistical mean value $\langle m(0)\rangle$ is equal to its variance $\sigma_m^2(0)$.

Once here, we consider that a commentary about expression (11) is important to make it now. From a strictly point of view, the value of P_{OUT} does not correspond exactly to the optical power of the amplified signal component. This is due to the unavoidable and inseparable components of spontaneous emission which are also present at the amplifier output. In consequence, expression (11) gives only a sufficiently precise estimation of the gain when the signal photon number is more higher than the spontaneous emission photon number involved in the amplification process. This really occurs when the mean value of the incident photon number is more larger than the spontaneous emission parameter of the amplifier. This parameter can be expressed as

$$\rho = \frac{A}{A-B-C} \tag{12}$$

which takes typical values between 2 and 3. For the set of values of Table 1 the resultant value is ρ=2.56 and, in consequence, we can consider that (11) gives very precise values when the incident optical power is above -13dBm (50μW) which approximately corresponds to 30 incident photons.

4. Small signal gain

When the mean number of the incident photons is low (low level of the optical power), expression (11) becomes inexact and it is then necessary to state an alternative formula. Solving equation (6) for s=0, it is easy to show that

$$\langle m(\mathcal{L})\rangle = G_s^o\langle m(0)\rangle + \rho(G_s^o-1) \tag{13}$$

where G_s^o and ρ have been respectively defined in (5) and (12). The expression for G_s^o is then

$$G_s^o = \frac{\langle m(\mathcal{L})\rangle + \rho}{\langle m(0)\rangle + \rho} \tag{14}$$

This expression is strictly valid only in the case of s=0 (linear amplification). Nevertheless, the authors of this paper have been able to prove that for appreciable values of s ($\simeq 10^{-4}$), even though with low values of the incident optical power, the expression

$$G_s = \frac{\langle m(\mathcal{L})\rangle + \rho}{\langle m(0)\rangle + \rho} \tag{15}$$

gives very accurate results whenever the value of $\langle m(\mathcal{L})\rangle$ has been precisely calculated by solving numerically the non-linear differential equations (6) and (7 or 9). As it is logical, if the optical power at the amplifier input is very high and the saturation is considerable, formula (15) has no sense since it has been formally deduced starting from a linear equation. On the other hand, it must be pointed out that in this case $\langle m(0)\rangle \gg \rho$ and formula (15) tends to expression (11). Definitively, it is a question of solving the problems that expression (11) presents for low levels of P_{IN}. This has been realized considering a linear formula, (15), but using exact values of $\langle m(\mathcal{L})\rangle$ in it.

5. Noise figure

When the signal photon number involved in the amplification process is high, we can define as a figure of merit of the amplified light on the z coordinate the ratio

$$Q(z) = \frac{\langle m(z)\rangle^2}{\sigma_m^2(z)}, \qquad 0 \le z \le L \tag{16}$$

Due to the non-linear behaviour shown by the optical amplifier when the saturation is appreciable, the Q ratio for every z will depend on the mean number of the incident photons $\langle m(0)\rangle$. We define the noise figure of the amplifier as

$$NF = \frac{Q(0)}{Q(z)} = \frac{\langle m(0)\rangle^2}{\sigma_m^2(0)} \cdot \frac{\sigma_m^2(\mathcal{L})}{\langle m(\mathcal{L})\rangle^2} \tag{17}$$

For those cases in which the incident light is coherent, the statistical distribution at the amplifier input follows a Poisson statistics and then $\sigma_m^2(0)=\langle m(0)\rangle$. This allows us to express the noise figure in the form

$$NF = \langle m(0)\rangle \frac{\sigma_m^2(\mathcal{L})}{\langle m(\mathcal{L})\rangle^2} \tag{18}$$

For the case where we have an approximately linear behaviour of the amplifier, a gain more larger than the unity and an adequate spectral filtering of the spontaneous emission at the amplifier output, then relation (18) can be approximated by 2ρ, [11]. This is a typical result when the incident optical power is lower than the saturation power of the amplifier [1], [6]. Nevertheless, when the optical power reaches certain values (10μW or more), the amplifier behaviour is nonlinear and the noise figure depends considerably on the input power. In these conditions it is necessary to calculate numerically the $\langle m(\mathcal{L})\rangle$ and $\sigma_m^2(\mathcal{L})$ values in order to found the NF value given by (18).

Concept	Not.	Values & Units	
Active layer length	$\mathcal{L}$	785	μm
Confinement factor	Γ	0.18	
Parabolic coefficient	γ	$6.2\cdot10^{18}$	cm^{-3}
Stimulated emission coeff.	A	$1.17\cdot10^{12}$	s^{-1}
Stimulated absorption coeff.	B	$3.375\cdot10^{11}$	s^{-1}
Linear losses	C	$3.75\cdot10^{11}$	s^{-1}
speed of light	v	$7.5\cdot10^{7}$	$m\cdot s^{-1}$
Peak wavelength	λ_p	1.53	μm
Peak net gain	g_p	61	cm^{-1}
Saturation parameter	s	$8.97\cdot10^{-5}$	
Optical Bandwidth	Δf	12.63 THz	

Table 1

6. Results and discussion

The results we present here have been obtained using computer simulations of a NTWOA which has been experimentally characterized by

[6]. The values of the device parameters are shown in Table 1. Thanks to the antireflecting techniques employed, the amplifier reflectivities are of the order of 10^{-4}, which guarantees a behaviour too close to the travelling wave.

The amplifier single pass gain (in dB) has been obtained by solving numerically equation (11). The results are shown in Fig.1 as a function of the incident optical power expressed in dBm. As it is natural, the maximum value of the gain (20.59dB) is obtained for the lower value of P_{IN} (-35dBm) we have used but, even in this case, the gain is slightly lower than the theoretical non saturated value G_s^o=20.8dB. This confirms that, in fact, the saturation always exists to a certain extent. From Fig.1 one can deduce that G_s has a 3dB decay for an input power of nearly -10dBm, i.e., $P_{IN}\simeq 100\mu W$. In consequence, the resulting output saturation power is about 7.75dBm, which corresponds, approximately, to a 6mW power (this result has been proved experimentally).

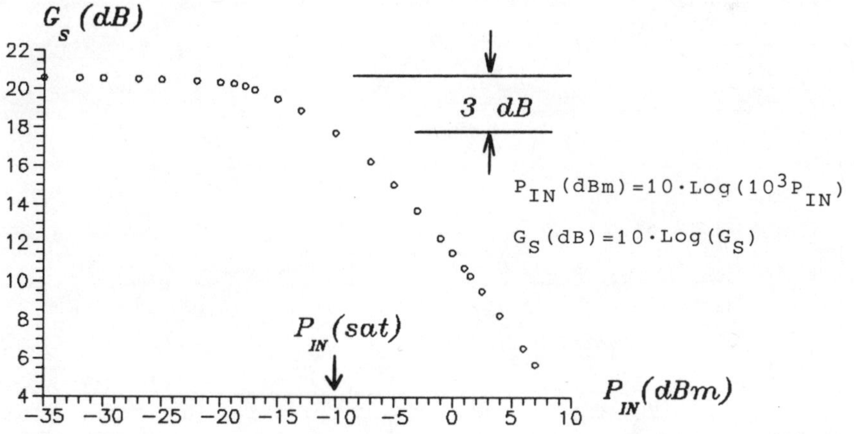

Figure 1

The NF noise figure has been calculated starting from the results given by the numerical solving of expression (18). In order to have calculations in accordance with experimental results, we have taken into account a spectral filtering (2.5GHz) of the spontaneous emission at the amplifier output. The noise figure is shown in Fig.2 as a function of the incident optical power. By observing it one can verify that for lower values of P_{IN}, the NF value is very close to 2ρ=5.12 (linear behaviour of the device). The results obtained predict a decreasing behaviour of the NF when P_{IN} increases over a large range of values till $P_{IN}\simeq$0dBm (=1mW). At this point we find a theoretical minimum of the noise figure (≃0.6). If the incident optical power increases over 0dBm, the noise figure turns to be rising. The theoretical limit when P_{IN} increases indefinitely is fixed by the attenuator behaviour it would present if the amplifier was totally saturated. Evidently, the practical limit is determined by the maximum power that the active layer of the device can support.

Taking into account the results obtained, it seems reasonable to presume that an amplifier behaviour beyond its linear characteristics can be convenient although, logically, with a lower gain. For instance, for powers close to 10dBm, the gain would be only of 18dB, approximately, but the noise figure could be considered excellent (≃2) since this value is lower, even, than the so-called quantum limit (NF=3). In the matter, it must be stated that this limit has been defined in the literature (see, e.g., [11]) for a linear behaviour of the optical amplifier.

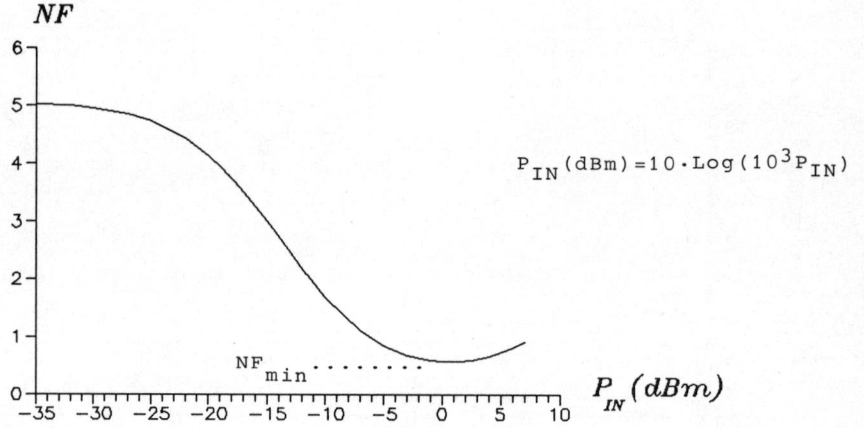

Figure 2

7. Conclusion

The simulations realized show that the performances of a travelling wave optical amplifier considerably differ from those predicted by its linear behaviour. Actually, from a strict point of view, an optical amplifier always shows a certain degree of saturation which is the more important the higher the incident optical power is.

The amplifier used in our simulations shows an incident power of saturation (half gain) of about -10dBm (0.1mW). The gain is then 17.75dB. This result, which has been confirmed experimentally, can be considered satisfactory if it is compared to those usually obtained [1], [2].

On the other hand, it has been confirmed that the noise figure can be reduced, with regard to its linear value, by a non-linear amplification process based on the saturation of the amplifier. Naturally, in such situation, the gain obtained will be lower than its non saturated value. An optimum situation is found for an incident optical power of about 0dBm (=1mW). In such case, the noise figure reaches a minimum value of 0.6 with a gain of 11.5 dB. This is an interesting theoretical result, but is this really feasible? The authors of this paper will try to confirm experimentally this behaviour of the optical amplifier during the following months.

Note: This work has been supported by the WP.4, RACE PROJECT No.1027.

References

[1] M.J.O'Mahony, "Semiconductor laser optical amplifiers for use in future fiber systems", J. Lightwave Technology, Vol.6, No.4, pp.531-544, 1988.

[2] G.Eisenstein and R.M.Jopson, "Measurements of the gain spectrum of near-travelling-wave and Fabry-Perot semiconductor optical amplifier at 1.5 μm", Int.J. Electronics, Vol.60, No.1, pp.113-121, 1986.

[3] Ch.Bendjaballah and G.Oliver, "Comparison of statistical properties of two models for saturated laser-light amplifier", Physical Rev. A, Vol.22, No.6, pp.2726-2731, 1980.

[4] S.Ruiz-Moreno, M.J.Soneira and J.Guitart, "Model of effective bandwith applied to a saturated near travelling wave optical amplifier", SIOE'90, Cardiff.

[5] S.Ruiz-Moreno, G.Junyent, J.R.Usandizaga and A.Calzada, "Resolution of moment equations in a nonlinear optical amplifier", Electronics Lett., Vol.23, No.1, pp.15-17, 1987.

[6] J.C.Simon and P.Doussiere, "Gain and Noise Measurement of near travelling wave semiconductor laser amplifiers", RACE 1027 WP 4, Deriverable 017, 1988.

[7] S.Ruiz-Moreno, M.J.Soneira, J.Guitart and G.Junyent, "Non-linear aspects in travelling wave semiconductor laser amplifier", RACE 1027 WP 4, Intermediate Report, 1990.

[8] J.L.Gimlett, "Progress in 10+ Gb/s lightwave systems technology", invited paper ECOC'90, Amsterdam.

[9] Y.Yamamoto, "AM and FM quantum noise in semiconductor lasers-Part I: theoretical analysis", IEEE J. of Quantum Electronics, Vol.QE-19, No.1, pp. 34-46, 1983.

[10] S.Ruiz-Moreno, G.Junyent, M.J.Soneira and J.R.Usandizaga, "Statistical analysis of nonlinear optical amplifier in high saturation", IEE Proceedings, Vol.135, Pt.J, No.1, pp.34-38, 1988.

[11] J.C.Simon, "Semiconductor optical amplifier for single mode optical fiber communications", J. of optical communications, Vol.4, No.2, pp.87-98, 1983.

Oral presentation

1.– Quantum coherences and dissipation.
C. Cohen–Tannoudji (Collège de France - Paris)

2.– Interferometric detection of gravitational waves : the fundamental limitations.
A. Brillet (GROG – Université de Paris-Sud – Orsay)

Index

Lecture Notes in Mathematics

Vol. 1236: Stochastic Partial Differential Equations and Applications. Proceedings, 1985. Edited by G. Da Prato and L. Tubaro. V, 257 pages. 1987.

Vol. 1237: Rational Approximation and its Applications in Mathematics and Physics. Proceedings, 1985. Edited by J. Gilewicz, M. Pindor and W. Siemaszko. XII, 350 pages. 1987.

Vol. 1250: Stochastic Processes – Mathematics and Physics II. Proceedings 1985. Edited by S. Albeverio, Ph. Blanchard and L. Streit. VI, 359 pages. 1987.

Vol. 1251: Differential Geometric Methods in Mathematical Physics. Proceedings, 1985. Edited by P.L. García and A. Pérez-Rendón. VII, 300 pages. 1987.

Vol. 1255: Differential Geometry and Differential Equations. Proceedings, 1985. Edited by C. Gu, M. Berger and R.L. Bryant. XII, 243 pages. 1987.

Vol. 1256: Pseudo-Differential Operators. Proceedings, 1986. Edited by H.O. Cordes, B. Gramsch and H. Widom. X, 479 pages. 1987.

Vol. 1258: J. Weidmann, Spectral Theory of Ordinary Differential Operators. VI, 303 pages. 1987.

Vol. 1260: N.H. Pavel, Nonlinear Evolution Operators and Semigroups. VI, 285 pages. 1987.

Vol. 1263: V.L. Hansen (Ed.), Differential Geometry. Proceedings, 1985. XI, 288 pages. 1987.

Vol. 1265: W. Van Assche, Asymptotics for Orthogonal Polynomials. VI, 201 pages. 1987.

Vol. 1267: J. Lindenstrauss, V.D. Milman (Eds.), Geometrical Aspects of Functional Analysis. Seminar. VII, 212 pages. 1987.

Vol. 1269: M. Shiota, Nash Manifolds. VI, 223 pages. 1987.

Vol. 1270: C. Carasso, P.-A. Raviart, D. Serre (Eds.), Nonlinear Hyperbolic Problems. Proceedings, 1986. XV, 341 pages. 1987.

Vol. 1272: M.S. Livšic, L.L. Waksman, Commuting Nonselfadjoint Operators in Hilbert Space. III, 115 pages. 1987.

Vol. 1273: G.-M. Greuel, G. Trautmann (Eds.), Singularities, Representation of Algebras, and Vector Bundles. Proceedings, 1985. XIV, 383 pages. 1987.

Vol. 1275: C.A. Berenstein (Ed.), Complex Analysis I. Proceedings, 1985–86. XV, 331 pages. 1987.

Vol. 1276: C.A. Berenstein (Ed.), Complex Analysis II. Proceedings, 1985–86. IX, 320 pages. 1987.

Vol. 1277: C.A. Berenstein (Ed.), Complex Analysis III. Proceedings, 1985–86. X, 350 pages. 1987.

Vol. 1283: S. Mardešić, J. Segal (Eds.), Geometric Topology and Shape Theory. Proceedings, 1986. V, 261 pages. 1987.

Vol. 1285: I.W. Knowles, Y. Saitō (Eds.), Differential Equations and Mathematical Physics. Proceedings, 1986. XVI, 499 pages. 1987.

Vol. 1287: E.B. Saff (Ed.), Approximation Theory, Tampa. Proceedings, 1985–1986. V, 228 pages. 1987.

Vol. 1288: Yu. L. Rodin, Generalized Analytic Functions on Riemann Surfaces. V, 128 pages, 1987.

Vol. 1294: M. Queffélec, Substitution Dynamical Systems – Spectral Analysis. XIII, 240 pages. 1987.

Vol. 1299: S. Watanabe, Yu.V. Prokhorov (Eds.), Probability Theory and Mathematical Statistics. Proceedings, 1986. VIII, 589 pages. 1988.

Vol. 1300: G.B. Seligman, Constructions of Lie Algebras and their Modules. VI, 190 pages. 1988.

Vol. 1302: M. Cwikel, J. Peetre, Y. Sagher, H. Wallin (Eds.), Function Spaces and Applications. Proceedings, 1986. VI, 445 pages. 1988.

Vol. 1303: L. Accardi, W. von Waldenfels (Eds.), Quantum Probability and Applications III. Proceedings, 1987. VI, 373 pages. 1988.

Lecture Notes in Physics

Vol. 356: R. Wehrse (Ed.), Accuracy of Element Abundances from Stellar Atmospheres. Proceedings, 1988. V, 109 pages. 1990.

Vol. 357: J.L. Lumley (Ed.), Whither Turbulence? Turbulence at the Crossroads. Proceedings, 1989. V, 535 pages. 1990.

Vol. 358: J. Kovalevsky, Astrométrie moderne. VI, 294 pages. 1990.

Vol. 359: K. Kirchgässner (Ed.), Problems Involving Change of Type. Proceedings, 1988. XII, 207 pages. 1990.

Vol. 360: Y. Mellier, B. Fort, G. Soucail (Eds.), Gravitational Lensing. Proceedings 1989. XV, 315 pages. 1990.

Vol. 361: P. Gaigg, W. Kummer, M. Schweda (Eds.), Physical and Nonstandard Ganges. Proceedings, 1989. IX, 310 pages. 1990.

Vol. 362: N.E. Kassim, K.W. Weiler (Eds.), Low Frequency Astrophysics from Space. Proceedings. XII, 280 pages. 1990.

Vol. 363: V. Ruždjak, E. Tandberg-Hanssen (Eds.), Dynamics of Quiescent Prominences. Proceedings, 1989. XI, 304 pages. 1990.

Vol. 364: T.T.S. Kuo, E. Osnes, Folded-Diagram Theory of the Effective Interaction in Nuclei, Atoms and Molecules. V, 175 pages. 1990.

Vol. 365: M. Schumacher, G. Tamas (Eds.), Perspectives of Photon Interactions with Hadrons and Nuclei. Proceedings. IX, 251 pages. 1990.

Vol. 366: M.-J. Goupil, J.-P. Zahn (Eds.), Rotation and Mixing in Stellar Interiors. Proceedings, 1989. XIII, 183 pages. 1990.

Vol. 367: Y. Osaki, H. Shibahashi (Eds.), Progress of Seismology of the Sun and Stars. Proceedings, 1989. XIII, 467 pages. 1990.

Vol. 368: L. Garrido (Ed.), Statistical Mechanics of Neural Networks. Proceedings. VI, 477 pages. 1990.

Vol. 369: A. Cassatella, R. Viotti (Eds.), Physics of Classical Novae. Proceedings, 1989. XII, 462 pages. 1990.

Vol. 370: H.-D. Doebner, J.-D. Hennig (Eds.), Quantum Groups. Proceedings, 1989. X, 434 pages. 1990.

Vol. 371: K.W. Morton (Ed.), Twelfth International Conference on Numerical Methods in Fluid Dynamics. Proceedings, 1990. XIV, 562 pages. 1990.

Vol. 372: F. Dobran, Theory of Structured Multiphase Mixtures. IX, 223 pages. 1991.

Vol. 373: C.B. de Loore (Ed.), Late Stages of Stellar Evolution. Computational Methods in Astrophysical Hydrodynamics. Proceedings, 1989. VIII, 390 pages. 1991.

Vol. 374: L. Ting, R. Klein, Viscous Nortical Flows. V, 222 pages. 1991.

Vol. 375: C. Bartocci, U. Bruzzo, R. Cianci (Eds.), Differential Geometric Methods in Theoretical Physics. Proceedings, 1990. XIX, 401 pages. 1991.

Vol. 376: D. Berényi, G. Hock (Eds.), High-Energy Ion-Atom Collisions. Proceedings, 1990. IX, 364 pages. 1991.

Vol. 377: W.J. Duschl, S.J. Wagner, M. Camenzind (Eds.), Variability of Active Galaxies. Proceedings, 1990. XII, 312 pages. 1991.

Vol. 378: C. Bendjaballah, O. Hirota, S. Reynaud (Eds.), Quantum Aspects of Optical Communications. Proceedings 1990. VII, 389 pages. 1991.